Drug Repurposing: Exploring Newer Therapeutic Potentials

Ravinder K. Kaundal
Rahul Shukla • Shubhini A. Saraf
Editors

Drug Repurposing: Exploring Newer Therapeutic Potentials

Springer

Editors
Ravinder K. Kaundal
Pharmacology and Toxicology
NIPER-Raebareli
Lucknow, Uttar Pradesh, India

Rahul Shukla
Department of Pharmaceutics
NIPER-Raebareli
Lucknow, Uttar Pradesh, India

Shubhini A. Saraf
NIPER-Raebareli
Lucknow, Uttar Pradesh, India

ISBN 978-981-95-4521-6 ISBN 978-981-95-4522-3 (eBook)
https://doi.org/10.1007/978-981-95-4522-3

This Springer imprint is published by the registered company Springer Nature Singapore Pte Ltd.
The registered company address is: 152 Beach Road, #21-01/04 Gateway East, Singapore 189721, Singapore

Preface

Scientific progress is often a story of rediscovery. Drugs that once held promise for one condition are now being repurposed to fight entirely different diseases, breathing new life into modern medicine. *Drug repurposing* is not just a strategy—it is the demonstration of human inventiveness, a means of expediting medical interventions, and a link between previous and upcoming discoveries.

The idea for this book emerged from a deep-seated passion for pharmacology and the need to highlight the immense potential of *repurposing existing drugs* in today's fast-paced world of biomedical research. In an era where developing a new drug can take over a decade and billions of dollars, repurposing offers a smarter, faster, and more efficient alternative. We have encountered difficult situations, such as the pandemic, which was a stark reminder of why we need quicker solutions, and drug repurposing played a crucial role in identifying treatments at an unprecedented pace.

This book results from *collaborative efforts* by researchers and educators who have spent years studying drug action, disease mechanisms, and therapeutic innovations. It brings together their collective wisdom to explore the promise and challenges of drug repurposing—from its scientific foundations to the regulatory, clinical, and patent-related roadblocks that must be navigated for success.

This book is a valuable resource for pharmacology enthusiasts as it provides motivation and understanding for scholars, medical professionals, and students. It aims to spark new ideas and appreciation for repurposed drugs.

Lucknow, Uttar Pradesh, India

Ravinder K. Kaundal
Rahul Shukla
Shubhini A. Saraf

Contents

Editors and Contributors

About the Editors

Ravinder K. Kaundal is working as an Assistant Professor at NIPER Raebareli. He has done his MS (Pharm.) and PhD in pharmacology and toxicology at NIPER-SAS Nagar, India. He received his postdoctoral training at Yale School of Medicine, New Haven, and Icahn School of Medicine at Mount Sinai, New York. He has made an outstanding contribution to the field of pharmacological research and has elucidated the involvement of various pathogenic pathways like oxidative stress, PARP overactivation, inflammation, and apoptosis in cerebral ischemic-reperfusion injury and diabetic neuropathy. Gene regulation is another of his interests, and he has developed various tools/methods to study the mechanisms of epigenetic memory. His work is published in journals of high repute and is highly cited. He is a reviewer and editorial board member of several international journals. His research interests are neuropharmacology, drug repurposing, molecular biology, ischemic-reperfusion injuries, and gene regulation.

Rahul Shukla working as an Assistant Professor at the Department of Pharmaceutics, NIPER Raebareli, India, did his PhD in pharmaceutical sciences from CSIR CDRI and JNU, New Delhi, India; MPharm from IIT BHU, Varanasi, India; and BPharm from Jamia Hamdard, New Delhi. He has the experience of as research scientist at Dr. Reddy's Laboratories, India, and D.S. Kothari Post-Doctoral Fellow at Panjab University, India. He has more than ten years of research and academic experience. He has more than 150 publications including 60 book chapters to his credit. He has filed 12 Patent and 1 Copyright. He was awarded SERB International Research Experience (SIRE) Fellow for the year, 2022–2023 fellowship at the School of Pharmacy and Biomedical Sciences, University of Central Lancashire, United Kingdom. He also received Science and Engineering Research Board (SERB) Research Grant. His current research interests include targeted drug delivery systems, particle engineering, controlled and targeted delivery for neurodegenerative diseases and cancer, and works on repurposing, particle engineering, and bioavailability enhancements.

Shubhini A. Saraf is working as director of the National Institute of Pharmaceutical Education and Research (NIPER), Raebareli (UP). Prior to joining NIPER, Raebareli, she was working as a Professor in the Department of Pharmaceutical Sciences and Director (R&D) at Babasaheb Bhimrao Ambedkar University (A Central University), Lucknow. She is an alumnus of the Department of Pharmaceutical Sciences, Dr. H.S. Gour University, Sagar, India. She has more than 27 years of teaching and research experience. Her research interests include nanotechnology and drug targeting as well as technology policy interventions for health care. Her work extensively discusses surface-decorated lipid nanoparticles for cancer therapy. She has 165 research papers published in journals of repute. She has authored 7 books and 22 chapters. She has attended many national and international conferences and delivered many invited lectures. Her administrative and research work has been recognized at various national and international forums. She is a sincere administrator, academician, and researcher and is known for her inclusive involvement and untiring hard work. Her contributions to the pharmacy profession are enormous and praiseworthy.

Contributors

S. Ali Delhi Institute of Pharmaceutical Sciences and Research, DPSRU, New Delhi, India

Alka Department of Pharmaceutical Sciences, Babasaheb Bhimrao Ambedkar University Lucknow (A Central University), Uttar Pradesh, Lucknow, Uttar Pradesh, India

Harapriya Baral Department of Pharmacology and Toxicology, National Institute of Pharmaceutical Education and Research, Raebareli (NIPER-R), Transit Campus, Lucknow, Uttar Pradesh, India

B. Behera Delhi Institute of Pharmaceutical Sciences and Research, DPSRU, New Delhi, India

Archana Bhatta Department of Pharmacology and Toxicology, National Institute of Pharmaceutical Education and Research, Raebareli (NIPER-R), Transit Campus, Lucknow, Uttar Pradesh, India

Shurabhi Chaurasia Department of Pharmaceutical Sciences, Birla Institute of Technology, Mesra, Ranchi, India

John Disouza SYBES's Bombay Institute of Pharmacy and Research, Dombivli, India

Kamal Dua Discipline of Pharmacy, Graduate School of Health, University of Technology Sydney, Ultimo, NSW, Australia

Faculty of Health, Australian Research Centre in Complementary and Integrative Medicine, University of Technology Sydney, Ultimo, NSW, Australia

Uttaranchal Institute of Pharmaceutical Sciences, Uttaranchal University, Dehradun, India

Monika Dwivedi Department of Pharmaceutical Sciences, Birla Institute of Technology, Mesra, Ranchi, India

Gul Naz Fatima Division of Pharmaceutical Chemistry, Faculty of Pharmacy, Babu Banarasi Das Northern India Institute of Technology, Ayodhya Road, Chinhut, Lucknow, Uttar Pradesh, India

Hera Fatma Division of Pharmaceutical Chemistry, Faculty of Pharmacy, Babu Banarasi Das Northern India Institute of Technology, Ayodhya Road, Chinhut, Lucknow, UP, India

Debapriya Garabadu Department of Pharmacology, School of Health Sciences, Central University of Punjab, Bathinda, India

S. Ghosh Delhi Institute of Pharmaceutical Sciences and Research, DPSRU, New Delhi, India

Gaurav Gupta Centre for Research Impact and Outcome, Chitkara College of Pharmacy, Chitkara University, Rajpura, Punjab, India

Centre of Medical and Bio-Allied Health Sciences Research, Ajman University, Ajman, UAE

Saurabh Gupta Chameli Devi Institute of Pharmacy, Department of Pharmacology, Indore, Madhya Pradesh, India

Krishna Jadhav Chemical Biology Unit, Institute of Nano Science and Technology (INST), Knowledge City, Mohali, Punjab, India

Agrim Jhilta Chemical Biology Unit, Institute of Nano Science and Technology (INST), Knowledge City, Mohali, Punjab, India

Vikas Kamble Department of Pharmaceutical Sciences and Technology, Institute of Chemical Technology, Matunga, Mumbai, India

Kunal Agam Kanaujia Department of Pharmaceutical Sciences, Babasaheb Bhimrao Ambedkar University Lucknow (A Central University), Uttar Pradesh, Lucknow, Uttar Pradesh, India

Ravinder K. Kaundal Department of Pharmacology and Toxicology, National Institute of Pharmaceutical Education and Research, Raebareli (NIPER-R), Transit Campus, Lucknow, Uttar Pradesh, India

Kaustubh Ajit Kolekar School of Pharmaceutical Sciences, Lovely Professional University, Phagwara, Punjab, India

Anoop Kumar Department of Pharmacology, Delhi Institute of Pharmaceutical Sciences and Research (DIPSAR), Delhi Pharmaceutical Science & Research University (DPSRU), New Delhi, India

Madhu Kumari Department of Pharmaceutical Sciences, Birla Institute of Technology, Mesra, Ranchi, India

Popat S. Kumbhar Department of Pharmaceutics, Tatyasaheb Kore College of Pharmacy, Warana University, Warananagar, Warananagar, Maharashtra, India

Pranav S. Kumbhar Tatyasaheb Kore College of Pharmacy, Panhala, Kolhapur, Maharashtra, India

S. Latha Department of Pharmacology, Delhi Institute of Pharmaceutical Sciences and Research (DIPSAR), Delhi Pharmaceutical Science & Research University (DPSRU), New Delhi, India

Priyanka Mahale Department of Pharmacology and Toxicology, National Institute of Pharmaceutical Education and Research, Raebareli (NIPER-R), Transit Campus, Lucknow, Uttar Pradesh, India

M. Maheshwari Delhi Institute of Pharmaceutical Sciences and Research, DPSRU, New Delhi, India

M. Pal Delhi Institute of Pharmaceutical Sciences and Research, DPSRU, New Delhi, India

Nidhi Mishra Department of Pharmaceutical Sciences, Babasaheb Bhimrao Ambedkar University Lucknow (A Central University), Uttar Pradesh, Lucknow, Uttar Pradesh, India

Rozhan Arif Muhammed Tishk International University, Department of Pharmacy, Sulaymaniyah, Iraq

Swarnima Negi Chemical Biology Unit, Institute of Nano Science and Technology (INST), Knowledge City, Mohali, Punjab, India

Raquibun Nisha Department of Pharmaceutical Sciences, Babasaheb Bhimrao Ambedkar University Lucknow (A Central University), Uttar Pradesh, Lucknow, Uttar Pradesh, India

Bhaskaranand Pancholi Department of Pharmacology, School of Health Sciences, Central University of Punjab, Bathinda, India

Anima Pandey Department of Pharmaceutical Sciences, Birla Institute of Technology, Mesra, Ranchi, India

Archita Patel Ramanbhai Patel College of Pharmacy, Charotar University of Science and Technology, Anand, Gujarat, India

Samir Patel Ramanbhai Patel College of Pharmacy, Charotar University of Science and Technology, Anand, Gujarat, India

Vandana Patravale Department of Pharmaceutical Sciences and Technology, Institute of Chemical Technology, Mumbai, Maharashtra, India

Omji Porwal Qaiwan International University, Department of Pharmacy, Sulaymaniyah, Iraq

Amarjeet Rao Department of Pharmacology, School of Health Sciences, Central University of Punjab, Bathinda, India

P. K. Sahoo Delhi Institute of Pharmaceutical Sciences and Research, DPSRU, New Delhi, India

Saima Department of Pharmacology, Magadh Institute of Pharmacy, Patna, India

Shailendra K. Saraf Division of Pharmaceutical Chemistry, Faculty of Pharmacy, Babu Banarasi Das Northern India Institute of Technology, Ayodhya Road, Chinhut, Lucknow, Uttar Pradesh, India

Shubhini A. Saraf Department of Pharmaceutics, National Institute of Pharmaceutical Education and Research-Raebareli, Lucknow, Uttar Pradesh, India

Department of Pharmaceutical Sciences, Babasaheb Bhimrao Ambedkar University Lucknow (A Central University), Uttar Pradesh, Lucknow, Uttar Pradesh, India

National Institute of Pharmaceutical Education and Research, Raebareli (NIPER-R), Lucknow, Uttar Pradesh, India

Ruchika Sharma Centre for Precision Medicine and Pharmacy, Delhi Pharmaceutical Science & Research University (DPSRU), New Delhi, India

K. Shubham Delhi Institute of Pharmaceutical Sciences and Research, DPSRU, New Delhi, India

Rahul Shukla Department of Pharmaceutics, National Institute of Pharmaceutical Education and Research-Raebareli, Lucknow, Uttar Pradesh, India

Neelu Singh Department of Pharmaceutical Sciences, Babasaheb Bhimrao Ambedkar University Lucknow (A Central University), Uttar Pradesh, Lucknow, Uttar Pradesh, India

Raghuraj Singh Chemical Biology Unit, Institute of Nano Science and Technology (INST), Knowledge City, Mohali, Punjab, India

Academy of Scientific and Innovative Research (AcSIR), Ghaziabad, UP, India

Sachin Kumar Singh School of Pharmaceutical Sciences, Lovely Professional University, Phagwara, Punjab, India

Faculty of Health, Australian Research Centre in Complementary and Integrative Medicine, University of Technology Sydney, Ultimo, NSW, Australia

Samipta Singh Department of Pharmaceutical Sciences, Babasaheb Bhimrao Ambedkar University Lucknow (A Central University), Uttar Pradesh, Lucknow, Uttar Pradesh, India

Abhishek Verma Department of Pharmaceutical Sciences, Babasaheb Bhimrao Ambedkar University Lucknow (A Central University), Uttar Pradesh, Lucknow, Uttar Pradesh, India

Rahul K. Verma Chemical Biology Unit, Institute of Nano Science and Technology (INST), Knowledge City, Mohali, Punjab, India

Sharad Visht College of Pharmacy, Cihan University-Erbil, Erbil, Iraq

Sukriti Vishwas Chitkara College of Pharmacy, Chitkara University, Rajpura, Punjab, India

D. K. Yuvaraj Department of Pharmacology and Toxicology, National Institute of Pharmaceutical Education and Research, Raebareli (NIPER-R), Transit Campus, Lucknow, Uttar Pradesh, India

Amit Halder Dr. B.C. Roy College of Pharmacy & Allied Health Sciences, Durgapur, West Bengal, India

Ankit Kumar Department of Pharmaceutical Technology, Jadavpur University, Kolkata, West Bengal, India

Arun Baidya Department of Pharmaceutical Technology, Jadavpur University, Kolkata, West Bengal, India

Axata Lama Department of Pharmaceutical Technology, Jadavpur University, Kolkata, West Bengal, India

Nilanjan Ghosh Department of Pharmaceutical Technology, Jadavpur University, Kolkata, West Bengal, India

Rituparna Chaki Dr. B.C. Roy College of Pharmacy & Allied Health Sciences, Durgapur, West Bengal, India

Saptapadi Saha Department of Pharmaceutical Technology, Jadavpur University, Kolkata, West Bengal, India

Shuvam Sar Department of Pharmaceutical Technology, Jadavpur University, Kolkata, West Bengal, India

Tanmoy Banerjee Department of Pharmaceutical Technology, Jadavpur University, Kolkata, West Bengal, India

Role of Industry–Academia Collaboration in the Translational Success of Drug Repurposing

1

Rahul Shukla and Shubhini A. Saraf

Abstract

Drug repurposing is the process of developing an existing approved drug to identify new medical indications. This process involves drug identification and efficacy in the preclinical phase for moving to Phase II of clinical trials. This can be done to optimize new doses, novel formulations, new combinations, and methods intended for patient benefits. The pharmaceutical industry and academic collaborations in the research and development of drug repurposing are actively exploring this as a driver for cost- and time-efficient pharmaceutical product development. The other paybacks of collaborative drug development include fewer risks, lower costs, and shorter timelines. Under industry and academic partnerships, each partner can contribute a unique set of technical and skill expertise that may be beneficial for both collaborators and end users. Further, with domain-specific expertise, industry scientists can contribute through highly product-focused process strategies, and high-risk, manpower-enabled academic research with low financial rewards can present a product-specific avenue for innovations. The very first example in the category of drug repurposing is a phosphodiesterase-5 (sildenafil) inhibitor, which was initially explored for angina pectoris and hypertension therapy by Pfizer but was successfully used for other indications. However, drug repurposing under the umbrella of industry and academic collaborations may face some challenges; for example, many academic institutions may not manage to engage a research consortium and establish an open innovation ecosystem. This book chapter review emphasized the current trends that lead us to have an essence in industry–academia collaborations for

R. Shukla · S. A. Saraf (✉)
Department of Pharmaceutics, National Institute of Pharmaceutical Education and Research-Raebareli, Lucknow, Uttar Pradesh, India
e-mail: director@niperraebareli.edu.in

R. K. Kaundal et al. (eds.), *Drug Repurposing: Exploring Newer Therapeutic Potentials*,
https://doi.org/10.1007/978-981-95-4522-3_1

drug repurposing for various therapeutic indications, driving a time- and cost-effective product development for patients.

Keywords

Industry · Academia · Collaborations · Drug repurposing · Pharmaceutical

Abbreviations

AD	Alzheimer's Disease
ALS	Amyotrophic Lateral Sclerosis
ANDA	Abbreviated New Drug Applications
ARCs	Academic Research Centres
ASIC1	Acid-Sensing Ion Channel Subunit 1
BRIEF	Biofilm Research–Industrial Engagement Framework
CHF	Congestive Heart Failure
CHIKV	Chikungunya Virus
CIBA	Central Institute of Brackishwater Aquaculture
CNN	Convolutional Neural Network
COPD	Chronic Obstructive Pulmonary Disease
COVID-19	Coronavirus Disease 2019
DHFR	Dihydrofolate Reductase
DNA	Deoxyribonucleic Acid
DNN	Deep Neural Networks
DPCO	Drug Price Control Order
EMA	European Medicines Agency
FDA	Food and Drug Administration
FNN	Feedforward Neural Network
GCN	Graph Convolutional Network
HD	Huntington's Disease
HGA	Human Granulocytic Anaplasmosis
HUVECs	Human Vascular Endothelial Cells
IBD	Inflammatory Bowel Disease
ICMR	Indian Council of Medical Research
IPR	Intellectual Property Rights
LSTM	Long Short-Term Memory
MS	Multiple Sclerosis
NCATS	National Center for Advancing Translational Sciences
NCE	New Chemical Entity
NIH	National Institutes of Health
NTD	Neglected Tropical Diseases
PheWAS	Phenome-Wide Association Study
R&D	Research and Development
RdRp	RNA-Dependent RNA Polymerase

RNN	Recurrent Neural Network
SVM	Support Vector Machine
THS	Thymidylate Synthase
UK	United Kingdom
VEGFR2	Vascular Endothelial Growth Factor Receptor 2
WHO	World Health Organization

1.1 Introduction

1.1.1 Drug Repurposing: Lessons Learnt from Drug Repurposing

Drug repurposing (drug repositioning) is the progression of reassessing a medication that has passed rigorous safety and efficacy testing. Redirecting the medication for a different or supplementary medical indication is the main goal. This complex strategy consists of three primary ideas. First, investigating medications that were initially approved for one indication for possible usage for other purposes is the first step in queue extension approvals. Second, the repurposing of compounds that have already been licensed, put on the market, or are in clinical trials for off-label use yields important information to support their repurposing for novel indications. The final step in drug rescue is to repurpose molecules that have reached a certain stage of clinical development but were dropped for various reasons, such as toxicity, poor effectiveness, or reprioritization, and reroute them to other purposes.

The development of medicinal compounds has yielded valuable insights, especially in the management of COVID-19. Interestingly, several medications that were previously approved for the treatment of other viral infections are now being used to treat COVID-19 patients. These medications include dexamethasone, ivermectin, hydroxychloroquine, ritonavir, and favipiravir, among others. These examples of repurposing already approved drugs highlight how adaptive and resourceful people can be when faced with new health issues, and they provide crucial insights for further research and improvement of therapeutic interventions. One noteworthy development is the repurposing of favipiravir (T-705) as an anti-influenza medicine for the treatment of COVID-19. This medication was created in Japan and was first licensed in 2014 to treat influenza. It works by preventing viral RNA-dependent RNA polymerase (RdRp) (Furuta et al. 2017). China made the first encouraging report on favipiravir's clinical activity against COVID-19, while other investigations offered different insights. Favipiravir was evaluated in a recent study at the School of Medicine, Zhejiang University, along with other compounds such as baloxavir and marboxil. When treating COVID-19 patients, researchers could not find any additional benefits from favipiravir at the study dosages (Lou et al. 2021).

These divergent results underscore the need for continued investigation and assessment when repurposing current drugs for newly developing infectious illnesses. The treatment of rheumatoid arthritis with baricitinib, a Janus kinase inhibitor that targets JAK1/2 specifically, has been approved. According to Richardson's research, baricitinib shows promise as a possible COVID-19 treatment option and may lower the risk of death and intensive care unit hospitalization. Based on these observations, individuals treated with this therapeutic molecule have faster viral clearance and greater discharge rates than those treated with standard care for COVID-19. Notably, baricitinib has been used as a therapy for patients with moderate pneumonia linked to COVID-19. The goal of ongoing randomized clinical trials (NCT04373044, NCT04421027, and NCT04640168) is to gain a thorough understanding of the benefits of baricitinib in the treatment of COVID-19. In hospitalized COVID-19 patients, a recent randomized clinical trial with 1033 patients presented improved therapeutic results with combination baricitinib and remdesivir therapy than with remdesivir alone (ACTT-2 Study Group) (NCT04401579) (Kalil et al. 2021).

These results add to our growing knowledge of possible COVID-19 therapeutic approaches. Computational drug repurposing, which involves examining current medications for new use, provides a quick way to find alternate applications with known pharmacological and safety profiles. This strategy involves evaluating a large number of compounds through the use of methods such as in silico and high-throughput screening, which offer quick and affordable screening approaches. Computational drug repurposing is positioned as an initial phase and involves important filtering stages. Scientists have been aggressively searching for medications that have been repurposed for COVID-19, highlighting the importance of computational methods during this pandemic. This cooperative effort offers insightful information about possible COVID-19 therapeutics and emphasizes the significance of applying computational techniques in drug discovery.

1.1.2 Translational Research and Development

With the primary objective of bridging the gap between basic and clinical development, the area of translational research has recently evolved as a scientific discipline. An expert in both fields who can translate the results of one branch into the language of the other is known as a translational researcher, who acts as a vital connection. This expert is skilled at recognizing the practical implications of novel findings for doctors in basic research and tying them in. Currently, the process of methodically translating research ideas into new health goods, diagnostic tools, and therapeutic approaches is known as translational research. This kind of research is usually carried out in academic institutions; its laboratory-based activities are similar to those of fundamental science, and its discoveries and developments are used in the clinical context of medical practice (Kalil et al. 2021).

Translational research has attracted a lot of interest, both nationally and in Washington, DC. Inspired by the US Congress's demand for concrete results, the National Institutes of Health (NIH) in Bethesda, MD, set forth a "Roadmap", which is essentially a roadmap for translational research. Numerous organizations, including the National Academies of Science in Washington, DC, scientific professional societies, medical advocacy groups, and some private foundations, have embraced this movement in large numbers. The dynamics of how science functions in society have changed as a result of the growing emphasis on translation. Scientists are increasingly expected to make explicit commitments to certain outcomes up front, rather than depending on implied guarantees of later results. In addition, there is a greater focus on specialized goals rather than broad advantages and a higher desire for faster outcomes. Furthermore, more specific guidance is being given by public investors. Ultimately, this leads to the development of a translation-focused mentality (Maienschein et al. 2008).

1.1.3 Drug Repurposing: Lab-to-Clinical Translational

The proliferation of chemical databases on the Internet has led to the development of cheminformatics, a sophisticated field rich in biological data. The application of computational and mathematical techniques offers special insights into biochemical data that could otherwise be missed when creating pharmacological and therapeutic relationships. Maximizing the efficacy of drug repurposing initiatives requires an understanding of how these factors interact, since pharmacodynamics cannot be fully explained by complementary binding in a single dimension, and the totality of pharmacodynamic modalities is not always additive. Drug repurposing relies heavily on molecular and chemical mechanics, but the endpoint perspective of disease phenotypes is becoming more and more important. Using PREDICT, a technique for similarity-based inference of new medication indications based on drug–drug and disease–disease similarities, Gottlieb et al. illustrated this (Gottlieb et al. 2011) by employing gene expression patterns like those found in CMap. PREDICT advances personalized medicine, which is a fundamental component of pharmacogenomics and next-generation pharmacogenetics.

Combining genetic/phenotypic markers with biochemical/biophysical and network modalities could lead to more precise drug repurposing as well as patient-specific, fine-tuned pharmacotherapy at a resolution that exceeds the current understanding. Just as "one gene, one phenotype" has changed over time, so too has the idea of "one drug, one protein". From the fundamental level of a single protein to the vast phenotypic network arising from the interaction of various pathways, functions, and expression patterns occurring at precisely the same time, the effects of therapeutic medications on the human body are intricately linked. Looking ahead, future attempts at drug repurposing need to take this complexity into account while simultaneously maintaining patient-centredness (Issa et al. 2013).

1.1.4 Industry and Academic Collaborations: Working at the Academia–Industry Interface

There are many reasons to support collaborations between academia and industry to work together, such as academic research being essential for basic science, finding and validating new targets, and conducting clinical studies to assess the safety and efficacy of treatment options. Academic institutions initiate research on rare diseases, industry players collaborate in development, and the industry oversees commercial launches (Bloom 2022). Even after submitting clinical trial results to regulatory bodies, pharmaceutical corporations maintain that the data are proprietary or confidential. On the other hand, there is pressure on academics to publish more often. Academic institutions are renowned for having a "publish or perish" mentality. Emphasizing the differences in cultures and procedures between academia and industry has become a popular practice. Academic researchers lack the resources and ability to develop drug candidates prior to the commercialization of the final product. From important nonclinical investigations, Phase I–III clinical trials, chemistry, manufacturing, and controls (CMC), and formulation operations to regulatory submissions, commercial launch activities, and life-cycle management, pharmaceutical businesses operate in an extremely regulated industry.

First and foremost, academia must comprehend industry needs when addressing real-world problems to which they can apply their knowledge. Academic research may be closely linked to practical implications in an industrial environment. Both have their respective roles to play in the drug development process; there is complementarity and not duplication, and there may be mutual support in learning from one another. Additionally, it is critical that the industry understand the critical role that publications, grant approvals, and worldwide networking play in academia, particularly for PhD students and young researchers, and that academia be aware of the pharmaceutical development of a drug. A mutual understanding of the expectations and limitations of academia and industry is crucial for promoting interactions and fostering productive collaboration. Each person must gain knowledge from the other person (Fetro 2023).

In his analysis of business-university partnership, Richard Lambert stressed the importance of working together to overcome the obstacles present in the "valley of death". The evaluation conducted by Lambert showed "clear evidence of the advantages brought to both the university system and British business by collaboration". Numerous organizations and government agencies have directed their attention towards this domain since the publication of Lambert's report in 2003; over 14 papers specifically highlight the significance of industry-university collaboration as a key driver of UK growth. The interest in this topic is said to be fuelled by "the significance of research and innovation as drivers of a knowledge-based economy, coupled with longstanding concerns regarding the UK's performance in converting research excellence into commercial success, and the need to boost UK productivity", according to the most recent of these reports, The Dowling Review of Business-University Research Collaborations. Indeed, a study team looking at the most novel or "transformational" medications created in the last 25 years discovered that just four of the 19 medications were fully studied and created by a single industry. This emphasizes how complementary the roles of the public and commercial sectors are when it comes to basic science and the various phases of drug discovery (Chakravarthy et al. 2016).

The importance of academia–industry collaboration in translational medicine, however, has received far less attention despite a wealth of studies on collaboration as a strategy for fostering innovation and growth. Furthermore, research has shown that university-industry ties are highly influenced by sector-specific dynamics. As a result, even while the interface between academia and business may be vital for fostering innovation, its complex and varied nature has not yet been fully investigated and understood. Presumably, the "valley of death" and other domain-specific obstacles could lead to different demands for industry–academia cooperation in this area. As such, the goal of this study is to pinpoint, investigate, and comprehend these demands, considering what is currently known about the cooperation between academia and industry.

1.1.5 Industrial Engagement

Industry and academia work together on a variety of projects and fields, creating a mutually beneficial partnership. The extent of this involvement can vary from smaller-scale initiatives to more extensive "strategic engagement", in which academic institutions and business associates work together on a variety of projects, including workforce development, clinical trials, curriculum development, student recruitment, and sponsored research. Universities are creating more and more specialized departments for business partnerships or industry involvement as a result of realizing the benefits of these partnerships. These offices actively work to draw in new businesses to support campus programmes, including student internships and funded research. Their goal is to foster relationships that are comprehensive and cover the whole "strategic" spectrum of cooperative endeavours. Universities stand to gain a great deal from these collaborations in terms of research funding, charitable contributions, and the wider effects they have on job creation, skill development, and regional development.

In addition, participating organizations stand to earn significantly from having access to innovative inventions, hiring competent individuals, and attracting new talent for their workforce—especially in high-demand sectors like computer science and engineering. The most fruitful collaborations between academic institutions and business sectors are those that are based on shared interests, enduring dedication, and essential support for economic growth and research. In addition to financial gains, these partnerships provide highly qualified graduates, improved employment prospects, creative solutions that benefit society, and a team effort that strengthens communities. In the end, these collaborations produce more prepared children, encouraging teachers, prosperous businesses, and stronger communities, which promote a constructive cycle of development and progress (Hobin and Galbraith 2012).

In the realm of biofilms, the differences between industrial activities and scholarly studies are evident. One such example is the widespread application of planktonic microorganisms in numerous industry-used standardized efficacy tests. Regrettably, planktonic assays are not applicable to sessile microorganisms that are frequently found in different industries. Although there is promise for translational research to close the gap between academia and industry, there are a number of obstacles in the way. While recognizing and removing these obstacles can be difficult, doing so has the potential to promote habits and research that benefit both parties. The institutes have created a two-dimensional framework called the Biofilm Research-Industrial Engagement Framework (BRIEF) in response to these difficulties.

This framework is intended to arrange current biofilm technologies, such as models, procedures, and tests, that are pertinent to biofilms in diverse industries. These technologies are categorized by the framework according to how well they match the existing scientific understanding and how well they perform in real-world industrial environments. Our approach focuses on the following sectors: wastewater treatment, agriculture, healthcare, the food industry, and marine habitats. By taking a holistic approach, we hope to promote collaboration and alignment between academic research and industry practices by facilitating a more nuanced understanding of biofilm-related technologies (Highmore et al. 2022).

1.1.6 Academic Contributions

Novel molecular entities are explored and advanced through the use of academic studies. Academic scientists are actively adjusting to a dynamic R&D world that demands both efficiency and innovation, much like their peers in industry and regulation. The definition and nature of translational research have changed over the last 10 years. Applying fundamental scientific findings to clinically relevant findings while concurrently formulating scientific enquiries based on clinical data is a fair definition of translational research. Examining the use of translational research approaches in certain therapeutic domains like pain management, nephrology, oncology, and ageing-related cognitive decline leads to a better understanding.

It is crucial to take into account the parallel and historical environment in order to understand the origins of translational research. While large and small pharmaceutical companies (PhRMA) struggled with shrinking pipelines in spite of high-throughput screening and other discovery-based innovations, academic medical centres within the post-genomic era struggled to connect basic sciences with clinical researchers who were familiar with the "bench" but rarely visited. Concurrently, regulatory agencies, such as the National Institute of Health (NIH) and the Food and Drug Administration (FDA), have developed critical path and roadmap programmes to tackle these issues and suggest remedies. The creation of interdisciplinary teams with the ability to establish quantitative connections between developmental sciences and discovery is fundamental to these approaches. These teams are responsible for testing basic and clinical pharmacology hypotheses, encouraging creative experimental designs, and participating in overall drug development decision-making. This work is essentially what is known as "quantitative pharmacology" for want of a better phrase (Barrett 2008).

Several noteworthy instances of medication repurposing have been reported by academia. One such example is the medication thalidomide, which was first created by CIBA and made available as a sedative in 1956. Even though it was abandoned because of its failure, data from academic researchers have helped thalidomide be repurposed, resulting in a novel medication that can be used to treat multiple myeloma and leprosy. This groundbreaking effort created a new drug class, in addition to saving previously unsuccessful drugs.

Once thought to be harmless in preclinical research and recommended for depression and anxiety, thalidomide has gained notoriety in other areas for generating teratogenic adverse effects during pregnancy. By refusing to approve thalidomide in the US, Frances Kelsey of the Food and Drug Administration (FDA) played a crucial role and ultimately improved the FDA's standing. Tecfidera, also known as dimethyl fumarate, is another effective example of a medication that has been repurposed by academia. In 1959, a German chemist reported dimethyl fumarate's first medicinal usage as a topical psoriasis treatment. Originally, it was employed as a fungicide to stop mould growth on furniture. Following the demonstration of its effectiveness in randomized clinical tests, an oral formulation for psoriasis was approved in Germany in 1994. Preclinical multiple sclerosis (MS) models were used by academic researchers to demonstrate dimethyl fumarate's immunosuppressive effects.

Clinical trials have verified its effectiveness in patients with MS, which is in line with these findings. Building on these findings, the European Medicines Agency approved dimethyl fumarate in 2014 after the FDA approved it in 2013 for the relapse of multiple sclerosis. These instances demonstrate the important roles that academics play in repurposing medications and transforming underutilized or abandoned substances into potent remedies for a range of illnesses.

1.1.7 Intellectual Property Considerations

Novel innovation models are emerging in response to declining product development, increasing costs, and duplication of effort in medication research. Collaboration among academia, government, business, non-governmental organizations, and patient organizations is emphasized in many of these strategies. The fundamental idea is that many research and development organizations required for a novel medication or treatment cannot be independently carried out by a single organization. Dynamic authority, including that of intellectual property rights (IPRs), is necessary for effective collaboration in such initiatives that use the strengths of several industries and enable the exchange of both formal and informal knowledge, data, and reagents.

While it is commonly believed that IPRs encourage the introduction of novel and valuable products and methods, there is mounting evidence that, in some situations, they may obstruct collaborations that are essential to accomplishing those goals. This necessitates a re-evaluation of the use of intellectual property rights based on data that reflects the intricacy of an intensely networked environment for research and development. This conversation emphasizes how innovative IPR governance is necessary to facilitate cross-sectoral cooperation. This effort entails adopting more open patent licencing, using more selective patent acquisition, and putting improved metrics into practice. Once accepted by funding organizations and research institutions, these metrics should not only assess how successful collaborations are but also act as a motivator for initiating and maintaining successful partnerships.

Innovative medication and therapy development is greatly aided by imaginative IPR management and shared knowledge production, while also appreciating the critical role other incentives play in the larger regulatory and financing environments. Instead of fruitless discussions about expanding IPR protection on the grounds that it will unavoidably spur pharmaceutical innovation, the proposal by the US National Centre for Advancing Translational Sciences (NCATS), through the National Institutes of Health (NIH), is more likely to advance translational science. Creative intellectual property rights management enables the creation of flexible structures that can respond to changes in the industry, such as the outsourcing of research and development to nations like China, India, and Brazil; the development of those nations' domestic R&D capacities; the shrinking of R&D in large pharmaceutical companies; and the rise in translational activity in academia (Bubela et al. 2012).

1.1.8 Drug Discovery Is a Risky and Expensive Journey

Research and development (R&D) expenditures have increased significantly in recent years, according to aggregate industry data. This illustrates the growth in inflation-adjusted R&D spending for PhRMA member companies in the US pharmaceutical business from 1963. Since 1963, the trend has been steadily rising, with a noticeable pick-up in the rate of increase after 1979 that has continued to this day. Since 1979, the pharmaceutical industry's R&D spending has grown at a compound annual growth rate of about 10%, even after taking inflation into account.

Nevertheless, the United States' new medicine approval rate has not kept up with this explosive growth. It is clear that there is a substantial delay between R&D costs and marketing approval; the majority of costs at the end of the period are probably related to upcoming approvals. However, it is difficult to ignore the suspicion—and possibly the conclusion—that research and development expenses have increased over time. Several studies have used extensive disaggregated data to estimate the cost of new drug development in order to solve the difficulties associated with connecting industry-level R&D spending with new chemical entity (NCE) approvals (DiMasi 1995).

The percentage of medications researched that are finally approved for commercialization serves as a proxy for the technical risks associated with new drug development, and these hurdles are significant. It is believed that thousands of molecules are synthesized for every medicine that is commercialized. The fact that these dangers continue until the very end of clinical testing may not be well known. By analysing development timeframes, success rates, attrition rates, and expenditures associated with trial periods for each therapeutic class, DiMasi et al. provided information about the likelihood that the FDA will approve new chemical entities (NCEs) at various stages of clinical testing. For example, there is a 20% chance that an anti-infective NCE will be approved and put into clinical testing, compared to a 30% chance for neuropharmacological NCEs. Approximately one-third of the medications that advance to Phase III testing may not receive marketing approval.

Notably, there is a 50/50 chance that NCEs that advance to Phase III trials will be approved, making neuropharmacological development very dangerous throughout the clinical development process. The number $802 million was first mentioned in a paper that was published in the Journal of Health Economics in March 2003. This amount has attracted considerable interest. Figure 1.1 shows how much pharmaceutical companies typically have to spend to get a new medication to the market. This number, though, was criticized for being exaggerated or called a

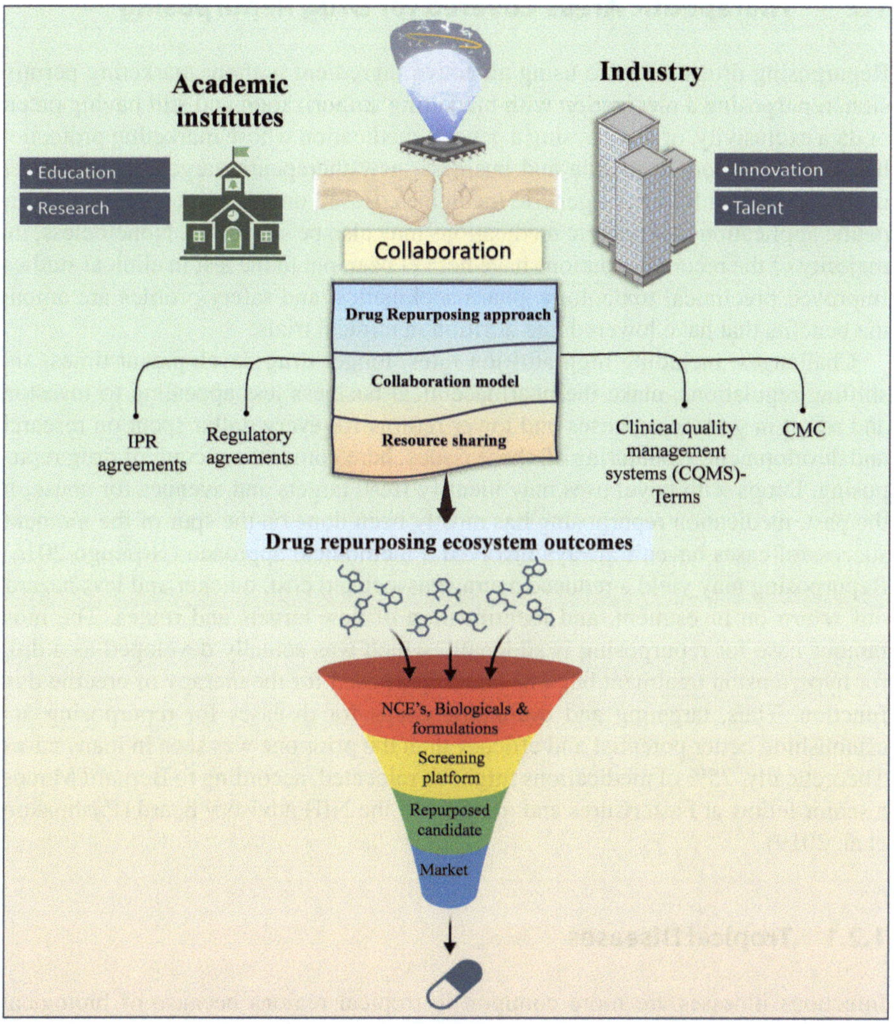

Fig. 1.1 Schematic representation of academia–industry collaboration for drug repurposing for pharmaceutical product development

"9-digit fairy tale". New estimates of the cost of developing new drugs have since surfaced, with health economists putting the current range between \$1.3 and \$1.7 billion. Experts agree that there is a noticeable increase in the expense of research and development, especially clinical trials, in the pharmaceutical business. Pharmaceutical firms frequently fund the majority of drug-related clinical trials, pointing to the costly product development process as the main cause of the high prices of their goods. This idea was emphasized in a 2004 GlaxoSmithKline (GSK) advertisement that said, "Today's medicines finance tomorrow's miracles" (Collier 2009).

1.2 Therapeutic Areas Covered for Drug Repurposing

Repurposing drugs refers to using an active ingredient without marketing permission, repurposing a medication with marketing authorization and still having patent or data exclusivity, or repurposing a generic medication whose marketing protection has lapsed. In both academia and business, new therapeutic targets are evaluated using a chemical library of generic medicines. Based on this screening, new therapeutic applications for generic medications may also be suggested. Nonetheless, the majority of the recommendations have not yet been put to the test in clinical studies. Improved preclinical toxicology, pharmacokinetics, and safety profiles are among the benefits that have lowered late attrition in clinical trials.

Challenges, including high attrition rates, longer drug development times, and shifting regulations, make the pharmaceutical business less appealing to investors and result in greater expenses and lower returns for every dollar spent on research and development. Countering all these issues, here comes the picture of drug repurposing. Drugs with novel uses may identify fresh targets and avenues for abuse. In the past, medication repurposing has mostly been done on the spur of the moment; successful cases haven't always involved a methodical approach (Nosengo 2016). Repurposing may yield a reduced average associated cost, quicker and less hazardous return on investment, and identification of new targets and routes. The most famous case for repurposing is sildenafil, which was actually developed as a drug for hypertension treatment but was later repurposed for the therapy of erectile dysfunction. Thus, targeting and evaluating drugs for diseases for repurposing and establishing better potential and efficacy than the prior one was seen in many cases. Theoretically, 75% of medications might be relocated, according to Bernard Munos, a senior fellow at FasterCures and member of the NIH advisory board (Pushpakom et al. 2019).

1.2.1 Tropical Diseases

Infectious illnesses are more common in tropical regions because of biological, environmental, and societal factors. Due to the variety of infections, hosts, and symptoms that these illnesses exhibit, especially febrile illnesses, early diagnosis and death rates are vital for medical professionals, healthcare facilities, and the general public. Neglected tropical diseases (NTDs) are communicable and affect the poorest one-sixth of the global population, known as the "bottom billion". Caused by vector-borne protozoa, bacteria, and helminths, they primarily affect tropical areas and are primarily caused by poverty. The World Health Organization has a dedicated department for NTD control (van der Pol et al. 2023; Feasey et al. 2010).

Repositioning of drugs according to the suspicion of potential upon incidence brought out more success in the confirmation of enhanced activity than that of the control drug of the prior treatment category. Tinidazole, a nitroimidazole compound, has been used to treat anaerobic protozoa like amoeba and bacteria. Later, it was tested for its efficacy in treating *Plasmodium vivax* infection, and one experiment showed that the medicine might treat the infection caused by *P. vivax* (Pradhan et al. 2022; Nzila et al. 2011). The Walter Reed Army Institute of Research is now conducting a Phase II study to determine the effectiveness of tinidazole for a radical cure, including against *P. vivax* hypnozoites and latent forms.

Next is the Chikungunya, and Tanzania was the place where the alphavirus known as the Chikungunya virus (CHIKV) was initially identified in 1952–1953. It bears Makonde's name and leaves victims with excruciating joint agony accompanied by a severe fever that lasts for 3–10 days, along with myalgia, rash, and nausea, and still, the actual symptoms are not clear. The early steps of the CHIKV replication cycle have been demonstrated to be inhibited by suramin, an FDA-approved medication for trypanosomiasis. Its direct interaction with viral particles was discovered by Albulescu et al. during their investigation of its mode of action. Another FDA-approved medication for candidiasis, micafungin, has shown a wide range of antagonistic effects against many alphaviruses, such as SFV, CHIKV, and SINV. While a time-of-addition assay revealed late-stage CHIKV infection inhibition, a molecular docking study suggested a conceivable interface with the CHIKV envelope glycoprotein.

Following the above, sofosbuvir, a drug for the treatment of hepatitis C, has demonstrated encouraging results against CHIKV in several cell lines, shielding adult Swiss mice from CHIKV-induced illness and raising the survival rates of neonatal mice (Battisti et al. 2021). Next NTD, in terms of disability-adjusted life years, visceral leishmaniasis (VL) ranks seventh in terms of disability and is the second most deadly tropical and subtropical illness. It is spread by the biting of infected sand flies and brought on by the protozoa *Leishmania infantum* and *L. donovani*. Miltefosine, pentamidine, and amphotericin are the drugs of choice for treatment, but several drugs screened under repurposing came out to be more effective than miltefosine, as evaluation of the IC50 values of lansoprazole and posaconazole revealed that they are just as effective as miltefosine and pentamidine in inhibiting *L. donovani*.

Posaconazole has a greater IC50 than typical anti-leishmanial medications and lansoprazole. However, it might be used in other therapeutic contexts due to its manageable toxicity. Likewise, thalidomide, which caused a major tragedy when used for treating morning sickness in pregnant women in 1962, has now been repurposed for treating tuberculosis, followed by several antibacterial drugs, such as linezolid and clofazimine, repurposed and even approved by the WHO for MDR-TB treatment. Furthermore, benzimidazole, when combined with miltefosine (a drug for treating leishmaniasis), provided better effectiveness in preventing intense parasitemia (Sharma et al. 2023; Porta et al. 2023).

1.2.2 Rare Diseases

Any heterogeneous disorder that affects a small fraction of the population (less than 200,000 persons in the USA) can be classified as a rare disease (RD). Around 30% of children with RD die before the age of five. RDs are frequently chronic and cause permanent impairment or premature mortality. Many RDs have a paediatric beginning (Bloom 2016). Somatic and germline gene mutations account for 70% of all RDs, which are genetic disorders. Among recessive disorders originating from genetics, a large number are monogenic in nature; they result from a single gene deficiency and are inherited according to a Mendelian pattern. Furthermore, RDs exhibit non-Mendelian inheritance, which includes epigenetic modification and mitochondrial diseases resulting from maternal transfer of mitochondrial DNA variations.

Alkaptonuria (AKU) is a rare disease that is brought on by a lack of homogentisic dioxygenase, which prevents the body from breaking down the harmful acid HGA. This ultimately causes bone loss and discolouration of the bone and cartilage. Nitisinone is being repurposed for the AKU. Currently, hereditary tyrosinemia type 1 is another uncommon condition for which nitisinone is permitted for treatment. Even though nitisinone is not authorized for use in AKU at this time, prior animal studies and tests have demonstrated that nitisinone could work well to treat it. US clinical research has demonstrated that nitisinone could cut HGA levels by up to 95%. There are presently two clinical studies investigating nitisinone in AKU.

It is taking place in Europe, with funding of more than $8 million for research obtained through government grants and crowdsourcing. Next, there are now several rare disorders for which eculizumab is indicated. Atypical haemolytic uraemic syndrome is an extremely uncommon and potentially fatal hereditary condition characterized by uncontrollably prolonged complement activation, which can cause severe end-organ damage and thrombotic microangiopathy. In adult patients with atypical haemolytic uraemic syndrome, an open-label single-arm Phase II trial demonstrated the advantages of eculizumab, reporting improvements in haematologic, renal, and quality-of-life parameters, as well as the ability to stop dialysis and protect organ transplant recipients (Ranganath et al. 2020).

1.2.3 Chronic Ailments

Chronic illnesses are often described as those that are chronic or long-lasting and need continuing medical care. Chronic illnesses include cancer, chronic obstructive pulmonary disease (COPD), inflammatory bowel disease (IBD), and congestive heart failure. IBD is characterized as a chronic disease with severe symptoms and gut inflammation. Thalidomide is a contentious medication due to the serious side effects of polyneuropathy. Medication is not included in the IBD treatment plan. However, thalidomide is used when individuals do not react to steroids or other proven IBD treatments. In addition, thalidomide was found to be beneficial in individuals with chronic illness in a small open-label clinical trial. Therefore, intensive research needs to be done to scrutinize the genetic information that coordinates the use of thalidomide in the mitigation of IBD (Bramuzzo et al. 2016).

Congestive heart failure (CHF), often known as cardiac failure, is a chronic medical illness in which the heart fails to properly pump blood to satisfy the demands of the body. This causes fluid to accumulate in numerous regions of the body, mainly in the lungs and legs. Preclinical research has revealed that inflammatory pathways and mediators, including tumour necrosis factor (TNF), IL-6, and IL-1, play an important part in the pathomechanism of disorders such as stroke, myocardial infarction, atherosclerosis, and cardiac failure. As a result, during the last decade, there has been a surge in interest in researching the repurposing of anti-inflammatory drugs for cardiovascular purposes. Despite promising preclinical results, the bulk of clinical trials have produced poor and contradictory results, as shown with TNF inhibitors and the IL-1 receptor antagonist anakinra. This shows that new techniques and a more extensive investigation of inflammatory pathways are required to properly treat heart failure (Onódi et al. 2023).

Chronic obstructive pulmonary disease, or COPD, is a set of disorders that produce airflow obstruction and breathing difficulties. Emphysema and chronic bronchitis are examples. Albuterol was initially marketed as Ventolin® by Allen and Hanbury in London, UK, in 1968. In 1981, it was brought to the US under the names Proventil® (Schering, Kenilworth, NJ) and Ventolin® (GlaxoSmithKline, Research Triangle Park, NC). As "rescue" medications, short-acting beta-agonists (SABAs) have proven to be incredibly helpful, providing rapid relief from acute bronchospasm. Beginning in 1990, advances in molecular structure manipulation resulted in the successful production of inhalational outcomes containing medicines with a duration of action of 12 hours, illustrated by salmeterol and formoterol. Following that, medications having a 24-hour duration of action, such as indacaterol and vilanterol in the twenty-first century, were launched. Long-acting agonists with action periods of 12 hours or more are now routinely used in the mitigation of COPD and asthma (Newman 2018).

1.2.4 Neurodegenerative Disorders

Neurodegenerative disorders are a group of conditions that cause progressive loss of neuronal function in the central nervous system and affect a large percentage of the population (Nieoullon 2011). People with neurodegenerative diseases have considerably shorter life expectancy and quality of life, imposing an enormous financial load on the healthcare system. The most common neurodegenerative illnesses are

Parkinson's disease (PD), amyotrophic lateral sclerosis (ALS), Huntington's disease (HD), multiple sclerosis (MS), and Alzheimer's disease (AD) (Wyss-Coray 2016). There are two approaches to drug repurposing: (1) Activity-based medication repurposing, which includes phenotypic and target-based screening methods. (2) Drug repurposing by computation. AD is a well-known neurodegenerative condition that accounts for 80% of dementia cases in the elderly. It causes gradual memory loss, reduced learning capacity, and a reduction in both behaviour and functionality (Kumar et al. 2015).

Bexarotene, a retinoid X receptor antagonist used to treat cutaneous T-cell lymphomas, has shown promise in reversing neurodegeneration, improving cognition, and lowering amyloid levels in mice carrying familial AD genes (Tousi 2015). PD, the second most common neurodegenerative ailment after AD, affects people all over the world. Sildenafil, a typical erectile dysfunction medication, acts as a phosphodiesterase-5 inhibitor that controls cGMP. This cGMP control has been linked to AD, where it has a dual function, adversely affecting learning, memory, and neuroplasticity, while perhaps having a favourable effect in alleviating memory deficits caused by amyloid. Sildenafil was found to be efficacious in suppressing neuroinflammation and lowering amyloid levels in aged mouse models (Zhang et al. 2013).

Amantadine was originally developed to treat influenza but was later repurposed to treat PD. It acts as a moderate glutamate receptor antagonist, increasing dopamine levels while inhibiting reuptake (Antony et al. 2013). Amantadine was initially developed to mitigate influenza, but it was later repurposed to treat Parkinson's disease (PD). It functions as a mild glutamate receptor antagonist, boosting dopaminergic levels while inhibiting reuptake (Lee and Kim 2016).

Methylphenidate, a central nervous system stimulant, works by inhibiting dopamine and noradrenaline absorption in the striatum and prefrontal cortex by blocking presynaptic dopamine and noradrenaline transporters. Originally used to treat attention deficit hyperactivity disorder, multiple studies show that it is also useful in treating gait problems and non-motor symptoms associated with PD (Devos et al. 2013).

In the industrialized world, HD is the most common monogenic neurological condition. It is an autosomal dominant disorder. Involuntary choreatic movements, behavioural and mental abnormalities, and dementia are among its symptoms. Clozapine, a neuroleptic medicine used to treat schizophrenia, has a high affinity for dopamine D1 and D4 receptors while having little antagonistic action for D2 dopaminergic receptors. Even though it has a lower risk of eliciting extrapyramidal side effects, which has led to its consideration as a possible symptomatic medication for chorea, clinical trials have shown mixed findings (Bonuccelli et al. 1994). Memantine, an adamantane derivative, is used in the mitigation of Alzheimer's disease (AD) as a non-competitive N-methyl-D-aspartate (NMDA) inhibitor. When the NMDA receptor is excessively stimulated, a large amount of calcium enters the cell, eventually leading to cell death. Memantine prevents brain cell death by inhibiting harmful calcium influx into neuronal cells. According to a study on its utility in Huntington's disease (HD), memantine can reduce the vulnerability of neurones to glutamate-induced excitotoxicity (Beister et al. 2004). Multiple sclerosis (MS) is a central nervous system autoimmune disease. It is characterized by persistent inflammation and, to varying degrees, causes the breakdown of myelin and axons. Several anticancer drugs have been repurposed to treat multiple sclerosis and its associated symptoms. Because of its immunosuppressive properties, mitoxantrone has been licensed for the mitigation of multiple sclerosis (MS).

It is linked to changes in the central nervous system T- and B-cell responses to antigens, a reduction in macrophage-mediated myelin degradation, and a reduction in axonal damage. Mitoxantrone inhibits T-cell activation, limits T- and B-cell proliferation, decreases antibody production, and inhibits macrophage activation. It also has a high tolerance for therapeutic applications (Hartung et al. 2002). Amiloride, a diuretic used for the management of hypertension and oedema associated with heart failure or liver illness, has been studied for its neuroprotective potential in MS. This medicine has the ability to suppress the neuronal proton-gated acid-sensing ion channel 1 (ASIC1), which is overexpressed in the axons and oligodendrocytes of people with MS. As a result, amiloride exhibits neuroprotective and myeloprotective effects in the context of MS. Furthermore, the beneficial element of amiloride's defensive function in the cascade of inflammation is notable, since it makes the treatment effective even when inflammation is just beginning (Arun et al. 2013).

Amyotrophic lateral sclerosis (ALS) is defined by the degeneration of both the upper and lower motor nerve cells, which regulate voluntary muscular movement. Masitinib, which was first used to treat canine cancer as a tyrosine kinase inhibitor, has been investigated for ALS due to its possible influence on aberrant glial cell proliferation in this illness. Masitinib has been demonstrated to decrease glial cell activation in relevant rat models, resulting in enhanced survival. Triumeq®, an antiretroviral drug used in HIV therapy, has been examined for possible applications in the treatment of amyotrophic lateral sclerosis (ALS). The finding that humans have endogenous retrovirus expression in the brains of patients with ALS, as well as the fact that patients with ALS have blood levels of reverse transcriptase, which is equivalent to HIV-infected individuals, is the basis for this investigation. Given these findings, there is a case to be made for anti-HIV medication being useful for ALS (Trias et al. 2016). In patients with ALS, Triumeq®, an antiretroviral amalgamation of dolutegravir (an integrase blocker), abacavir, and lamivudine, is safe and tolerated.

1.2.5 Oncology and Cancer

Cancer is a category of diseases characterized by uncontrolled cell division and proliferation in the body. These cells can infiltrate neighbouring tissues and spread to other areas of the body via the circulation and lymphatic system. There are several forms of cancer, each with its own set of characteristics and behaviours. Breast cancer, lung cancer, colorectal cancer, prostate cancer, and skin cancer are all common types (Trias et al. 2016).

Repurposing non-cancer drugs for precision oncology significantly broadens the range of actionable molecular targets, hence expanding the potential pool of patients who can benefit from precision cancer treatment. The precision oncology method provides individual patients with access to innovative treatment choices in cancer types with substantial genetic heterogeneity, where finding suitable repurposed medication candidates for conventional therapy is challenging.

Metformin, a biguanide chemical often given for the treatment of type II diabetes, is thought to work by inhibiting oxidative phosphorylation. This causes energetic stress and gluconeogenesis inhibition, notably in liver cells. Several epidemiological studies, as discussed in the context of cancer prevention, have discovered a relationship between metformin use and a decreased risk of cancer. Notably, metformin usage has been associated with a favourable response to therapy and higher survival rates in a wide range of malignancies, including hepatocellular carcinoma, colorectal cancer, prostate cancer, HER2+ breast cancer, ovarian, pancreatic, oesophageal, and rectal cancer (Larsson et al. 2007).

Gene and protein targets for biological pathway analysis may identify novel molecular targets (MOAs) as well as novel therapeutic uses for already approved medications. In this regard, alpelisib offers a useful example of innovative DR since it shows how examining a biological pathway can result in the therapy of illnesses not originally targeted by the research. Alpelisib, an inhibitor of PIK3CA (phosphatidylinositol-4,5-bisphosphate 3-kinase catalytic subunit alpha), was primarily created for the treatment of PIK3CA-mutated malignancies and then repurposed for the treatment of PIK3CA-related overgrowth spectrum (PROS). Individuals whose malignancies are hormone receptor (HR)-positive/HER2-negative and with PIK3CA mutations have poor prognoses and treatment resistance (Nagabukuro and Uemura 2020).

The US Food and Drug Administration (FDA) approved alpelisib, developed by Novartis, as Piqray® when used in conjunction with fulvestrant to treat advanced or metastatic breast cancer with a PIK3CA mutation. A specific antagonist of the oestrogen receptor. Piqray® is used with fulvestrant after hormone treatment alone has failed.

PIK3CA-related overgrowth spectrum (PROS), a group of overgrowth syndromes, also includes activating PIK3CA mutations. PROS and PIK3CA-related tumours have nearly identical PIK3CA mutational profiles and the same pathogenetic pathway. In a Phase III randomized double-blind, placebo-controlled study (SOLAR-1) with acceptable safety, alpelisib (BYL719) was administered to patients with metastatic breast cancer that was HR-positive, PIK3CA-mutated, and had no receptor for human epidermal growth factor receptor 2. Canaud's laboratory was given permission to give BYL719 to 17 more patients with PROS after making remarkable progress first in PROS mouse models and then in two patients with highly severe and potentially fatal clinical signs of PROS/CLOVES syndrome (Venot et al. 2018).

The 2018 study's publication supported PIK3CA inhibition as a potential treatment approach for patients with PROS. After 4 years, in 2022, the Food and Drug Administration (FDA) approved alpelisib—sold under the brand name

Vijoice®—for the treatment of severe signs of PROS in patients aged 2 years and older. Based on the real-world response rate and duration of response data from 57 patients in the single-arm EPIK-P1 study, this indication for Vijoice® has been authorized. Based on real-world response rate and duration of response data from 57 patients in the single-arm EPIK-P1study, a retrospective chart review study, this indication for Vijoice® has been authorized. Regulation-making in rare and ultra-rare diseases may benefit from the use of real-world data (RWD) and real-world evidence (RWE) (Fetro 2023).

Penfluridol (PF), a first-generation antipsychotic first recommended for schizophrenia, has anti-proliferative effects that are thought to be related to interference with integrin signalling. In vivo investigations have shown that it has anti-tumour properties that are beneficial in localized disorders as well as experimental models of breast cancer spread to the brain. Notably, in pancreatic xenograft tumours, penfluridol has been shown to inhibit tumour progression (Ranjan et al. 2016).

Lithium (LiCl), which was formerly used to treat depression, is now recognized to have anticancer action, owing mostly to the control of glycogen synthase kinase 3 (GSK3). GSK3 inhibits and phosphorylates glycogen synthase, which is an important participant in glycogen production and a negative regulator of WNT signalling. Lithium influences neurotransmitter release, changes the activity of several phosphoproteins, and directly suppresses inositol monophosphatase (IMPase), and it is required for G-protein coupled receptor signalling (Phiel and Klein 2001). Itraconazole, on the other hand, is an antifungal medicine that works by blocking 14-lanosterol demethylase (14LDM), an enzyme required for cholesterol production. Itraconazole has anti-angiogenic properties as well, since it inhibits endothelial cell proliferation and lowers VEGFR2 expression, as well as multiple downstream applications in endothelial cells. Itraconazole lowered the phosphorylation of growth factor receptors in human vascular endothelial cells (HUVECs), thereby reducing cell migration and tube formation. Another study found that itraconazole medication reduced angiogenesis and increased regression in non-small cell lung cancer xenografts (Nacev et al. 2011).

Doxycycline, a tetracyclic antibiotic recognized for its efficiency against a variety of infectious ailments, was shown to have anti-angiogenic characteristics in the early 1990s. Doxycycline also inhibits growth in osteosarcoma, prostate cancer, and mesothelioma cells, according to further research. Doxycycline has also been shown to have pro-apoptotic effects on pancreatic and leukaemic cells (Fife et al. 1997).

Several preclinical studies have focused on leflunomide, a frequently used immunomodulatory medication for treating rheumatic illnesses such as rheumatoid arthritis and psoriatic arthritis. Leflunomide suppresses growth, alters cell cycle control, and promotes apoptosis in a variety of cancer cells, including those linked with thyroid cancer, neuroblastoma, chronic lymphocytic leukaemia, and glioma (Alhefdhi et al. 2013).

1.3 Current Status of Molecules Repurposed Under Industry–Academia Collaborations

The concept of drug repurposing was ideated by the Food and Drug Administration (FDA) to minimize the unexpected adverse effects of new molecules during their drug discovery and development (Dalwadi et al. 2023). It also has the advantages of being less time-consuming and economical for discovering novel therapeutic interventions and drug development. The recent strategy of public-private collaboration brings institutes, government, funding agencies, and pharmaceutical industries together (Nagabukuro and Uemura 2020).

In this context, the collaboration of academia and industry favours better outcomes because the academic provides access to trial participants and methodology expertise, whereas the industry provides expertise and funding (Rasmussen et al. 2018). An example of a pharmaceutical company, Bayer, has introduced an international funding plan (Grants4Indication) for 2 years (Verbaanderd et al. 2021). Clinical trials are phases of drug development that explore the safety and efficacy of therapeutic interventions. Clinical trials are the decisive phase for several molecules to establish them as potent therapeutic molecules and to introduce them as a new medicinal product (Akhondzadeh 2016).

The drug molecules that enter clinical trials are not necessarily eliminated due to their therapeutic potency; they can be dropped due to their severe toxicity. In the case of drug repurposing, therapeutic efficacy is the segment to be explored, whereas toxicity and safety remain unchallenged (Chakraborty et al. 2021). So, various institutes are in the exploration of new indications of the existing drugs and also conducting clinical trials to establish new therapeutic indications. For example, the University of New Mexico has explored new indications of the existing drug and is further taking it for pre-clinical and clinical trials. Drugs like raltegravir, ketorolac, astemizole, phenothiazines, and beta-adrenergic receptor drugs are in various clinical and preclinical phases. If the result provides the potential repurpose of the drug, it may give the possible outcomes of collaboration between academia and industry (Buranda et al. 2011). Some academic institutes explored the new indication of the old drug, took patents for repurposing, and further collaborated or transferred the licensing to pharmaceutical industries for production and marketing. Also, some drug repurposing studies were conducted in collaboration with institutes and industries. Examples of these collaborations are listed in Table 1.1.

Industry has long collaborated with outside parties to gain access to specialized knowledge and technology. However, it now has to meet a wider range of criteria, including new targets, disease knowledge, signalling pathways, human cells and tissues, patients for clinical trials for target validation studies, platform technologies, new chemical approaches, disease and biomarker discovery, safety and efficacy investigations, and biotechnology refining. Conventional academic research supported by research councils and grant agencies keeps generating a stream of innovations and product concepts that can be licenced. Even though universities have had an alliance with the pharmaceutical industry for many years, they have not been recognized as significant stakeholders in the sector. Lack of funding for

Table 1.1 Summary of formulation, NCE, and biologics developed through academia–industry collaboration

Drug/formulation	Initial clinical use	Repurposed clinical use	Academic/ Institute	Industry/ Company	National/ International	Current status	Reference
Minoxidil/Rogaine®	Hypertension	Alopecia	University of Colorado School of Medicine	Johnson & Johnson	International	Patent expired	Goldfarb (2008); Zhou et al. (2023)
Propranolol/ Hemangiol®	Hypertension	Infantile haemangioma	Bordeaux University	Pierre Fabre Dermatologie Laboratories	International	Patent expired	Fetro (2023)
Alpelisib/*VIJOICE*®	Breast cancer	PIK3CA-related overgrowth spectrum (PROS)	University of Paris, France	Novartis	International	Patented and still in force	Canaud; Pagliazzi et al. (2021); Morin et al. (2022)
Amikacin/Arikayce®	Infections with Gram-negative and Gram-positive organisms	Non-tuberculous mycobacterial lung infection	Karolinska Institute	PARI Pharma GmbH	International	Patented and still in force	Weers et al. (2009); Eagle et al (2019); Forslöw et al. (2003)
Mercaptamine/ Crysatdrops	Cystinosis	Corneal cystine crystal deposits	Hacettepe University	Recordati Rare	International	Patent expired	Bozdağ et al. (2008); Dufier et al. (1987); National Institute for Health and Care Excellence (2019)
Zidovudine	Anticancer	Antiviral	National Cancer Institute (NIH) and Duke University	Burroughs Wellcome	International	Patent expired	Kesselheim et al. (2015)

commercially promising research hinders university-industry collaboration on less profitable technology.

But currently, this situation has changed, as both industrial and academic scientists are working together to bridge the translational gap and are more interested in providing effective treatment and care for patients. The recent COVID-19 pandemic highlighted the importance of academia–industry collaboration. India's indigenous vaccine, COVAXIN®, was commercially available in the market through the collaboration between Bharat Biotech and the Indian Council of Medical Research (ICMR) at the National Institute of Virology (NIV). This partnership reshaped the landscape, producing the vaccine in Bharat Biotech's advanced BSL-3 (Bio-Safety Level 3) facility, showcasing the fusion of academic and industry capabilities. Similarly, the Oxford-AstraZeneca vaccine was developed in the United Kingdom through the collaboration between Oxford University and AstraZeneca (a British-Swedish company).

The research prowess of Oxford University, combined with the marketing capabilities of AstraZeneca, led to the successful commercialization of the vaccine. This vaccine is sold under the brand names Covishield® and Vaxzevria® However, industry and academic collaborations have proved fruitful, even during the COVID-19 pandemic, by addressing the global health crisis.

A well-known professor at Princeton University, Edward C. Taylor, who was an advisor to Eli Lilly, was able to collaborate with Eli Lilly scientists and indulge in drug discovery. He worked on different folic acid derivatives called antifolates that can hinder cell growth by impeding the synthesis of DNA and RNA. They achieve this by targeting crucial enzymes like thymidylate synthase, dihydrofolate reductase (DHFR), and glycinamide ribonucleotide formyltransferase (GARFT). As a result of his collaboration with Eli Lilly Company, Alimta (pemetrexed), a new drug candidate, was discovered in 1989. The US FDA granted approval for the use of Alimta in 2004 to treat malignant pleural mesothelioma (a form of lung tumour) in combination with Platinol (cisplatin). Additionally, Alimta gained approval as a second-line treatment for non-small cell lung cancer (Gray 2006).

Another accomplished collaboration between researchers at Abbott Labs, the University of Wisconsin, and Johns Hopkins University led to the identification of a novel application for selective antagonists of endothelin receptor A (ET_A). Originally, there was considerable pharmaceutical interest in developing pharmacological methods to inhibit endothelin-1, a potent vasoconstrictive peptide hormone, for various cardiovascular applications. Efforts have led to the discovery of selective and potent ET_A antagonists with suitable pharmacological properties for in vivo studies. Despite promising preclinical data, ET_A antagonists were overshadowed by successful ACE inhibitors. Rather than storing the drug leads, Abbott Labs chose to provide access to the drugs for academic researchers, who in turn found that these compounds demonstrated an inhibitory effect on the proliferation of cultured prostate cancer cells. Through these academic collaborations, Abbott has successfully progressed to one of the drugs, Xinlay (astrasentan), for hormone-refractory prostate cancer that arrived into Phase 3 trials (Gray 2006).

Apart from discovering new applications for existing drugs, academic researchers have achieved success in creating innovative compounds to combat targets that pharmaceutical companies have seemingly already explored, such as human immunodeficiency virus (HIV) protease inhibitors. The nonpeptide protease inhibitor Prezista (darunavir), with enhanced efficacy, takes down a range of mutant foes and was identified by Professor Arun Ghosh, a medicinal chemistry expert at Purdue University. In partnership with Tibotec, a pharmaceutical and development company, accelerated FDA approval for a novel drug, Prezista, was granted in June 2006. Specifically, for adults who haven't responded well to earlier HIV drugs, Prezista and Norvir join forces to fight the virus (Kovalevsky et al. 2006).

Another promising avenue for collaborative partnerships lies in the realm of "large-scale" scientific endeavours, particularly in the fields of sequencing and high-throughput screening involving complementary DNA, small interfering RNA (siRNA), and small-molecule libraries. A noteworthy example is the RNAi Consortium, a collaborative initiative involving researchers from prestigious institutions such as the Broad Institute, Harvard Medical School, Massachusetts Institute of Technology, Dana Farber Cancer Institute, and Whitehead Institute for Biomedical Research, along with industry players like Novartis, Eli Lilly, Bristol-Myers Squibb, Sigma-Aldrich, and Academia Sinica Research Institute in Taiwan. A consortium was formed with the aim of creating short hairpin lentiviral RNA interference (shRNA) libraries designed to target 15,000 genes in mice and 15,000 genes in humans. The consortium recently released its inaugural publication utilizing RNA interference (RNAi) strategies to identify novel human cell cycle-regulated genes (Moffat et al. 2006).

1.4 The Indian Scenario of the Pharmaceutical Industry and Research

The pharmaceutical sector in India has played a crucial role in improving healthcare outcomes not only within the country but also in various developing and developed nations. Indian pharmaceutical firms have leveraged their expertise in synthetic organic chemistry and reverse engineering to effectively create numerous affordable generic medications. In 2017, Indian pharmaceutical companies secured 304 approvals from the USFDA for Abbreviated New Drug Applications (ANDA). In 2017, the USFDA approved 46 new drugs, including 11 novel biologics. In 2018, 42 new drugs were approved, out of which 17 were classified as biologics. While this presents a significant growth opportunity for the pharmaceutical industry, there are specific and unique entry barriers. These encompass evolving regulatory demands and development expectations, combined with ongoing patent disputes. This results in research and development efforts that are resource-intensive when compared to generic research.

The Indian pharmaceutical industry stands as one of the leading global producers, meeting more than 50% of the worldwide demand for various vaccines, 40% of the demand for generics in the USA, and contributing to 25% of all medicines in the

UK. Currently, India hosts a staggering number of manufacturing units, exceeding 11,000, and is home to over 3000 pharmaceutical companies (Festa et al. 2022). With the reintroduction of product patents in 2005, small and mid-level Indian corporations shifted from generic drugs to drug discovery research, technology licensing, and contract manufacturing. Prominent players in the Indian pharmaceutical sector comprise Lupin, Biocon, Sun Pharma, Dr. Reddy's Lab, Torrent, Ipca, Cipla, Zydus, and Glenmark. These companies have established a strong presence in both Indian and international markets, achieving growth through both organic expansion and strategic acquisitions.

The primary reasons behind these acquisitions include the desire to expand market share, respond to the rise in the Indian population, address the increase in chronic diseases, extend product portfolios and brands, the absence of substantial research and development, and the gradual expiration of patented products in the global market. In 2017, Aurobindo Pharma acquired the Portuguese company Generis Farmaceutica for $142.5 million, and Lupin acquired Symbiomix Therapeutics for $150 million (Chakraborty 2020).

There is limited emphasis on research and development (R&D) in the pharmaceutical sector, primarily centred around generics. This is due to insufficient R&D infrastructure in India and a weaker industry–academia connection for research. The Drug Price Control Order (DPCO) sets forth diverse pricing parameters that determine the pricing structure. This policy diminishes the profitability of companies investing in innovative drugs, thus requiring significant capital. The absence of product patents has impeded this sector, dissuading foreign companies from introducing new drugs into the Indian market, thereby hindering innovation and drug discovery (Mahajan et al. 2018). Nevertheless, the government's investment in healthcare is far from satisfactory, and Indian companies must explore innovative approaches to reach extensive rural areas.

1.5 Identifying Academic Research with Potential for Drug Development

The majority of fundamental scientific research leading to the development of groundbreaking medicines takes place within academic laboratories and research institutes. Over the past 15 years, numerous academic centres have expanded their participation in areas of drug discovery that were traditionally exclusive to the pharmaceutical industry. These laboratories and institutes aim to: (i) devise biochemical, cell, and other in vitro assays for screening small molecules; (ii) generate libraries of peptides, fragments, small molecules, and other compounds to explore hits, leads, and chemical scaffolds; (iii) establish moderate- to high-throughput screening platforms for drug discovery; (iv) comprehend Structure-Activity Relationships (SARs) by examining analogues obtained from chemical catalogues and through synthetic chemistry; (v) conduct studies on drug metabolism, pharmacokinetics, drug delivery, formulation, and toxicology, both in vitro and in vivo, to pinpoint experimental compounds with favourable drug-like properties; (vi) create animal models of disease for screening pharmacologic and pharmacodynamic effects in vivo; and (vii) discover and assess clinical candidates for Phase I and II trials (Silber 2010).

Numerous academic laboratories are presently engaged in uncovering genes and striving to establish connections between genes, proteins, and various diseases. Additionally, some of these laboratories are actively participating in the exploration and application of chemical biology methods for drug discovery initiatives. The University of Wisconsin is creating novel structures using beta-peptides endowed with antimicrobial characteristics. Similarly, the University of California, San Francisco (UCSF), is engaged in the exploration of gene discovery, disease biology, chemical biology, and the identification of targets for numerous diseases. Academic institutes have assets of skilled and inexpensive manpower. Ultimately, academic investigators have the potential to significantly contribute to justifying the development of new therapeutic agents, even when the intellectual property rights for the drug are owned by external entities (Shamas-Din and Schimmer 2015).

1.6 Recommendations from a Recent Pharmacology and Translational Research Round Table

1.6.1 Optimizing Access to and Use of Databases

In recent years, tremendous growth in technological advancements across various domains of biomedical research has resulted in a substantial surge in the accessibility of data, thereby amplifying data dimensionality and interconnectivity. The availability of omics data led to the validation of novel data categories, which in turn created new frameworks for translational research and personalized therapy. The vast amount of data that must be assembled, sorted, and assessed using appropriate tools, artificial intelligence, and informatics is now essential to help humans with these activities (De Maria Marchiano et al. 2021).

Systems for storing electronic health records (EHRs) have quickly emerged as essential tools, and research shows that physicians devote 2 hours of their time on back-office EHR tasks for every hour of patient care (Sinsky et al. 2016). Apart from conventional data sources, there is a growing fascination with data obtained from real-world scenarios, also referred to as "real-world data" (RWD), which is revolutionizing approaches to data analysis and interpretation. A viable substitute for the conventional evidence-based medicine (EBM) methods, which rely on randomized clinical trials to evaluate a specific hypothesis, is RWD-based translational applications. It is crucial to recognize that the RWD approach is not to be seen as an alternative to conventional EBM, but rather as a unique approach with different dimensions, collection methods, and interpretations of data.

The broad adoption of sensor technologies and omics data analysis, combined with the ability to gather massive amounts of clinical, behavioural, and environmental data, is driving the digitization of health and disease monitoring. To improve the understanding, interpretation, and use of omics data, we need to better characterise tissue and systemic dysfunction at the molecular level. This study explores several different fields, including radiomics, microbiomics, transcriptomics, proteomics, interactomics, metabolomics, genomics, and epigenomics, each of which evaluates unique environmental and biological aspects. Among the various omics platforms, those dedicated to nucleic acid analysis are the most advanced and cost-effective, boasting well-established validation practices.

Sequencing of the human genome has tremendously advanced the understanding of biological processes, although much of the acquired knowledge still requires clarification, particularly regarding the functions of biomolecule classes, particularly proteins. Health systems are using biobanks containing human germline DNA samples to produce genomic data that are connected from Electronic Health Records (EHRs) to clinical data. For genome research, translation, and medical applications, these biobanks are an invaluable resource. Hereditary-linked biobanks, which are well-equipped with extensive and long-term clinical data, are essential for examining how illnesses naturally develop. This, in turn, facilitates the adoption of individualized strategies for early detection, prevention, and disease management. National biobanks are starting to appear in several nations, including the UK (Collins 2012), Japan (Nagai et al. 2017), and China (Chen et al. 2011), marking a significant global trend towards leveraging genomic data for advancing medical research and healthcare practices (Stark et al. 2019).

An expansion of the existing Web, the Semantic Web (Sharma et al. 2023; Porta et al. 2023), allows automated processes to facilitate navigation and valuable use of digital information. This extension relies on standardized formats that facilitate data consolidation and integration from various sources. In the existing web landscape, links on web pages lack characterization. The mRNA on the Entrez page of the human TP53 gene is directly related to TP53, but no explicit information is provided to a computer to explain this relationship. On the other hand, the relationship between a gene and its transcribed mRNA product is made evident in the context of the Semantic Web by a statement that describes both entities and their nature.

Statements of this type are referred to as "triples" since they consist of three components: the subject, predicate, and object. For example, the human TP53 gene might be considered the subject. GeneProduct is the predicate, and the object is human TP53 mRNA. Similar to how Uniform Resource Identifiers (URIs) identify the subject and object, the relationship also possesses a URI. This site may provide a definition of the connection that is legible by humans when viewed via a web browser. Given that URIs have the capability to represent names, all of the Web's current content can be included in statements that are part of the Semantic Web framework. When two statements have the same URI, it means that they are discussing the same subjects. This alignment allows for the consolidation of data references, forming the foundation for integrating knowledge and data on the Semantic Web.

1.6.2 Anticipating Preclinical and Clinical Requirements

Despite the identification of numerous possible new drug targets by molecular biology methodology and the Human Genome Project, the rates of success of medications during clinical testing continue to be low. The primary cause of these failures is the insufficient efficacy observed during Phases II and III of the clinical trials. Implementing stricter success criteria in the non-clinical phases of drug development has the potential to decrease the risk of failures. This is particularly true for a given development candidate throughout the preclinical proof-of-concept and target validation phases. The use of animal models that enable evaluation of the target validity and prediction of the clinical effectiveness of a particular chemical is crucial for success during these phases.

This model is a simplified depiction of a complex system. Animal models play a critical role in enhancing our comprehension of medicine and biology, including the discovery and development of novel medications when they are well-designed and carried out. Safety assessments typically involve healthy animals, who may be less susceptible to any negative effects than ill individuals. The disease in humans and the animal model's biology and symptoms are similar. While evaluating face validity, target validity and the mechanism of action are crucial for the validation of a disease model. These standards are usually applied to offer a model's general validation. However, as animal models can be employed for a wide range of goals (such as investigating pathomechanisms, comparing drugs to the standard of care, or demonstrating the viability of a novel target), the validation offered for a particular model must be appropriate for its intended use.

A single, even extremely sophisticated, model may not always accurately represent the actual scenario; rather, a group of models may eventually do so. The model validity score method introduced by Sams-Dodd aims to define an ideal combination of models (Sams-Dodd 2006). Five distinct factors, such as species, complexity, disease simulation, predictivity, and face validity, are used in this scoring system. This scoring system can assist in building up a model combination or screening cascade that has the highest possible validity overall.

Humanization Models

Humanized mouse models, which allow for the engrafting of human cells or tissues into immunodeficient mice, are thought to be very helpful for functional research studies in vivo, which in turn supports clinical translation. Various humanized models and mouse strains are used, depending on the human ailment and the subject being studied (Ito et al. 2012). The most widely used models for studying cancer are the human cancer xenograft models and humanized mouse models, which imitate the human immune system. A highly informative preclinical resource for analysing pharmacological action, including new medication combinations and identifying predictive biomarkers, is well-characterized patient-derived xenograft (PDX) models (Tentler et al. 2012). PDX models are useful for predicting the effectiveness of cytotoxic therapies in humans when these chemotherapeutic medications are evaluated in mice at pharmacokinetically and clinically equivalent drug dosages (Fiebig et al. 2004).

Tumour graft models created for melanoma, ovarian cancer, adenocarcinoma, and sarcoma showed a good association with clinical success when the response of the tumour graft was utilized to direct patient treatment, serving as proof of principle for the latter. The creation of a suitable humanized mouse model to support preclinical translational research on the in vivo human immune system has advanced significantly within the past 10 years. Numerous strains have been developed since the initial immunodeficient strains (SCID and nude mice) were discovered (Ito et al. 2012). In clinical trials of various chronic diseases, quality of life is a crucial outcome, mostly to facilitate reimbursement. The most common approach used to assess this clinical evaluation is questionnaires, which are mainly inappropriate for use with experimental animals. The predicted human PK characteristics include clearance, bioavailability, and plasma concentration-time profiles.

1.6.3 Improving the Regulatory Procedures for Market Access

Regulators and medical innovators in academia possess some knowledge about introducing disruptive medical innovations into the market, but new strategies are required to address some of the difficulties associated with modifying the current legal framework to accommodate the modern medications being created. Regarding this, the collaboration known as the Consolidation Training of Academia in Regulatory Science (STARS) seeks to gain a deeper comprehension of the regulatory obstacles impeding the translational drug development of innovative products or novel therapeutic paradigms that have been launched in academic settings. Therefore, improving communication among various stakeholders in a multifaceted manner is one goal. Regulatory agencies play a crucial role in serving as permitting and guiding bodies at the NCAs or European (EMA) levels. They also produce as well as disseminate guidelines for establishing effective drug development programmes by utilizing particular pathways to connect with key academic stakeholders, which are hospitals and university medical centres.

Different research institutes and individual scientists involved in drug research and development, and governmental as well as corporate financing agencies that facilitate these studies, are additional important parties. Assuring a superior clinical progress dossier and, consequently, the regulatory acceptance of both non-clinical and clinical study designs when stakeholders demand marketing authorization is the goal of this conversation with regulatory bodies. As a result, funding for drug development projects should produce additional scientific and societal benefits, increasing the likelihood that the approved research findings will be used in clinical applications. Significantly, there is two-way contact between recipients and regulatory authorities. For instance, creative research and conclusions may encourage changes to the regulations (Starokozhko et al. 2021).

Reforming domestic regulations can be made easier by opening up international markets. Trade agreements ratified at the regional or global level can advance broad regulatory reform ideas or particular components as well as serve as a catalyst for national regulatory reform initiatives. A wide range of instruments, including binding and non-binding principles, technical aid, and capacity building, are used in regional and bilateral trade and economic initiatives to support regulatory reforms. Important problems are ahead of the incorporation of regulatory issues into international accords, including how to secure suitable domestic capacities for regulatory reform and avoid interfering with national regulatory sovereignty.

1.6.4 Creating a Public-Private Partnership Support Structure on a European Scale

The partnership between academic institutions and pharmaceutical corporations is essential for this newly integrated paradigm of innovation (Kaitin 2012). Assisting reasons include the need for many academic research centres (ARCs) to identify new sources of income due to the lack of NIH funding and the requirement for several major pharmaceutical firms to significantly reduce their R&D expenditures. The United States and the European Union (EU) have created initiatives to support translational science over the last 10 years (Milne and Kaitin 2009).

The US NIH, for instance, published its NIH roadmap in 2001 with the goal of re-engineering the clinical research sector, supporting future research teams, and investing in new avenues for drug discovery. The Clinical and Translational Science Award programme was introduced in 2006 with the goal of promoting the exchange of knowledge between clinical medicine and basic science. At present, the NIH has awarded about 60 prizes in the field of clinical and translational science, making it abundantly evident that it supports academic institutions as active partners in bioinnovation (Kaitin 2012).

Similarly, the FDA launched the Critical Path Initiative in 2004 to enhance the translation of fundamental research discoveries into safe and efficacious medications. The EU Innovative Medicines Initiative, formed through collaboration between the European Community and the European Federation of Pharmaceutical Industries and Associations in 2007, shares goals with the Critical Path Initiative. Its objective is to enhance precompetitive research by combining the abilities of industry, academia, and the government to discover novel biomarkers and other methods that enhance drug candidate selection and increase pipeline success rates. Successful partnerships face substantial challenges despite the unwavering backing of the government and shared commitment from academia and industry.

These challenges include language barriers (industry speaks business, whereas academics speak science); misaligned reward systems (academics receive grants and promotions for research and publications, while industry employees receive bonuses, promotions, and company goals for meeting regulatory filings and pipeline success); intellectual property challenges (while industry requires enough rights to justify downstream investment, academics strive to maintain as much ownership as feasible); and increased awareness of academic conflicts of interest and resistance to integrate more closely with the private sector (Kaitin 2012). However, there are numerous reasons to be enthusiastic about the changes brought by collaborations between academia and business. Specifically, academics get the opportunity to acquire knowledge about drug development and a higher possibility that their research discoveries may eventually lead to novel treatments and medicines, while businesses obtain access to state-of-the-art science and innovative technology. Furthermore, collaborations between academia and business are essential to achieving translational science's ultimate goals (Lenselink et al. 2017).

1.7 Role of AI–ML in Drug Repurposing for Fast Approval

A drug should go through five stages of refinement before entering the market. Occasionally, the drug may fail in the last phase; hence, the time and cost invested are wasted. The average medication design process takes 10–15 years, and the procedure costs anything from 897 million to 1.9 billion dollars. As a result, the cost of these drugs is high. Drug repurposing has recently evolved to use artificial intelligence (AI) to address these problems. Programmes are made to deal with and organize problems. Drug repurposing only requires four steps, as opposed to the five phases of drug discovery, where the first step is disregarded and is thought to be the most time-consuming in the "discovery and development of a medicinal compound". Through medication repurposing, AI leveraged personified knowledge and obtained clarification to solve the problems brought on by a variety of disorders, including COVID-19, malignancies, and rheumatoid arthritis (Kim et al. 2019; Mali and Pratap 2021).

Therapeutic repurposing has several advantages over de novo synthesis, which involves the creation and manufacture of novel therapeutic compounds. Repurposing, on the other hand, results in drug manufacturing in a considerably shorter time than de novo synthesis, has a significantly lower rate of failure of drug molecules, and brings them to the clinical phase because its contents contain chemicals that have already been studied. Furthermore, drug repurposing allows for the elimination of preclinical testing steps in humans and other animal models, thereby lowering the overall cost of drug development (Das et al. 2008).

If safety testing has been done, it will show dose compatibility for the new disease. The repurposing of the drug was perceived by coincidence through random experimentation and study. For example, after it was discovered in the market that sildenafil citrate was a hypertensive drug, this medicine was repurposed by Pfizer, leading to the creation of Viagra, which contained sildenafil citrate, which was effective in erectile dysfunction treatment, clinically proven by direct evidence from its formulation approach, and achieved massive sales (Kumar et al. 2018).

Artificial intelligence (AI) is a significant advancement in technology. Artificial intelligence is the term for technology that uses various computing techniques to process text, images, and sound, and is predicted to think similarly to a human brain. A bountiful revolution in artificial intelligence has resulted from the recent expansion of extraordinary data about biomedicals like proteomics, genomics, drug libraries, metabolomics, and data of patient care; the advent of supercomputers; and the ongoing creation of novel algorithms. AI has been incorporated into nearly every scientific discipline since it is accurate, dependable, and easily accessible, especially when it comes to repurposing drugs (Selvaraj et al. 2021).

Repositioning a medication presents a few obstacles for AI. When a medication that is currently being used to treat one disease is repositioned to treat another, it can do so quickly and effectively if the drugs have comparable structures, side chains, or side effects. The FDA has provided a detailed list of medications' negative effects (Setoain et al. 2015).

Despite the promise, current machine learning (ML) systems fail to comprehend the data's heterogeneous, semi-structured feature sets. As a result, we argue that ML can augment—but not replace—the viewpoints of the experts in the drug repurposing domain. Human intellect and AI are becoming increasingly complementary and integrated. We believe that in the current state, machine learning (ML) is widely used to advise human "go or no-go" decision-making in drug repurposing activities. Relying on stringent rule-based criteria, as is often demanded by ML, by ignoring the fact that key information remains undefined properly to even apply a rule (Challa et al. 2021).

Phenome-wide association study (PheWAS) is an ML approach for generating hypotheses regarding novel medicinal uses for an already approved medicine by data mining from a public dataset. PheWAS, a logistic classification approach, is used to formalise the connections, which yield p-values and odds ratios to quantify the strength of each projected mutation-disease pair. Drug repurposing studies primarily use bipartite graph learning (BGL) and deep neural network (DNN), CNN, graph convolutional network (GCN), FNN, similarity measures (SM), support vector machine (SVM), long short-term memory (LSTM), and RNN models or algorithms (Aliper et al. 2016). The "knowledge graph" of the UK-based AI startup BenevolentAI is a sizable collection of organized medical data. They discovered the medication baricitinib, which has the therapeutic potential to treat COVID-19 and can prevent the virus from entering host lung cells. In addition to overcoming the limitations of experimentally mapped DTI-based methodologies, accurate drug target identification can be used to repurpose existing therapeutic compounds that have already received approval (Zhavoronkov et al. 2020).

DTI mappings can be used in a variety of silico-based techniques to repurpose medicinal molecules. Mei et al., for example, suggested a multiple learning approach for drug repurposing that depends on previously identified pharmacological targets. Every drug molecule is regarded as a class label throughout this procedure, and the target protein serves as a training dataset for the I2 logistic regression model (Weinstein et al. 1994).

For one drug molecule, stratified multi-labelled cross-validation yields about 89% correct validation; the suggested framework demonstrated 84.5% accurate DTI predictions, in contrast to the other drug molecules. The data demonstrate that the suggested framework is typically compatible with large-frame medications and just needs the chemical structure and target identification information. DTI and drug repositioning are used in the recently developed iDrug technique to create a new cohesive model known as cross-network embedding. It offers a general framework for transferring knowledge between different drug-target interactions, and by doing so, it predicts the accuracy of drug-disease linkages as well as DTI relationships. Since iDrug's functionality has been evaluated on numerous datasets with information from a variety of disease categories, it can now be used for drug repurposing due to many reasons (Koromina et al. 2019).

Between 2000 and 2021, a total of 245 orphan drugs were approved in Europe for the rare diseases they treat. They covered a total of 148 rare diseases. About 20% of orphan drugs are repurposed drugs (Fetro 2023).

In the subject of repurposing, three important participants can be identified: academia and research institutions, pharmaceutical agencies, and companies that are based on repurposing technology. They all share a similar strategy for the development of a drug by assessing new indications for already-marketed medications. Academia is based on two approaches: generating a molecule and an indication, and concentrating on technologies like HTS or in silico. It is not bound for commercial or economic achievement, but it is more reliant on scientific innovations that recruit talent, collaborations with profit-making organizations, and government assistance. Conversely, repurposed technology firms are constrained by the company model they choose, which can change depending on the company's capabilities and goals.

A hybrid model is frequently utilized, which should improve both corporate performance and the development consequences of repurposing. Repurposed technology-based businesses frequently use cutting-edge technologies, experience, targeted research, and flexibility, but they might not have the resources required to conduct clinical and preclinical research on a medicine in a regulatory-focused setting efficiently and promptly. On the contrary, larger pharmaceutical organizations might be more concentrated on the life cycle management of a particular medication or chemical moiety, which is typically part of the post-marketing or late-stage development of a product. These factors make effective collaborations between larger pharmaceutical corporations and smaller repurposing technology businesses an appealing way to fuse superior repurposing skills with extensive knowledge regarding drug development. A small biotech company called Adamas Pharmaceuticals created a novel drug combination for the recently approved memantine. This drug combination was combined with donepezil to treat Alzheimer's disease (Cha et al. 2018).

1.8 Case Study of Successful Drug Repurposing

Sharma et al. 2023 recently reported a strategy that combines several datasets and methodologies. They separately developed disease and drug pair similarity ratings in genome data and in laboratory test data extracted from EMRs based on the fundamental idea that comparable pharmaceuticals can be used to treat similar ailments.

Given the parallels between ALS and Kawasaki syndrome, as well as between terbutaline sulfate and ursodeoxycholic acid, a β2-adrenoceptor agonist commonly used to treat asthma, it was suggested as a possible treatment for amyotrophic lateral sclerosis (ALS). The avoidance of axon abnormalities and neuromuscular junction degradation in a zebrafish model of ALS was subsequently used to establish and validate the potential therapeutic benefit of terbutaline sulfate for ALS.

One example of Parkinson's disease (PD) repurposing is Type II (diabetes mellitus), which is a potential risk for patients with PD. The UK Clinical Practice Research Datalink (CPRD) and Norwegian Prescription Database (NorPD) autonomously assessed the risk of PD in patients with diabetes. However, there was no significant effect of metformin or sulfonylureas on PD. But treatment with exenatide, which is a GLP-1 agonist, displayed a significant decrease in the progression of PD. The neuroprotective action of the GLP-1 receptor in a dose-dependent manner is proven by preclinical studies of exenatide. Clinical studies (Phases I and II) showed improvement in motor scores when treated with exenatide. Phase III trials are also moving on for the evaluation of the neuroprotective action of this disease.

1.9 Conclusions and Future Perspectives

The collaborative approach to developing a drug molecule—whether already approved for a therapeutic indication or at an investigational stage—has demonstrated success on many occasions. This opportunity represents advantages such as cheaper, faster, and safer drug development for newer treatments. Orphan diseases, rare diseases, cancer, or any other life-threatening disease can be targeted. Industry and academia may be different in some aspects, such as objectives, culture, and constants, although they build on each other's capabilities to promote research and development of repurposed medications for patients. This approach has already proven successful in several instances; however, its full potential still needs to be explored. The successful story of sildenafil has been widely publicized, but lacks full reflection. Collaborative success fully depends on proof of concept, robustness, intellectual property rights of the candidate drug, regulatory strategies, and commercialization. The potential vested in academic research institutions is a vital component of drug discovery, and a lot of educational projects are being explored with a wide variety of expertise, but a lack of industry standards and project-specific soft corners may lead to challenges which can be easily overcome by proper planning at the industry–academia level, and there is a need for a sustainable drug repurposing business model to address patient-specific queries. However, there are challenges and misconceptions like regulatory procedures, patent rights, royalty, intellectual property rights, and market access, and without thorough research, there are no innovations, even without incentives associated with drug repurposing under industry–academia collaborations. A few well-known success stories serve as excellent examples of academia–industry drug repurposing cooperation that have been effective; yet, they may only represent the tip of the iceberg. There are also difficulties in sour and private partnerships.

Acknowledgements The authors acknowledge support from the Department of Pharmaceuticals, Ministry of Chemicals and Fertilizers, Government of India. The NIPER-R communication number is NIPER-R/Communication.

Funding Statement The authors declare that no funds, grants, or other support were received during the preparation of this manuscript.

Data Availability Statement The reviewed article-related raw data files/figures/tables are available from the corresponding author.

Ethics Approval and Consent to Participate No animal or human studies were carried out for this article by the authors.

Competing Interests The authors do not declare any conflicts of interest among themselves.

Consent for Publication All authors who have been involved in completing this manuscript have given approval to the final version of the manuscript for publication.

References

Akhondzadeh S (2016) The importance of clinical trials in drug development. Avicenna J Med Biotechnol 8(4):151

Alhefdhi A, Burke JF, Redlich A, Kunnimalaiyaan M, Chen H (2013) Leflunomide suppresses growth in human medullary thyroid cancer cells. J Surg Res 185(1):212–216

Aliper A, Plis S, Artemov A, Ulloa A, Mamoshina P, Zhavoronkov A (2016) Deep learning applications for predicting pharmacological properties of drugs and drug repurposing using transcriptomic data. Mol Pharm 13(7):2524–2530

Antony PMA, Diederich NJ, Krüger R, Balling R (2013) The hallmarks of Parkinson's disease. FEBS J 280(23):5981–5993

Arun T, Tomassini V, Sbardella E, de Ruiter MB, Matthews L, Leite MI et al (2013) Targeting ASIC1 in primary progressive multiple sclerosis: evidence of neuroprotection with amiloride. Brain 136(1):106–115

Barrett JS (2008) The role of quantitative pharmacology in an academic translational research environment. AAPS J 10(1):9–14

Battisti V, Urban E, Langer T (2021) Antivirals against the Chikungunya virus. Viruses 13(7):1307

Beister A, Kraus P, Kuhn W, Dose M, Weindl A, Gerlach M (2004) The N-methyl-D-aspartate antagonist memantine retards progression of Huntington's disease. J Neural Transm Suppl 68:117–122

Bloom BE (2016) Recent successes and future predictions on drug repurposing for rare diseases. Expert Opin Orphan Drugs 4(1):1–4

Bloom B (2022) Collaboration models for repurposing. In: Drug repurposing. The Royal Society of Chemistry, pp 74–100

Bonuccelli U, Ceravolo R, Maremmani C, Nuti A, Rossi G, Muratorio A (1994) Clozapine in Huntington's chorea. Neurology 44(5):821–821

Bozdağ S, Gümüş K, Gümüş Ö, Ünlü N (2008) Formulation and in vitro evaluation of cysteamine hydrochloride viscous solutions for the treatment of corneal cystinosis. Eur J Pharm Biopharm 70(1):260–269

Bramuzzo M, Ventura A, Martelossi S, Lazzerini M (2016) Thalidomide for inflammatory bowel disease. Medicine 95(30):e4239

Bubela T, FitzGerald GA, Gold ER (2012) Recalibrating intellectual property rights to enhance translational research collaborations. Sci Transl Med 4(122):122cm3

Canad, Guillaume, inventor. Byl719 (alpelisib) for use in the treatment of pik3ca-related overgrowth spectrum. World Intellectual Property Organization patent application WO 2017/140828 A1. Published August 24, 2017
Cha Y, Erez T, Reynolds IJ, Kumar D, Ross J, Koytiger G et al (2018) Drug repurposing from the perspective of pharmaceutical companies. Br J Pharmacol 175(2):168–180
Chakraborty P (2020) Indian pharmaceuticals industry in global scenario: an appraisal. J Health Manag 22(3):424–429
Chakraborty C, Sharma AR, Bhattacharya M, Agoramoorthy G, Lee SS (2021) The drug repurposing for COVID-19 clinical trials provide very effective therapeutic combinations: lessons learned from major clinical studies. Front Pharmacol 12:1–15
Chakravarthy R, Cotter K, DiMasi J, Milne CP, Wendel N (2016) Public- and private-sector contributions to the research and development of the most transformational drugs in the past 25 years: from theory to therapy. Ther Innov Regul Sci 50(6):759–768
Challa AP, Zaleski NM, Jerome RN, Lavieri RR, Shirey-Rice JK, Barnado A et al (2021) Human and machine intelligence together drive drug repurposing in rare diseases. Front Genet 12:707836
Chen Z, Chen J, Collins R, Guo Y, Peto R, Wu F et al (2011) China Kadoorie Biobank of 0.5 million people: survey methods, baseline characteristics and long-term follow-up. Int J Epidemiol 40(6):1652–1666
Collier R (2009) Drug development cost estimates hard to swallow. Can Med Assoc J 180(3):279–280
Collins R (2012) What makes UK Biobank special? Lancet 379(9822):1173–1174
Dalwadi SM, Hunt A, Bonnen MD, Ghebre YT (2023) Computational approaches for drug repurposing in oncology: untapped opportunity for high value innovation. Front Oncol 13:1–5
Das S, Abraham A, Konar A (2008) Swarm intelligence algorithms in bioinformatics [Internet]. In: Computational intelligence in bioinformatics. Studies in computational intelligence, vol 94. Springer, Berlin, Heidelberg. Available from: http://www.springerlink.com/
De Maria Marchiano R, Di Sante G, Piro G, Carbone C, Tortora G, Boldrini L et al (2021) Translational research in the era of precision medicine: where we are and where we will go. J Pers Med 11(3):216
Devos D, Moreau C, Delval A, Dujardin K, Defebvre L, Bordet R (2013) Methylphenidate. CNS Drugs 27(1):1–14
DiMasi JA (1995) Trends in drug development costs, times, and risks. Drug Inf J 29(2):375–384
Dufier JL, Dhermy P, Gubler MC, Gagnadoux MF, Broyer M (1987) Ocular changes in long-term evolution of infantile cystinosis. Ophthalmic Genet 8(2):131–137
Feasey N, Wansbrough-Jones M, Mabey DCW, Solomon AW (2010) Neglected tropical diseases. Br Med Bull 93(1):179–200
Festa G, Kolte A, Carli MR, Rossi M (2022) Envisioning the challenges of the pharmaceutical sector in the Indian health-care industry: a scenario analysis. J Bus Ind Mark 37(8):1662–1674
Fetro C (2023) Connecting academia and industry for innovative drug repurposing in rare diseases: it is worth a try. Rare Dis Orphan Drug J 2(2):7
Fiebig HH, Maier A, Burger AM (2004) Clonogenic assay with established human tumour xenografts. Eur J Cancer 40(6):802–820
Fife RS, Rougraff BT, Proctor C, Sledge GW (1997) Inhibition of proliferation and induction of apoptosis by doxycycline in cultured human osteosarcoma cells. J Lab Clin Med 130(5):530–534
Forslöw U, Geborek A, Hjelte L, Petrini B, Heurlin N (2003) Early chemotherapy for non-tuberculous mycobacterial infections in patients with cystic fibrosis. Acta Paediatr 92(8):910–915
Furuta Y, Komeno T, Nakamura T (2017) Favipiravir (T-705), a broad spectrum inhibitor of viral RNA polymerase. Proc Jpn Acad Ser B 93(7):449–463
Goldfarb BNM (2008) Can you handle the truth? HIPAA Compl Clin Res 4(2):4–6
Gottlieb A, Stein GY, Ruppin E, Sharan R (2011) PREDICT: a method for inferring novel drug indications with application to personalized medicine. Mol Syst Biol 7(1):496

Gray NS (2006) Drug discovery through industry-academic partnerships. Nat Chem Biol 2(12):649–653

Eagle G, and gupta R (2019) 'Methods for treating pulmonary non-tuberculous mycobacterial infections'

Hartung HP, Gonsette R, Konig N, Kwiecinski H, Guseo A, Morrissey SP et al (2002) Mitoxantrone in progressive multiple sclerosis: a placebo-controlled, double-blind, randomised, multicentre trial. Lancet 360(9350):2018–2025

Highmore CJ, Melaugh G, Morris RJ, Parker J, Direito SOL, Romero M et al (2022) Translational challenges and opportunities in biofilm science: a BRIEF for the future. NPJ Biofilms Microbiomes 8(1):68

Hobin JA, Galbraith RA (2012) Engaging basic scientists in translational research. FASEB J 26(6):2227–2230

Issa N, Byers S, Dakshanamurthy S (2013) Drug repurposing: translational pharmacology, chemistry, computers and the clinic. Curr Top Med Chem 13(18):2328–2336

Ito R, Takahashi T, Katano I, Ito M (2012) Current advances in humanized mouse models. Cell Mol Immunol 9(3):208–214

Kaitin KI (2012) Translational research and the evolving landscape for biomedical innovation. J Investig Med 60(7):995–998

Kalil AC, Patterson TF, Mehta AK, Tomashek KM, Wolfe CR, Ghazaryan V et al (2021) Baricitinib plus remdesivir for hospitalized adults with Covid-19. N Engl J Med 384(9):795–807

Kesselheim AS, Tan YT, Avorn J (2015) The roles of academia, rare diseases, and repurposing in the development of the most transformative drugs. Health Aff 34(2):286–293

Kim E, Choi A-S, Nam H (2019) Drug repositioning of herbal compounds via a machine-learning approach. BMC Bioinformatics 20(S10):247

Koromina M, Pandi MT, Patrinos GP (2019) Rethinking drug repositioning and development with artificial intelligence, machine learning, and omics. OMICS 23(11):539–548

Kovalevsky AY, Tie Y, Liu F, Boross PI, Wang YF, Leshchenko S et al (2006) Effectiveness of nonpeptide clinical inhibitor TMC-114 on HIV-1 protease with highly drug resistant mutations D30N, I50V, and L90M. J Med Chem 49(4):1379–1387

Kumar A, Singh A, Ekavali. (2015) A review on Alzheimer's disease pathophysiology and its management: an update. Pharmacol Rep 67(2):195–203

Kumar R, Sharma A, Siddiqui MH, Tiwari RK (2018) Prediction of drug-plasma protein binding using artificial intelligence based algorithms. Comb Chem High Throughput Screen 21(1):57–64

Larsson SC, Mantzoros CS, Wolk A (2007) Diabetes mellitus and risk of breast cancer: a meta-analysis. Int J Cancer 121(4):856–862

Lee HM, Kim Y (2016) Drug repurposing is a new opportunity for developing drugs against neuropsychiatric disorders. Schizophr Res Treatment 2016:1–12

Lenselink EB, ten Dijke N, Bongers B, Papadatos G, van Vlijmen HWT, Kowalczyk W et al (2017) Beyond the hype: deep neural networks outperform established methods using a ChEMBL bioactivity benchmark set. J Cheminform 9(1):45

Lou Y, Liu L, Yao H, Hu X, Su J, Xu K et al (2021) Clinical outcomes and plasma concentrations of baloxavir marboxil and favipiravir in COVID-19 patients: an exploratory randomized, controlled trial. Eur J Pharm Sci 157:105631

Mahajan V, Nauriyal DK, Singh SP (2018) Efficiency and its determinants: panel data evidence from the Indian pharmaceutical industry. Margin J Appl Econ Res 12(1):19–40

Maienschein J, Sunderland M, Ankeny RA, Robert JS (2008) The ethos and ethics of translational research. Am J Bioeth 8(3):43–51

Mali SN, Pratap AP (2021) Targeting infectious coronavirus disease 2019 (COVID-19) with artificial intelligence (AI) applications: evidence based opinion. Infect Disord Drug Targets 21(4):475–477

Milne CP, Kaitin KI (2009) Translational medicine: an engine of change for bringing new technology to community health. Sci Transl Med 1(5):5cm5

Moffat J, Grueneberg DA, Yang X, Kim SY, Kloepfer AM, Hinkle G et al (2006) A lentiviral RNAi library for human and mouse genes applied to an arrayed viral high-content screen. Cell 124(6):1283–1298

Morin G, Degrugillier-Chopinet C, Vincent M, Fraissenon A, Aubert H, Chapelle C et al (2022) Treatment of two infants with PIK3CA-related overgrowth spectrum by alpelisib. J Exp Med 219(3):1–8

Nacev BA, Grassi P, Dell A, Haslam SM, Liu JO (2011) The antifungal drug itraconazole inhibits vascular endothelial growth factor receptor 2 (VEGFR2) glycosylation, trafficking, and signaling in endothelial cells. J Biol Chem 286(51):44045–44056

Nagabukuro H, Uemura N (2020) Drug reprofiling and repurposing: a state-of-the-art public-private collaboration model. Transl Regul Sci 2(2):47–50

Nagai A, Hirata M, Kamatani Y, Muto K, Matsuda K, Kiyohara Y et al (2017) Overview of the BioBank Japan Project: study design and profile. J Epidemiol 27(3):S2–S8

National Institute for Health and Care Excellence (2019) Clinical evidence review of mercaptamine hydrochloride for treating corneal cystine deposits in people with cystinosis aged older than 2 years, pp 1–65

Newman SP (2018) Delivering drugs to the lungs: the history of repurposing in the treatment of respiratory diseases. Adv Drug Deliv Rev 133:5–18

Nieoullon A (2011) Neurodegenerative diseases and neuroprotection: current views and prospects. J Appl Biomed 9(4):173–183

Nosengo N (2016) Can you teach old drugs new tricks? Nature 534(7607):314–316

Nzila A, Ma Z, Chibale K (2011) Drug repositioning in the treatment of malaria and TB. Future Med Chem 3(11):1413–1426

Onódi Z, Koch S, Rubinstein J, Ferdinandy P, Varga ZV (2023) Drug repurposing for cardiovascular diseases: new targets and indications for probenecid. Br J Pharmacol 180(6):685–700

Oprea TI, Bauman JE, Bologa CG, Buranda T, Chigaev A, Edwards BS et al (2011) Drug repurposing funding projects. Drug Discov Today Ther Strateg 8(3–4):61–69

Pagliazzi A, Oranges T, Traficante G, Trapani C, Facchini F, Martin A et al (2021) PIK3CA-related overgrowth spectrum from diagnosis to targeted therapy: a case of CLOVES syndrome treated with alpelisib. Front Pediatr 9:1–6

Phiel CJ, Klein PS (2001) Molecular targets of lithium action. Annu Rev Pharmacol Toxicol 41(1):789–813

Porta EOJ, Kalesh K, Steel PG (2023) Navigating drug repurposing for Chagas disease: advances, challenges, and opportunities. Front Pharmacol 14:1233253

Pradhan AU, Uwishema O, Wellington J, Nisingizwe P, Thambi VD, Onyeaka CVP et al (2022) Challenges of addressing neglected tropical diseases amidst the COVID-19 pandemic in Africa: a case of Chagas disease. Ann Med Surg 81:104414

Pushpakom S, Iorio F, Eyers PA, Escott KJ, Hopper S, Wells A et al (2019) Drug repurposing: progress, challenges and recommendations. Nat Rev Drug Discov 18(1):41–58

Ranganath LR, Psarelli EE, Arnoux JB, Braconi D, Briggs M, Bröijersén A et al (2020) Efficacy and safety of once-daily nitisinone for patients with alkaptonuria (SONIA 2): an international, multicentre, open-label, randomised controlled trial. Lancet Diabetes Endocrinol 8(9):762–772

Ranjan A, Gupta P, Srivastava SK (2016) Penfluridol: an antipsychotic agent suppresses metastatic tumor growth in triple-negative breast cancer by inhibiting integrin signaling axis. Cancer Res 76(4):877–890

Rasmussen K, Bero L, Redberg R, Gøtzsche PC, Lundh A (2018) Collaboration between academics and industry in clinical trials: cross sectional study of publications and survey of lead academic authors. BMJ 363:1–10

Sams-Dodd F (2006) Strategies to optimize the validity of disease models in the drug discovery process. Drug Discov Today 11(7–8):355–363

Selvaraj G, Kaliamurthi S, Peslherbe GH, Wei DQ (2021) Application of artificial intelligence in drug repurposing: a mini-review. Curr Chin Sci 1(3):333–345

Setoain J, Franch M, Martínez M, Tabas-Madrid D, Sorzano COS, Bakker A et al (2015) NFFinder: an online bioinformatics tool for searching similar transcriptomics experiments in the context of drug repositioning. Nucleic Acids Res 43(W1):W193–W199

Shamas-Din A, Schimmer AD (2015) Drug discovery in academia. Exp Hematol 43(8):713–717

Sharma K, Ahmed F, Sharma T, Grover A, Agarwal M, Grover S (2023) Potential repurposed drug candidates for tuberculosis treatment: progress and update of drugs identified in over a decade. ACS Omega 8(20):17362–17380

Silber BM (2010) Driving drug discovery: the fundamental role of academic labs. Sci Transl Med 2(30):30cm16

Sinsky C, Colligan L, Li L, Prgomet M, Reynolds S, Goeders L et al (2016) Allocation of physician time in ambulatory practice: a time and motion study in 4 specialties. Ann Intern Med 165(11):753

Stark Z, Dolman L, Manolio TA, Ozenberger B, Hill SL, Caulfied MJ et al (2019) Integrating genomics into healthcare: a global responsibility. Am J Hum Genet 104(1):13–20

Starokozhko V, Kallio M, Kumlin Howell Å, Mäkinen Salmi A, Andrew-Nielsen G, Goldammer M et al (2021) Strengthening regulatory science in academia: STARS, an EU initiative to bridge the translational gap. Drug Discov Today 26(2):283–288

Tentler JJ, Tan AC, Weekes CD, Jimeno A, Leong S, Pitts TM et al (2012) Patient-derived tumour xenografts as models for oncology drug development. Nat Rev Clin Oncol 9(6):338–350

Tousi B (2015) The emerging role of bexarotene in the treatment of Alzheimer's disease: current evidence. Neuropsychiatr Dis Treat 11:311

Trias E, Ibarburu S, Barreto-Núñez R, Babdor J, Maciel TT, Guillo M et al (2016) Post-paralysis tyrosine kinase inhibition with masitinib abrogates neuroinflammation and slows disease progression in inherited amyotrophic lateral sclerosis. J Neuroinflammation 13(1):177

van der Pol KH, Aljofan M, Blin O, Cornel JH, Rongen GA, Woestelandt AG et al (2023) Drug repurposing of generic drugs: challenges and the potential role for government. Appl Health Econ Health Policy 21(6):831–840

Venot Q, Blanc T, Rabia SH, Berteloot L, Ladraa S, Duong JP et al (2018) Targeted therapy in patients with PIK3CA-related overgrowth syndrome. Nature 558(7711):540–546

Verbaanderd C, Rooman I, Huys I (2021) Exploring new uses for existing drugs: innovative mechanisms to fund independent clinical research. Trials 22(1):1–13

Weers J, Metzheiser B, Taylor G, Warren S, Meers P, Perkins WR (2009) A gamma scintigraphy study to investigate lung deposition and clearance of inhaled amikacin-loaded liposomes in healthy male volunteers. J Aerosol Med Pulm Drug Deliv 22(2):131–138

Weinstein JN, Myers T, Buolamwini J, Raghavan K, Van Osdol W, Licht J et al (1994) Predictive statistics and artificial intelligence in the U.S. National Cancer Institute's drug discovery program for cancer and AIDS. Stem Cells 12(1):13–22

Wyss-Coray T (2016) Ageing, neurodegeneration and brain rejuvenation. Nature 539(7628):180–186

Zhang J, Guo J, Zhao X, Chen Z, Wang G, Liu A et al (2013) Phosphodiesterase-5 inhibitor sildenafil prevents neuroinflammation, lowers beta-amyloid levels and improves cognitive performance in APP/PS1 transgenic mice. Behav Brain Res 250:230–237

Zhavoronkov A, Vanhaelen Q, Oprea TI (2020) Will artificial intelligence for drug discovery impact clinical pharmacology? Clin Pharmacol Ther 107(4):780–785

Zhou C, Fan W, Zou J, Zhang J, Fang H, Lv Z et al (2023) Comparison of a new 5% minoxidil foam and Rogaine® in the treatment of androgenetic alopecia in Chinese men: a randomized, double-blind, phase III, equivalence trial. Dermatol Ther 2023:1–10

Drug Repurposing: Exploring Newer Therapeutic Potentials

2

Popat S. Kumbhar, Kaustubh Ajit Kolekar, Vikas Kamble, Omji Porwal, Pranav Kumbhar, Sukriti Vishwas, John Disouza, Gaurav Gupta, Rozhan Arif Muhammed, Saurabh Gupta, Vandana Patravale, Archita Patel, Samir Patel, Kamal Dua, Sachin Kumar Singh, and Sharad Visht

Abstract

Drug repurposing is becoming a more captivating idea in treating sorts of diseases owing to the elevated mortality, exorbitant costs, and very poor success rate of new drug discovery. A range of computational and experimental avenues is employed in the identification of repurposed drug candidates against diverse diseases. A large number of therapeutics from different pharmacological classes have been reported to be effective in managing diseases. Moreover, phytochemicals have also shown great promise as repurposed drug candidates in managing

P. S. Kumbhar
Department of Pharmaceutics, Tatyasaheb Kore College of Pharmacy, Warana University, Warananagar, Maharashtra, Warananagar, India

K. A. Kolekar
School of Pharmaceutical Sciences, Lovely Professional University, Phagwara, Punjab, India

V. Kamble
Department of Pharmaceutical Sciences and Technology, Institute of Chemical Technology, Matunga, Mumbai, India

O. Porwal
Qaiwan International University, Department of Pharmacy, Sulaymaniyah, Iraq

P. Kumbhar
Tatyasaheb Kore College of Pharmacy, Panhala, Kolhapur, Maharashtra, India

S. Vishwas
Chitkara College of Pharmacy, Chitkara University, Rajpura, Punjab, India

J. Disouza
SYBES's Bombay Institute of Pharmacy and Research, Dombivli, India

R. K. Kaundal et al. (eds.), *Drug Repurposing: Exploring Newer Therapeutic Potentials*,
https://doi.org/10.1007/978-981-95-4522-3_2

diverse diseases. However, only limited repurposed drug candidates are observed to be available in the market or in clinical trials. Patents, regulatory concerns, and organizational obstacles are the key bottlenecks in the path of drug repurposing. The current chapter discusses drug repurposing and its significance and the avenues for drug repurposing. This article offers insights into drug repurposing for various diseases, including cancer, cardiovascular, neurological, and infectious diseases. Additionally, repurposed phytochemicals for diverse disease management and challenges associated with repurposing drugs are also briefed. Furthermore, repurposed drugs that are on the market or currently undergoing clinical studies are covered.

Keywords

Drug repurposing · Computational approaches · Experimental approaches · Phytochemicals · Clinical trials

G. Gupta
Centre for Research Impact and Outcome, Chitkara College of Pharmacy, Chitkara University, Rajpura, Punjab, India

Centre of Medical and Bio-Allied Health Sciences Research, Ajman University, Ajman, UAE

R. A. Muhammed
Tishk International University, Department of Pharmacy, Sulaymaniyah, Iraq

S. Gupta
Chameli Devi Institute of Pharmacy, Department of pharmacology, Indore, Madhya Pradesh, India

V. Patravale
Department of Pharmaceutical Sciences and Technology, Institute of Chemical Technology, Mumbai, Maharashtra, India

A. Patel · S. Patel
Ramanbhai Patel College of Pharmacy, Charotar University of Science and Technology, Anand, Gujarat, India

K. Dua
Discipline of Pharmacy, Graduate School of Health, University of Technology Sydney, Ultimo, NSW, Australia

Faculty of Health, Australian Research Centre in Complementary and Integrative Medicine, University of Technology Sydney, Ultimo, NSW, Australia

Uttaranchal Institute of Pharmaceutical Sciences, Uttaranchal University, Dehradun, India

S. K. Singh (✉)
School of Pharmaceutical Sciences, Lovely Professional University, Phagwara, Punjab, India

Faculty of Health, Australian Research Centre in Complementary and Integrative Medicine, University of Technology Sydney, Ultimo, NSW, Australia
e-mail: Sachin.16030@lpu.co.in

S. Visht
College of Pharmacy, Cihan University-Erbil, Erbil, Iraq

2.1 Introduction

The path from a concept to an approved treatment is a protracted, expensive, and unpredictable one in the conventional drug discovery pathway. The financial and temporal investments are substantial, and the failure rate is considerable (De Rosa et al. 2023). This approach includes preclinical and clinical investigations in addition to lead selection and optimization. This conventional drug discovery process may cost approximately 1800 million dollars and takes between 10 and 17 years to complete. Furthermore, this avenue may not be able to control the mortality associated with various diseases, including cancer and cardiovascular diseases, in low- and middle-income countries. Also, this strategy could fail to satisfy the needs of therapeutics for sorts of diseases (Sleire et al. 2017).

A substitute for the conventional drug discovery technique is drug repurposing, which allows drugs that have been licensed for other purposes to be employed for different kinds of diseases. The benefits of drug repurposing comprise lower expenses and time to bring a medicine to market due to the widely recognized pharmacokinetics, pharmacodynamics, and toxicity characteristics of the cargo. Furthermore, there is little chance of disaster, considering that the safety of medications has already been established in preclinical and initial-stage clinical trials. As a result, this approach is being widely used by researchers and medical professionals to solve the issue of medicine scarcity while looking for novel cargos for treating various diseases (Kumbhar et al. 2023).

An emerging variety of computational and experimental approaches that exploit clinical and molecular data allows for various medication repurposing strategies. When these tools are integrated, drug repurposing can be quickly applied on a progressively larger scale. Nowadays, efficient techniques for systematic drug-repositioning strategies that exploit sizable databases of substances with biological activity are accessible. These techniques have led to the discovery of several intriguing therapeutics that could be used to treat various ailments; a handful of these drugs are currently undergoing clinical testing (Pushpakom et al. 2019). Nevertheless, significant organizational, regulatory, and technological obstacles still exist, limiting the progress in drug repurposing. In this chapter, we have discussed about drug repurposing and its significance and avenues in drug repurposing. Insight into synthetic drug repurposing and the repurposing of phytoceuticals for various diseases is also briefed. Furthermore, repurposed drugs in the market or in clinical trials are described as challenges in drug repurposing.

2.2 Drug Repurposing and Its Significance

Drug repurposing, also known as drug reprofiling or repositioning, is a strategy to discover fresh applications for authorized or experimental drugs outside of their original intended medical usage. This approach provides multiple advantages over creating entirely new drugs for specific purposes (Ashburn and Thor 2004). First and foremost, because repurposed medications have previously shown adequate safety in animal models and the initial stages of human studies, they lower the likelihood of disaster. This reduces the likelihood of safety-related failures in subsequent efficacy trials. Secondly, it accelerates the drug development timeline because much of the testing in animal models, safety evaluation, and, in some instances, formulation work is already accomplished. Thirdly, it requires less investment, albeit this varies depending on the development stage of the reprofiling drug (Breckenridge and Jacob 2019). When repurposed medications are contrasted with new therapeutics in an identical indication, regulatory and Phase III expenditures may stay essentially consistent, although preclinical and Phase I and II charges can still be significantly reduced. Taken together, these benefits can result in a quicker and less hazardous payback on investment for repurposed drug research, with lower average costs when mishaps are taken into account. In actuality, the typical expense of getting a repurposed medicine to the market is believed to be $300 million, while the total expense of an entirely novel chemical entity is projected to be between $2 and $3 billion (Nosengo 2016). The de novo drug development and drug repurposing processes timeline is represented in Fig. 2.1. Finally, repurposed medications can reveal new targets and avenues for more research (Pushpakom et al. 2019). The cost of drugs approved for treating a variety of diseases, including cancer and infectious diseases, is huge and causes high mortality in low- and middle-income countries. The repurposed drug candidates from different pharmacological classes have reasonable prices and display significant potential in treating sorts of diseases effectively. Thus, repurposed drug candidates at reasonable prices contribute to reducing the mortality allied with various diseases in low- and middle-income countries.

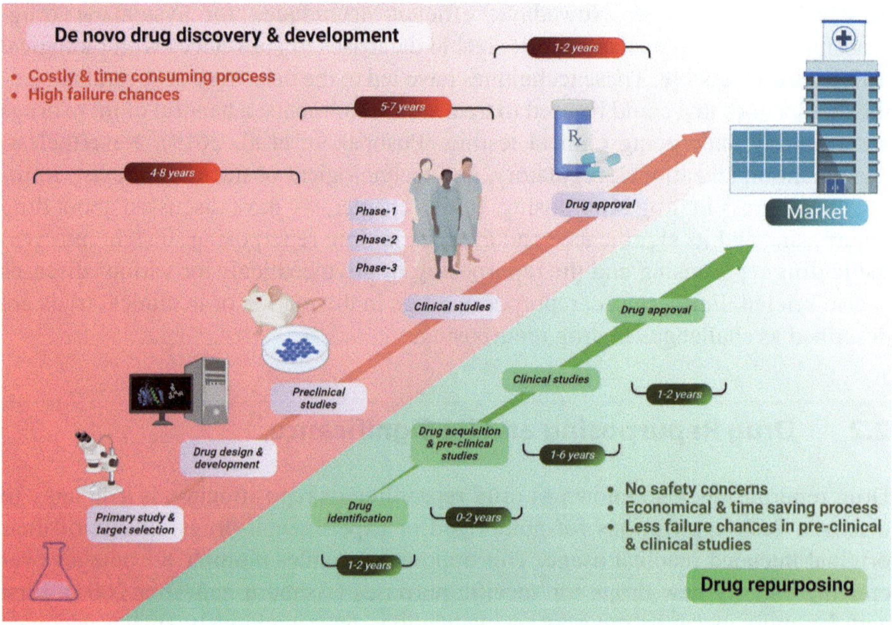

Fig. 2.1 Processes of de novo drug development and drug repurposing

2.3 Drug Repurposing Avenues

Usually, a drug repurposing approach comprises three main phases before progressing the candidate medication via a conduit for development: pinpointing a fitting candidate molecule for a particular indication, conducting an exhaustive examination of the drug's mechanism of action in preclinical models, and assessing its efficacy in Phase II clinical trials. In these three stages, the first phase involves determining with great precision which medication is best suited for a given reason, and this is extremely important. This is the phase where contemporary methods for hypothesis generation play a crucial role. These methodical strategies fall into two categories: computational and experimental, and they are gradually combined (Fig. 2.2). Leveraging clinical information for drug repurposing is covered by these two broad areas (Pushpakom et al. 2019).

2.3.1 Computational Approaches

Computational methods are predominantly driven by data, entailing the systematic analysis of various data types, including gene expression, genotype, chemical structure, electronic health records (EHRs), and proteomic data. This analysis can subsequently give rise to the generation of hypotheses for drug repurposing. The following section explores the most frequently employed computational methods and provides illustrative examples of drug repurposing (Hurle et al. 2013).

2.3.1.1 Matching Signatures

In drug repurposing, signature matching stands out as a strategy wherein the distinctive features, or 'signature,' of a drug are compared with those of a distinct medication, ailment, or clinical feature. This signature is extracted from three primary data types, including metabolomic or proteomic facts, transcriptomic (RNA) data, chemical structures, and adverse event characteristics.

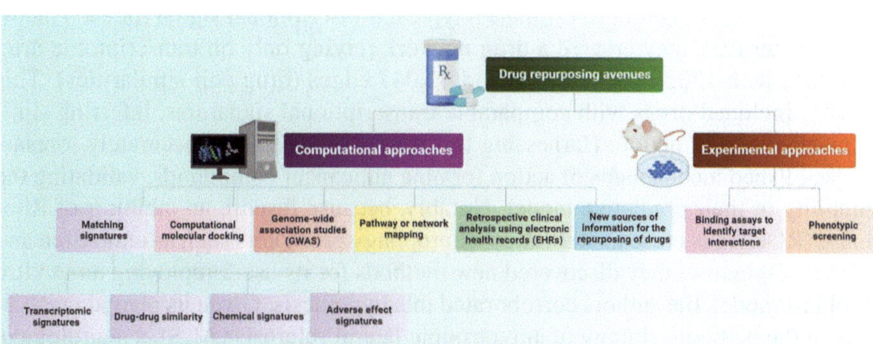

Fig. 2.2 Different types of drug repurposing avenues

Transcriptomic Signatures

Matching transcriptomic signatures facilitates comparisons between drugs and diseases. In the former scenario, a drug's transcriptomic signature is formulated by contrasting the gene expression profile pre- and post-treatment, subsequently comparing the results with a manifestation signature linked to a disease. The inverse connection between the drug's gene expression profile and that of the disease may suggest the potential efficacy of the drug against the disease. The foundation of this method is the signature reversion principle, which postulates that a medication might alleviate an illness by reversing the sequence of gene expression linked to the ailment. Dudley and collaborators conducted a comparison between the gene expression signatures of inflammatory bowel disease (IBD) extracted from publicly available data and the gene expression profiles of 164 drugs accessible from the Connectivity Map (cMap). The degree of negative association between the drug's gene expression signature and the disease's gene expression signature served as the foundation for their therapeutic predictions of drug-disease combinations. Prednisolone, a corticosteroid that is frequently used to reduce inflammation related to IBD, was surprisingly one of their most reliable treatment predictions for the disease. Moreover, a high association was found between the antiepileptic topiramate and ulcerative colitis and Crohn's disease, two important manifestations of IBD (Dudley et al. 2011a).

Drug-Drug Similarity

This method seeks to unveil common mechanisms of action among diverse drugs, irrespective of their chemical composition. It operates based on the 'guilt by association' principle, suggesting that drugs sharing a transcriptomic signature might have similar therapeutic applications. This notion has worked well for locating alternate targets for currently available medications and disclosing possible adverse outcomes. For instance, Iorio and associates used open-access transcriptome patterns of drugs to build a therapeutic network using the "guilt by connection" idea. Using this strategy, they were able to identify medications that had similar transcriptional signatures, signifying a shared mechanism of action. They applied cMap to investigate gene expression patterns spanning multiple treatments on numerous cell lines and dosages in order to estimate a typical transcriptional signal for each medication. After that, they created a drug network relying only on transcriptome drug identities, with 1302 nodes (drugs) and 41,047 edges (drug pair similarities). This network included drugs with comparable transcriptional signatures, inferring similar mechanisms of action. Harnessing this network, the authors accurately foresaw the established mechanisms of action for nine anticancer compounds, validating the reliability of their predictive model. Notably, because fasudil, an inhibitor of Rho-kinase (ROCK), is close to well-known autophagy promoters such as tamoxifen and 2-deoxy-d-glucose, they discovered new methods for its use. Employing an in vitro fibroblast model, the authors corroborated this hypothesis. Given its pivotal involvement in the pathophysiology of amyotrophic lateral sclerosis (ALS), a neurological disease for which there is currently no recognized treatment, fasudil's relationship with autophagy is remarkable. Fasudil improves motor neuron survival in transgenic mice by generating mutant superoxide dismutase 1 (SOD1), the gene liable for congenital ALS forms, according to research in ALS animal models. The favorable effects of fasudil in ALS preclinical models were further affirmed in other studies. As a result, a clinical study of open-label use is being conducted in China to examine the safety and effectiveness of fasudil in the treatment of ALS (NCT01935518) (Iorio et al. 2010a, b; Lee et al. 2015; Takata et al. 2013; Günther et al. 2017; Tönges et al. 2014). Both transcriptomic approaches heavily rely on openly available gene expression facts, including cMap and other repositories. These resources allow for the matching of transcriptomic signatures, facilitating the discovery of new drug-disease connections and potential repurposing opportunities.

Chemical Signatures

Examining the chemical fingerprints of one drug in contrast to another constitutes an alternative approach to signature matching. This technique seeks out common chemical attributes, suggesting a shared biological impact. Despite its promise, this method comes with constraints, including potential inaccuracies in chemical structures and the presence of physiological effects that extend beyond structural relationships (Dudley et al. 2011b).

Adverse Effect Signatures

The adverse reaction profile of a drug can function as a surrogate for its phenotype. Aligning adverse effect signatures operates on the premise that drugs inducing similar adverse effects might influence common targets or pathways. This method has the potential to unveil novel drug-target associations and repurposing opportunities for various diseases. However, identifying undesirable effects and obtaining adverse effect information from medication package inserts can be difficult. Text mining and processing natural languages are two instances of machine learning methods that offer intriguing answers to these problems. Signature matching in drug repurposing emerges as a hopeful path for discovering fresh therapeutic applications and broadening the scope of existing drugs (Yang and Agarwal 2011).

2.3.1.2 Computational Molecular Docking

This method predicts how well a drug fits into a target in the body. It analyzes the structures of molecules to see if they match (Kitchen et al. 2004). Conventional docking can be used to test numerous medications on a target that is known to be involved in a disease. Alternatively, we can use inverse docking to test drug libraries against different targets and discover novel interactions for potential repurposing. In a study, Dakshanamurthy and team used computational docking to analyze 3671 FDA-approved drugs and found that mebendazole, an anthelmintic drug, could inhibit a key player in angiogenesis (Dakshanamurthy et al. 2012). However, using molecular docking for repurposing drugs faces challenges. Some protein targets lack 3D structures, especially membrane proteins like G protein-coupled receptors (GPCRs), though progress is being made (Cooke et al. 2015). There's also a shortage of well-curated databases for accurate structural information, though this is improving (Kharkar et al. 2014; Huang et al. 2018). Lastly, questions arise about the reliability of docking algorithms in predicting how well drugs bind. While progress is happening, differences among software packages and limitations in predictability, such as how drugs bind and their effects, still exist (Warren et al. 2006; Pagadala et al. 2017).

2.3.1.3 Genome-Wide Association Studies (GWAS)

In the last decade, there's been a big increase in GWAS thanks to better technology and lower costs. GWAS aims to find genetic differences linked to common diseases, giving us insights into how diseases work. This information could lead to possible drug repurposing, as it may help identify new targets, some of which are shared by illnesses studied in GWAS and those treated with pharmaceuticals. Sanseau and other researchers found that genes associated with illness have a higher probability of encoding proteins that could be therapeutic targets. The pharmaceutical industry has shown a 2.7-fold increase in pursuing these targets in GWAS. Additionally, 92 genes with GWAS characteristics distinct from the intended application of the medicine were discovered, indicating the possibility of assessing medications for novel applications. However, there are difficulties in repositioning drugs using GWAS data. Tracing the origin of genes or variants might be challenging, as there are signals in regions rich in genes. It might be difficult to determine the impact of a gene mutation, and it requires more research (Sanseau et al. 2012). Predicting novel uses for medications should be done with caution because GWAS data do not provide comprehensive illness information (Wang and Zhang 2013). Furthermore, as our knowledge of the human genome is continually expanding, there might be additional genes yet to be found (Willyard 2018).

2.3.1.4 Pathway or Network Mapping

Scientists often use this technique to find new uses for existing drugs or identify potential drug targets (Smith et al. 2012). Sometimes, the genes found through methods like GWAS may not be the best option for drug targeting. In such studies, a pathway-focused approach helps identify genes related to the GWAS target, opening up new possibilities for repurposing (Greene and Voight 2016). Network analysis involves creating networks based on disease information, gene expression, GWAS data, or protein interactions to find other potential repositioning candidates (Iorio et al. 2013). An investigation by Greene and colleagues integrated genetic data with functional interaction networks, proving more accurate in identifying disease-gene associations than GWAS alone. They applied this approach to hypertension, finding more potential drug targets compared to GWAS (Greene et al. 2015). Another study looked at gene expression data from infections involving respiratory viruses in human hosts. They found 67 common pathways important in pulmonary viral infections (Smith et al. 2012). Exploring these routes against the DrugBank database revealed medications like pranlukast (anti-asthmatic) and amrinone (used for heart failure) that, by influencing the immune response, could prove helpful in managing viral infections.

2.3.1.5 Retrospective Clinical Analysis Using Electronic Health Records (EHRs)

Researchers often go back and analyze past clinical data to find new uses for drugs, like sildenafil (Ashburn and Thor 2004). They have also found opportunities for repurposing drugs, such as using aspirin for colorectal cancer (US Preventive Services Task Force 2018), raloxifene for breast cancer, and propranolol for osteoporosis (Cavalla and Singal 2012). However, these discoveries weren't from a systematic analysis of the data. It is now recommended to examine clinical data using a methodical way in order to find further prospects for repurposing (Jensen et al. 2012). Clinical data can come from different sources, like post-marketing surveillance, EHRs, and clinical trials. EHRs provide an abundance of patient data, both structured and unstructured, that can be used to repurpose medications. Extracting information from over 13 years of EHRs identified terbutaline sulfate as an anti-asthmatic and a possible treatment for ALS (Paik et al. 2015).

The WHO global database for adverse drug reactions (VigiBase), the FDA Adverse Event Reporting System (FAERS), EudraVigilance, the UK Clinical Practice Research Data Link (CPRD), and the Yellow Card Scheme are additional large data platforms. These contain crucial data for drug repurposing. Challenges like ethical and legal obstacles in accessing EHR data and hurdles while attempting to extract unstructured data need improvement. Using natural language processing and machine learning methods could make these databases more useful for opportunities like drug repurposing. Post-marketing surveillance and information from clinical trials are vital, but getting hold of them could be restricted due to financial or privacy concerns. Understanding the significance of providing accessibility to these data can encourage further studies on the development of drugs. The European Medicines Agency now gives information from clinical trials accessibility to the general public, allowing independent reanalysis, which could point to more opportunities for drug repositioning (Pushpakom et al. 2019).

2.3.1.6 New Sources of Information for the Repurposing of Drugs

Scientists use immortalized human cancer cell lines in extensive drug tests to see how different compounds affect the cells. They combine the results with detailed genetic information about the cell lines, helping them understand how the cell features interact with drugs (pharmacogenomic interactions). Analyzing publicly available datasets with both genetic and drug information from these cells is a new way to find opportunities for repurposing drugs (Barretina et al. 2012; Basu et al. 2013; Iorio et al. 2016; Seashore-Ludlow et al. 2015). Even though these cell lines are not perfect models due to certain biases, they can still help identify interactions similar to those seen in clinical settings. Recent studies combined these cell line tests with genetic information from large groups of primary tumors. This integration helps prioritize interactions based on how common genetic changes are clinically (Iorio et al. 2016). Interestingly, many of the new interactions were specific to certain cancers and involved drugs already used for other diseases. This research could be crucial for finding drug repurposing opportunities and tailoring cancer therapy based on specific genetic changes.

Using large DNA biobanks connected to electronic health records (EHRs) could speed up drug repurposing (Wei and Denny 2015). For example, GlaxoSmithKline used the China Kadoorie Biobank (CKB) (Chen et al. 2011) to investigate gene variants in major vascular diseases. This method of proving a medication ineffective might additionally be employed to verify gene targets for therapeutic repurposing. Accessing large biobanks linked to EHRs, like the UK Biobank, could help evaluate potential drug targets (Millwood et al. 2016; O'Donoghue et al. 2014; Investigators 2014). Advanced sequencing technologies, like Illumina's HiSeq X Ten, collect extensive genetic data, aiding drug repurposing. Projects like the UK's 100,000 Genome Project and all US research programs use these technologies to genotype individuals. Similar initiatives worldwide contribute to collecting large-scale genomic data. These projects offer insights into the genetic basis of disease, identifying targets for new drug discovery and repurposing (Pushpakom et al. 2019; Eisenstein 2015; Peplow 2016).

Yet, managing extensive data from these initiatives and implementing high-throughput technologies presents difficulties in analyzing and applying them for drug discovery and repurposing. Ultimately, online self-disclosed patient information is proposed as a prospective reservoir for drug repurposing. As an illustration, self-disclosed information regarding the use of lithium carbonate by ALS patients was employed to deduce insights into its efficacy (Wicks et al. 2011). Despite the potential acceleration of clinical discovery through this method, utilizing patient-reported data online entails risks such as potential bias and safety concerns when patients engage in self-prescription.

2.3.2 Experimental Approaches

2.3.2.1 Binding Assays to Identify Target Interactions

Scientists utilize techniques like affinity chromatography and mass spectrometry to identify the binding partners of drugs (Brehmer et al. 2005). In the realm of chemical biology, it's customary to investigate the targets and unintended targets of drugs for validation and repurposing purposes. The cellular thermo-stability assay (CETSA) method anticipates how drugs fortify target proteins within cells by delineating their interaction (Molina et al. 2013). Early triumphs encompass verifying cellular targets for the tyrosine kinase inhibitor crizotinib (Alshareef et al. 2016) and uncovering quinone reductase 2 as a cellular off-target of paracetamol (Miettinen and Björklund 2014).

In response to the extensive utilization of protein kinase inhibitors (Eyers et al. 1999), endeavors are intensifying to create improved compounds for preclinical investigation and guide clinical cargo progress and repurposing via an evidence-based pharmacological 'audit trail' in cells (Cohen 2002). The identification of 'errors' in kinase drug discovery holds paramount importance. Initial impartial affinity methods aid in comprehending compound impacts on cells, encompassing instances of paradoxical kinase activation by inhibitors (Blagg and Workman 2017), which can result in off-target tumor initiation in patients (Hall-Jackson et al. 1999). For instance, Brehmer and team employed a matrix with gefitinib to comprehend its effects, pinpointing over 20 protein kinases as potential targets for gefitinib (Brehmer et al. 2005). More comprehensive initiatives have utilized ‘kinobeads’ to assess kinase inhibitors, capturing proteins prior to quantification, frequently unveiling new off-targets for well-established drugs (Su et al. 2012; Duncan et al. 2012). Analogous methodologies have disclosed numerous on- and off-targets for clinical agents like the BCR–ABL inhibitor imatinib (Klaeger et al. 2016), repurposed for KIT-driven gastrointestinal stromal tumors (Troutman et al. 2016). Knowing the binding and effectiveness link in the biological environment is facilitated by chemical genetics (Wisniewski et al. 2002). These discoveries swiftly translate into new clinical domains or mitigate the consequences of drug resistance, a common response to kinase inhibitor therapy in cancer (Blanke et al. 2008; Sloane et al. 2010). Numerous studies originate from industry-driven high-throughput assays, examining small-molecule kinase-binding kinome-wide utilizing organism-based and in vitro experiments to create heat maps of interconnections that are of biological importance (Bago et al. 2016; Carter et al. 2005; Davis et al. 2011). In a specific investigation, Karaman and colleagues (Karaman et al. 2008) utilized a competition binding assay to evaluate 38 kinase inhibitors against 317 human protein kinases, identifying 3175 binding interactions. Certain kinase inhibitors, like dasatinib and sorafenib, have a stronger preference for secondary kinase targets than for their identified primary target, which may have an impact on how patients are treated with them. The identification of non-kinase targets for small compounds intended to impede protein kinases is becoming increasingly imperative (Munoz 2017), offering repurposing prospects in cancer, modulation of the Zika virus (Xu et al. 2016), and effective weapons against microbes resistant to antibiotics (Sun et al. 2016).

2.3.2.2 Phenotypic Screening

This technique is employed to identify compounds that influence illnesses in model systems, regardless of the lack of explicit information on precise goals (Moffat et al. 2017). When applied to drug repurposing, screening existing drugs can unveil new opportunities for repurposing. In conventional in vitro phenotypic screens, diverse cell-based assays are utilized frequently in a 96-well format. As an illustration, Iljin and collaborators (Iljin et al. 2009) scrutinized 4910 drug-like substances in prostate cancer and prostate cell lines that were not malignant, recognizing disulfiram, a medication for alcohol abuse, as a potential anticancer agent. This finding was validated through genome-wide gene expression studies. In the realm of drug repurposing, researchers also leverage whole-organism phenotypic assays. Applying a zebrafish model, Cousin and colleagues (Cousin et al. 2014) investigated 39 FDA-approved drugs for the management of tobacco dependency. Researchers noticed that substances such as apomorphine and topiramate influenced the model’s behavior when nicotine and ethanol were introduced.

2.4 Insight into Drug Repurposing for Various Diseases

Drugs from different pharmacological classes approved for specific indications have been reported to be effective in treating other indications, including cancer, neurological disorders, infectious diseases, and cardiovascular diseases. The drug candidates repurposed in treating various diseases are summarized in Table 2.1.

2.4.1 Cancer

Globally, cancer is the biggest reason for fatality. Non-oncological drugs from different pharmacological classes have shown great promise in treating various cancers, including breast, lung, skin, and brain cancers. Drugs from diverse pharmacological classes, such as antidiabetics, immunosuppressants, antimicrobials, anesthetics, antihypertensives, antimalarials, antipsychotics, antidepressants, anthelmintics, antihyperlipidemics, and non-steroidal anti-inflammatory drugs (NSAIDs), have demonstrated effectiveness against various cancers (Kumbhar et al. 2022a, b). These therapeutic anti-tumor efficacies have been shown through a variety of pathways, such as the destruction of DNA, apoptosis induction, cell cycle halting, and impairment of the mitochondrial membrane potential. For instance, antidiabetic metformin is noticed to be effective in the treatment of skin, breast, colorectal, and endometrial cancers. Metformin exhibited potent anti-melanoma activity against SKMEL-28 and A375 melanoma cells. Metformin displayed anti-melanoma activity through the induction of apoptosis and triggered autophagy. Similarly, lipid-lowering agents, such as simvastatin, lovastatin, and atorvastatin, have shown anti-tumor potential via the induction of apoptosis (Kumbhar et al. 2022b).

2.4.2 Neurological Disorders

Alzheimer's ailment ranks as the most prevalent neurodegenerative condition, trailed by Parkinson's disorder, with frontotemporal dementia, Huntington's malady, amyotrophic lateral sclerosis (Lou Gehrig's disease), progressive supranuclear palsy, spinocerebellar ataxia, Pick's disease, and lastly, prion disease appearing much less frequently (Scatena et al. 2007). Diverse classes of repurposed drugs, including antihypertensive agents, antiepileptic drugs, anticancer drugs, antidiabetic drugs, and antibiotic agents, are employed for treating these diseases (Cummings and Zhong 2014). These agents utilize various pathways to avert neurodegenerative diseases, including the reduction of amyloid-β, preservation of neuronal function, prevention of cell death, and mitigation of pro-inflammatory cytokine and chemokine secretion by brain cells. Imatinib, recognized for its cancer treatment, operates through two mechanisms: reducing amyloid-β and providing neuroprotection. Similarly, liraglutide, a potent antidiabetic drug, combats neurodegeneration by reducing amyloid-β formation (Durães et al. 2018).

Table 2.1 Drugs from different pharmacological classes repurposed for different indications

Pharmacological class	Drug	Repurposed indication	References
Antidiabetic	Phenformin (Biguanide)	Skin cancer, breast cancer	Yuan et al. (2013) and Liu et al. (2015)
	Metformin (Biguanide)	Skin cancer, breast cancer, cardiac hypertrophy	Tomic et al. (2011) and Shukla et al. (2019)
Antihypertensive	Telmisartan (Angiotensin receptor blocker)	Skin cancer	Grahovac et al. (2019)
	Propranolol (Beta blocker)	Skin cancer	Zhou et al. (2016)
Anthelmintic	Mebendazole	Skin cancer, breast cancer	Simbulan-Rosenthal et al. (2017)
	Flubendazole	Skin cancer	Čáňová et al. (2018)
	Niclosamide	Skin cancer, anti-viral	Zhu et al. (2019) and Zheng et al. (2018a)
Antirheumatic	Leflunomide	Skin cancer	Hanson et al. (2018)
	Auranofin	Skin cancer, antibacterial	Gelosa et al. (2020) and Goenka and Simon (2020)
Antiretroviral	Lopinavir	Skin cancer	Paskas et al. (2019)
	Nelfinavir	Breast cancer	Soprano et al. (2016)
	Ritonavir	Breast cancer	Srirangam et al. (2006)
Cardiac glycoside	Digitoxin	Skin cancer	Eskiocak et al. (2016)
Serotonin receptor agonist	Tegaserod	Skin cancer	Liu et al. (2020)
Antifungal	Itraconazole	Skin cancer	Liu et al. (2020)
	Clotrimazole	Breast cancer	Marcondes et al. (2015)
	Haloprogin	Skin cancer	Riedel et al. (2018)
NSAIDs	Piroxicam	Skin cancer	Khodaie et al. (2017)
	Ibuprofen	Breast cancer	Rico et al. (2017)
	Naproxen	Breast cancer	Rico et al. (2017)
	Aspirin	Breast cancer	Kanwal et al. (2022)
Corticosteroids	Dexamethasone	Skin cancer	Sau et al. (2017)
Antibiotic	Tigecycline	Skin cancer	Hu et al. (2016)
	Nifuroxazide	Skin cancer	Sarvi et al. (2018)
	Ciprofloxacin	Skin cancer	Beberok et al. (2018)
	Moxifloxacin	Skin cancer	Beberok et al. (2019)
	Ciprofloxacin	Breast cancer	Zhao et al. (2019)
	Azithromycin	Breast cancer	Abdel-Aal et al. (2021)
	Erythromycin	Alzheimer's disease	Agnati et al. (1992)

(continued)

Table 2.1 (continued)

Pharmacological class	Drug	Repurposed indication	References
Antidote	Methylene blue	Skin cancer	Del Grande et al. (2022)
Antipsychotics	Pimozide	Skin cancer	Jia et al. (2018)
	Chlorpromazine	Breast cancer	Yang et al. (2019)
	Tetrabenazine	Huntington's disease	Paleacu (2007)
Antimalarial	Chloroquine	Skin cancer	Lakhter et al. (2013)
	Amodiaquine	Breast cancer	Parvathaneni et al. (2022)
	Quinacrine	Breast cancer	Zhang et al. (2012)
Local anesthetics	Lidocaine	Skin cancer	Xu et al. (2019)
Antihistamines	Desloratadine and loratadine	Skin cancer	Fritz et al. (2020)
	Chlorcyclizine	Antiviral	Zheng et al. (2018a)
Alcohol antagonist	Disulfiram	Breast cancer	Fasehee et al. (2017)
Statin	Atorvastatin	Breast cancer	Gambhire et al. (2018)
	Simvastatin	Breast cancer	Elakkad et al. (2021)
Antidepressant	Fluoxetine	Breast cancer	Bowie et al. (2015)
	Imipramine	Breast cancer	Timilsina et al. (2022)
Protease inhibitor	Saquinavir	Breast cancer	Singh et al. (2015)
Lipase inhibitor	Orlistat	Breast cancer	Paulmurugan et al. (2016)
Anticonvulsant	Valproic acid	Breast cancer	Zhou et al. (2019)
Immunosuppressant	Rapamycin	Breast cancer	Katiyar et al. (2016)
	Mitoxantrone	Multiple sclerosis	Hartung et al. (2002)
Antiepileptic	Flunarizine	Breast cancer	Zheng et al. (2018b)
	Retigabine	Amyotrophic lateral sclerosis	Martinez et al. (2017)
	Zonisamide	Parkinson's disease	Bermejo and Anciones (2009)
Anticancer	Tamoxifen	Amyotrophic lateral sclerosis	Goodman (2005)
Antigout	Colchicine	Myocardial infarction	Gelosa et al. (2020)
	Allopurinol	Endothelial dysfunction	Gelosa et al. (2020)
Alzheimer's disease	Donepezil	Heart failure	Gelosa et al. (2020)

2.4.3 Infectious Diseases

Infectious diseases encompass a range of ailments, including pulmonary infections, diarrheal diseases, HIV/AIDS, malaria, tuberculosis, measles, pertussis, tetanus, sexually transmitted diseases, and meningitis (Fauci 2001). Various repurposed drug classes, such as anti-arthritics and antihistamines, are employed for their treatment. These drugs exert their effects through diverse mechanisms, including inhibition of host cell pathogenesis factors like thioredoxin reductases, glycine reductase, and trypanothione reductase. Auranofin, typically used for arthritis, operates by inhibiting host cell pathogenesis of thioredoxin reductases and demonstrates efficacy against HIV/AIDS, as well as multidrug-resistant bacteria such as methicillin-resistant *Staphylococcus aureus* and *Klebsiella pneumoniae* (Zheng et al. 2018a).

2.4.4 Cardiovascular Condition

Cardiovascular disease encompasses conditions like atherosclerosis, coronary artery disease, arrhythmias, myocardial infarction, stroke, endothelial dysfunction, and hyperlipidemia (Cantrell et al. 2021). Various classes of repurposed pharmacological drugs, including anti-inflammatory, antidiabetic, antigout, antibiotic, and antiviral drugs, are employed for the treatment of a variety of cardiovascular diseases. These drugs act through diverse mechanisms such as TNF-α inhibitors, IL-1β receptor antagonists, and microtubule polymerization inhibition. For instance, colchicine, initially approved for gout treatment, is repurposed for cardiovascular treatments due to its microtubule polymerization inhibition properties (Onódi et al. 2023).

2.5 Repurposing of Phytoceuticals

Various types of phytoceuticals, including resveratrol, quercetin, epigallocatechin-3-gallate, and many more, are utilized as repurposed candidates for the treatment of various diseases such as cancer, neurological disorders, infectious diseases, and cardiovascular diseases. For instance, resveratrol contains polyphenol, which is obtained from grapes, mulberries, tiger cane, peanuts, and various plants. Primarily, it is used for the treatment of aging and inflammation, and now it is repurposed for treating cancer and neurodegenerative diseases. Moreover, Yao and team investigated that curcumin's primarily approved anti-inflammatory use has shown its potential in treating central nervous diseases, such as Alzheimer's and epilepsy (Zeng et al. 2022). Quercetin flavonoids possess antioxidant and anti-inflammatory qualities that may help prevent heart disease, reduce edema, kill tumor cells, and control blood sugar. It has demonstrated anti-tumor efficaciousness toward brain and breast cancers in addition to the previously described actions. In human glioblastoma, quercetin caused cell death and inhibited matrix metallopeptidase-9 (MMP-9) (Pan et al. 2015). The number of phytoceuticals repurposed for treating various diseases is summarized in Table 2.2.

Table 2.2 Repurposed phytochemicals for various diseases

Phytochemical	Source	Primary indication	Repurposed indication	Reference
Resveratrol	Grapes, mulberries, peanuts, tiger cane, etc.	Anti-aging, anti-inflammatory	Neurodegenerative diseases, lung cancer, prostate cancer, breast cancer	Jin et al. (2019), Li et al. (2018a), and Singh et al. (2017)
Quercetin	Orange, lemon, tea, berries, etc.	Anti-inflammatory, antioxidant	Cancer, cardiovascular diseases	Pan et al. (2015 and Li et al. (2018b)
Epigallocatechin-3-Gallate	Green tea	Antioxidant, weight management,	Bladder cancer, skin cancer	Zhang et al. (2015), Sun et al. (2019), and Ellis et al. (2011)
Fisetin	Vegetables, fruits,	Anti-inflammatory, antioxidant	Cancer, cardiovascular disease, neurodegenerative diseases, diabetes	Khan et al. (2013)
Berberine	Berberis vulgaris and Berberis aristata	Infectious diseases, antioxidants, gastrointestinal diseases	Cancer, cardiovascular diseases, neurodegenerative diseases	Hamsa and Kuttan (2012) and Katiyar et al. (2009)
Sanguinarine	Bloodroot, celandine	Inflammation, skin treatment, respiratory treatment	Cancer, infectious diseases, neurodegenerative diseases	Wei et al. (2017)
Caffeic acid phenethyl ester	Propolis, honey, vegetables	Anti-inflammatory, antibacterial	Cancer, cardiovascular, neurodegenerative diseases	Kudugunti et al. (2011)
Capsaicin	Red pepper	Pain relief	Skin cancer, cardiovascular health	Shin et al. (2008)
Eugenol	Clove	Anti-mutagenic, anti-inflammatory, dental application	Melanoma	Pal et al. (2010)

2.6 Challenges in Drug Repurposing

Reprofiling is not successful every time, often failing selected drug candidates, typically during Phase III trials. It's reasonable to anticipate some late-stage failures, as is common with entirely new drug development. Nevertheless, these setbacks are expected to be less attributable to toxicity issues, considering the well-established safety records of the candidates. However, other factors can contribute to setbacks in the realm of repurposing, such as the inability to progress a promising candidate beyond preliminary studies. These challenges are specific to drug repurposing and involve issues such as patent considerations, regulatory factors, and organizational obstacles (Pushpakom et al. 2019).

2.6.1 Patent Consideration

Several legal and intellectual property obstacles hinder the process of drug repurposing (Ashburn and Thor 2004; Breckenridge and Jacob 2019). The primary challenges revolve around obtaining patents for newly repurposed use and enforcing patent rights. This significantly impacts the expected profitability of the repurposed product (Ashburn and Thor 2004). If a known medicinal molecule has a new medical application that is both novel and non-obvious, it can be obtained in most major pharmaceutical marketplaces. Nonetheless, a plethora of prospective repurposing uses has already been recorded in clinical practice and scholarly publications. Even without clinical validation, prior knowledge might hinder patent fortification, unless the patentee distinguishes their claims from publicly available information. Facts proving the drug's legitimacy for its novel purpose must be provided in the patent application submitted by the patentee in order to be granted a patent for a newly repurposed medicinal use (Pushpakom et al. 2019).

For off-patent drugs, acquiring a new method-of-use (MOU) patent for the repurposed use of an old generic drug is possible, as long as this use is novel, inventive, and supported by credible data. Enforcing these patents becomes challenging if the repurposed indication uses existing formulations and dosage forms of the generic drug (Kremer and Jones 2014). The generic manufacturer can legally label their product only for non-patented indications, making it challenging to claim infringement on the new MOU patent unless they actively promote its use for the patented indication. This situation may pose difficulties in restricting off-label use, potentially impacting the product's profitability (Kremer and Jones 2014). Nonetheless, if the repurposed indication necessitates a unique formulation or dosage regimen not easily achieved with the available generic versions, off-label use can be limited.

Market exclusivity presents another significant hurdle for repurposed drugs. The UK's 2015–16 Off-Patent Drugs Bill aimed to address situations where an expired patent drug proves effective for a new, unlicensed indication, but fails to become legislation. Afterward, the Association of Medical Research Charities collaborated with various stakeholders to make recommendations for facilitating drug repurposing. The report suggests ways for sponsors to collaborate with regulatory authorities, develop financial incentives for generic manufacturers, and create a framework for investigating drug repurposing. The OPEN ACT in the US prolongs a commercial drug's patent validity when it is employed for the cure of an uncommon illness, providing an additional 6 months of exclusivity. Creating novel compositions, dosage forms, or compounds with a comparable therapeutic impact and securing unique marketing authorization in new geographical areas are further tactics to improve patentability (Ashburn and Thor 2004).

2.6.2 Regulatory Factors

Regulatory elements play a pivotal role in shaping the path of repurposed drug development. Murteira et al. conducted a study that examined the regulatory framework for medications that were recently reprofiled and reformulated, offering insights within the EU, with a particular focus on the UK, France, and Germany. The research underscored the centralized procedure as the primary pathway for repurposed drug submissions and approvals. In the United States, as classified by Murteira and team (Breckenridge and Jacob 2019), new drug application (NDA) chemical type 1 (new molecular entity), NDA type 6 (new indication), and supplemental new drug application (sNDA; new indication) were solely employed for repurposing submissions. NDA types 3 (new dosage form) and 4 (new combination), on the other hand, were flexible and could be utilized for reformulation or drug repurposing. Notably, the study observed that a significant percentage of repurposing cases in both the EU (Germany: 88.9%, France: 83.3%, and the UK: 93.8%) and the US (69.6%) secured approval before the original product's patent expiration.

For repurposed medications labeled as orphan drugs, the EU/EEA offers a 10-year shield against market competition from similar drugs with comparable indications, coupled with an extra 2 years if they adhere to an agreed Pediatric Investigation Plan. All applications of orphan drugs must undergo a centralized procedure. Repurposed drugs lacking orphan designation can enjoy a 10-year period of data exclusivity for Article 8(3) (complete dossier) applications. A year of information exclusivity can be allowed for submissions under Article 10(5) that seek inventive use of well-known drugs. Nevertheless, these restrictions on data confidentiality cannot be applied to changes made to already issued marketing approvals.

In the United States, the FDA provides a 3-year span of data exclusivity for fresh applications of marketed drugs. However, firms often do not recoup their investment in repurposing certain pharmaceuticals during this 3-year timeframe. Moreover, as discussed above in the section on patent considerations, employing a repurposed generic medication off-label might also reduce its market worth.

2.6.3 Organizational Obstacles

Pharmaceutical giants are collaborating with smaller biotech startups and academic groups as they recognize the prospect of repurposing medications outside their core illness domain of emphasis. A prime example is the AstraZeneca Open Innovation Platform, created to encourage external partnerships for collaborative research in drug repurposing. This means getting possession of well-characterized drugs that can be used in Phase II clinical studies in addition to integrative and preclinical investigations that are appropriate for repurposing. Notable instances of such endeavors include GlaxoSmithKline's Center for Excellence for External Drug Discovery, Pfizer's Centers for Therapeutic Innovation, and Bayer's Grant4Indications program.

Drug repurposing in the pharmaceutical sector, however, may run into some organizational roadblocks. These impediments are especially likely to arise if the repurposed indication falls outside the company's main ailment prioritization or when a substance was shelved over development, leaving the R&D division without a current endeavor to support the new indication. In such instances, the company might face shortages in personnel, funding, and resources essential for advancing a potential drug repurposing project. To tackle these challenges in clinical studies, external resources, such as contract manufacturing organizations, can be employed for drug supply, along with seeking regulatory support and pharmacovigilance. Exploring alternative funding pathways, as elaborated in the subsequent section, can be considered to ensure the examination of robust scientific hypotheses when pursuing such projects internally may not have been feasible. Moreover, broadening the array of available compounds can generate more opportunities for repurposing, especially those targeting innovative mechanisms. However, most of these substances are typically actively undergoing development, raising concerns about sharing them beyond the pharmaceutical company, including possible new threats to public safety and concerns with intellectual property holdings. It might help allay these worries if productive instances of reprofiling chemicals that are still under research are shared (Pushpakom et al. 2019).

2.7 Repurposed Drug in Market or Clinical Trials

Several repurposed medications are currently available in the market or undergoing clinical trials for the treatment of various diseases. Initially, specific drugs received approval for particular treatments, but due to limited clinical efficacy and notable side effects, these medications find utility in the treatment of alternative diseases or disorders, showcasing outstanding pharmacological effects and improved patient compliance. As an illustration, minoxidil was initially used for antihypertensive treatment owing to its vasodilation properties. However, its notable side effect of promoting excessive hair growth led to its repurposed use in treating alopecia, or hair loss. Furthermore, various repurposed therapeutics are undergoing clinical trials at various phases in treating different diseases. Antihypertensive propranolol hydrochloride is being investigated in early Phase I research (NCT02013492) to treat individuals with solid tumors that are metastatic or regionally recurring and are unable to be surgically removed, including stage IV melanoma and breast tumors. The repurposed drug candidates approved in the market and undergoing clinical trials for the treatment of various diseases are summarized in Table 2.3.

2.8 Conclusion and Future Perspectives

Drug repurposing has emerged as an alternative and promising avenue to de novo drug discovery in treating different kinds of diseases. It is a quicker and less costly method of increasing the arsenal of approved treatments for a range of illnesses and

Table 2.3 Repurposed drugs in the market or clinical trials for various diseases

Drug name	Approved primary use	Repurposed use	Year of approval/ phase of clinical trial	Reference/ clinical trial identifier
Aspirin	Pain suppression	Colorectal cancer	2015	Drew et al. (2016)
Ketoconazole	Fungus-related infections	Cushing's syndrome	2014	Viecceli et al. (2022)
Topiramate	Seizure disorder	Obesity	2012	Smith et al. (2013)
Dapoxetine	Pain relief and depression	Premature ejaculation	2012	Sangkum et al. (2013)
Fingolimod	Organ rejection	Multiple sclerosis	2010	Fragoso et al. (2019)
Raloxifene	Low bone density	Breast cancer	2007	Gennari et al. (2008)
Rituximab	Cancers	Rheumatoid arthritis	2006	Lopez-Olivo et (al. 2015)
Duloxetine	Depression	Stress urinary incontinence	2004	Guay (2005)
Atomoxetine	Parkinsonism	Attention-deficit hyperactivity disorder	2002	Walker et al. (2015)
Celecoxib	Pain and inflammation	Familial adenomatous polyposis	2000	Lynch et al. (2010)
Thalidomide	Morning sickness	Erythema nodosum leprosum and multiple myeloma	1998	Sampaio et al. (2002)
Sildenafil	Chest pain	Erectile dysfunction	1998	Krishnappa et al. (2019)
Minoxidil	Hypertension	Alopecia	1988	Goren and Naccarato (2018)
Zidovudine	Cancer	HIV/AIDS	1987	Cihlar and Ray (2010)
Metformin	Diabetics	Age-related macular degeneration	Phase II	NCT02684578
Dimethyl fumarate	Psoriasis	Cutaneous T-cell lymphoma	Phase II	NCT02546440
Povidone-iodine	Antiseptic for wound care	COVID-19	Phase II	NCT04549376
Dexamethasone	Anti-inflammatory	Tuberculosis, meningitis, drug-induced liver injury, and HIV	Phase III	NCT03092817
Celecoxib	Painkiller	Breast cancer	Phase II	NCT00056082
Propranolol	Hypertension	Skin cancer	Phase II	NCT03384836

is therefore considered as significant when compared to the de novo drug discovery process. Various drug repurposing avenues, including computational and experimental, play a major role in the identification of drugs having the potential to be repurposed as drug candidates in treating diseases. However, further advances in these approaches are highly required.

Therapeutics from different pharmacological classes and phytoceuticals are found to be potential in treating diseases like cancer, cardiovascular, neurological, and infectious diseases. Since these medications can react with multiple targets, it is crucial to comprehend their polypharmacology. To obtain a greater therapeutic effect, it is therefore necessary to identify appropriate therapies with greater specificity for disease targets using various computational methods. Preclinical research is also necessary to determine the safety and effectiveness of repurposed medicines. In addition, numerous treatments are still being developed in laboratories; there are new opportunities for pharmaceutical scientists to advance them to the clinical and commercial stages. Drug repurposing is also allied with various bottlenecks like patents, regulations, and organizational challenges, and therefore, additional efforts are necessary to conquer these bottlenecks. Moreover, the main obstacle to the traditional method of repurposing drug delivery is its nonspecific dispersion and the toxicities associated with its initial use. Consequently, it is crucial to administer repurposed medicines via optimal nanocarriers through the right route of administration. Repurposed drug-loaded nanoformulations are not widely available in the market. Thus, there is a fantastic chance to create an efficient nanoformulation that is loaded with repurposed therapies to combat a range of malignancies and to commercialize those formulations.

References

Abdel-Aal MA, Shaykoon MS, Abuo-Rahma GE, Mohamed MF, Badr M, Abdel-Aziz SA (2021) Synthesis, antitumor, antibacterial and urease inhibitory evaluation of new piperazinyl N-4 carbamoyl functionalized ciprofloxacin derivatives. Pharmacol Rep 73:891–906

Agnati LF, Benfenati F, Solfrini V, Biagini G, Fuxe K, Guidolin D, Carani C, Zini I (1992) Brain aging and neuronal plasticity. Ann N Y Acad Sci 673:180–186

Alshareef A, Zhang HF, Huang YH, Wu C, Zhang JD, Wang P, El-Sehemy A, Fares M, Lai R (2016) The use of cellular thermal shift assay (CETSA) to study Crizotinib resistance in ALK-expressing human cancers. Sci Rep 6(1):33710

Ashburn TT, Thor KB (2004) Drug repositioning: identifying and developing new uses for existing drugs. Nat Rev Drug Discov 3(8):673–683

Bago R, Sommer E, Castel P, Crafter C, Bailey FP, Shpiro N, Baselga J, Cross D, Eyers PA, Alessi DR (2016) The hVps34-SGK 3 pathway alleviates sustained PI3K/Akt inhibition by stimulating mTORC 1 and tumour growth. EMBO J 35(17):1902–1922

Barretina J, Caponigro G, Stransky N, Venkatesan K, Margolin AA, Kim S, Wilson CJ, Lehár J, Kryukov GV, Sonkin D, Reddy A (2012) The cancer cell line encyclopedia enables predictive modelling of anticancer drug sensitivity. Nature 483(7391):603–607

Basu A, Bodycombe NE, Cheah JH, Price EV, Liu K, Schaefer GI, Ebright RY, Stewart ML, Ito D, Wang S, Bracha AL (2013) An interactive resource to identify cancer genetic and lineage dependencies targeted by small molecules. Cell 154(5):1151–1161

Beberok A, Wrześniok D, Minecka A, Rok J, Delijewski M, Rzepka Z, Respondek M, Buszman E (2018) Ciprofloxacin-mediated induction of S-phase cell cycle arrest and apoptosis in COLO829 melanoma cells. Pharmacol Rep 70(1):6–13
Beberok A, Rzepka Z, Respondek M, Rok J, Stradowski M, Wrześniok D (2019) Moxifloxacin as an inducer of apoptosis in melanoma cells: a study at the cellular and molecular level. Toxicol In Vitro 55:75–92
Bermejo PE, Anciones B (2009) A review of the use of zonisamide in Parkinson's disease. Ther Adv Neurol Disord 2(5):313–317
Blagg J, Workman P (2017) Choose and use your chemical probe wisely to explore cancer biology. Cancer Cell 32(1):9–25
Blanke CD, Demetri GD, Von Mehren M, Heinrich MC, Eisenberg B, Fletcher JA, Corless CL, Fletcher CD, Roberts PJ, Heinz D, Wehre E (2008) Long-term results from a randomized phase II trial of standard-versus higher-dose imatinib mesylate for patients with unresectable or metastatic gastrointestinal stromal tumors expressing KIT. J Clin Oncol 26(4):620–625
Bowie M, Pilie P, Wulfkuhle J, Lem S, Hoffman A, Desai S, Petricoin E, Carter A, Ambrose A, Seewaldt V, Yu D (2015) Fluoxetine induces cytotoxic endoplasmic reticulum stress and autophagy in triple negative breast cancer. World J Clin Oncol 6(6):299
Breckenridge A, Jacob R (2019) Overcoming the legal and regulatory barriers to drug repurposing. Nat Rev Drug Discov 18(1):1–2
Brehmer D, Greff Z, Godl K, Blencke S, Kurtenbach A, Weber M, Müller S, Klebl B, Cotten M, Kéri G, Wissing J (2005) Cellular targets of gefitinib. Cancer Res 65(2):379–382
Čáňová K, Rozkydalová L, Vokurková D, Rudolf E (2018) Flubendazole induces mitotic catastrophe and apoptosis in melanoma cells. Toxicol In Vitro 46:313–322
Cantrell MS, Soto-Avellaneda A, Wall JD, Ajeti AD, Morrison BE, Warner LR, McDougal OM (2021) Repurposing drugs to treat heart and brain illness. Pharmaceuticals 14(6):573
Carter TA, Wodicka LM, Shah NP, Velasco AM, Fabian MA, Treiber DK, Milanov ZV, Atteridge CE, Biggs WH III, Edeen PT, Floyd M (2005) Inhibition of drug-resistant mutants of ABL, KIT, and EGF receptor kinases. Proc Natl Acad Sci 102(31):11011–11016
Cavalla D, Singal C (2012) Retrospective clinical analysis for drug rescue: for new indications or stratified patient groups. Drug Discov Today 17(3–4):104–109
Chen Z, Chen J, Collins R, Guo Y, Peto R, Wu F, Li L (2011) China Kadoorie biobank of 0.5 million people: survey methods, baseline characteristics and long-term follow-up. Int J Epidemiol 40(6):1652–1666
Cihlar T, Ray AS (2010) Nucleoside and nucleotide HIV reverse transcriptase inhibitors: 25 years after zidovudine. Antivir Res 85(1):39–58
Cohen P (2002) Protein kinases—the major drug targets of the twenty-first century? Nat Rev Drug Discov 1(4):309–315
Cooke RM, Brown AJ, Marshall FH, Mason JS (2015) Structures of G protein-coupled receptors reveal new opportunities for drug discovery. Drug Discov Today 20(11):1355–1364
Cousin MA, Ebbert JO, Wiinamaki AR, Urban MD, Argue DP, Ekker SC, Klee EW (2014) Larval zebrafish model for FDA-approved drug repositioning for tobacco dependence treatment. PLoS One 9(3):e90467
Cummings JL, Zhong K (2014) Repackaging FDA-approved drugs for degenerative diseases: promises and challenges. Expert Rev Clin Pharmacol 7(2):161–165
Dakshanamurthy S, Issa NT, Assefnia S, Seshasayee A, Peters OJ, Madhavan S, Uren A, Brown ML, Byers SW (2012) Predicting new indications for approved drugs using a proteochemometric method. J Med Chem 55(15):6832–6848
Davis MI, Hunt JP, Herrgard S, Ciceri P, Wodicka LM, Pallares G, Hocker M, Treiber DK, Zarrinkar PP (2011) Comprehensive analysis of kinase inhibitor selectivity. Nat Biotechnol 29(11):1046–1051
De Rosa MC, Purohit R, García-Sosa AT (2023) Drug repurposing: a nexus of innovation, science, and potential. Sci Rep 13(1):17887
Del Grande MP, Miyake AM, Nagamine MK, Leite JV, da Fonseca II, de Oliveira Massoco C, Dagli ML (2022) Methylene blue and photodynamic therapy for melanomas: inducing differ-

ent rates of cell death (necrosis and apoptosis) in B16-F10 melanoma cells according to methylene blue concentration and energy dose. Photodiagn Photodyn Ther 37:102635
Drew DA, Cao Y, Chan AT (2016) Aspirin and colorectal cancer: the promise of precision chemoprevention. Nat Rev Cancer 16(3):173–186
Dudley JT, Sirota M, Shenoy M, Pai RK, Roedder S, Chiang AP, Morgan AA, Sarwal MM, Pasricha PJ, Butte AJ (2011a) Computational repositioning of the anticonvulsant topiramate for inflammatory bowel disease. Sci Transl Med 3(96):96ra76
Dudley JT, Deshpande T, Butte AJ (2011b) Exploiting drug–disease relationships for computational drug repositioning. Brief Bioinform 12(4):303–311
Duncan JS, Whittle MC, Nakamura K, Abell AN, Midland AA, Zawistowski JS, Johnson NL, Granger DA, Jordan NV, Darr DB, Usary J (2012) Dynamic reprogramming of the kinome in response to targeted MEK inhibition in triple-negative breast cancer. Cell 149(2):307–321
Durães F, Pinto M, Sousa E (2018) Old drugs as new treatments for neurodegenerative diseases. Pharmaceuticals 11(2):44
Eisenstein M (2015) Big data: the power of petabytes. Nature 527(7576):S2–S4
Elakkad YE, Mohamed SN, Abuelezz NZ (2021) Potentiating the cytotoxic activity of a novel simvastatin-loaded cubosome against breast cancer cells: insights on dual cell death via ferroptosis and apoptosis. Breast Cancer Targets Ther 13:675–689
Ellis LZ, Liu W, Luo Y, Okamoto M, Qu D, Dunn JH, Fujita M (2011) Green tea polyphenol epigallocatechin-3-gallate suppresses melanoma growth by inhibiting inflammasome and IL-1β secretion. Biochem Biophys Res Commun 414(3):551–556
Eskiocak U, Ramesh V, Gill JG, Zhao Z, Yuan SW, Wang M, Vandergriff T, Shackleton M, Quintana E, Frankel AE, Johnson TM (2016) Synergistic effects of ion transporter and MAP kinase pathway inhibitors in melanoma. Nat Commun 7(1):12336
Eyers PA, Van Den Ijssel P, Quinlan RA, Goedert M, Cohen P (1999) Use of a drug-resistant mutant of stress-activated protein kinase 2a/p38 to validate the in vivo specificity of SB 203580. FEBS Lett 451(2):191–196
Fasehee H, Ghavamzadeh A, Alimoghaddam K, Ghaffari SH, Faghihi S (2017) A comparative cytotoxic evaluation of disulfiram encapsulated PLGA nanoparticles on MCF-7 cells. Int J Hematol Oncol Stem Cell Res 11(2):102
Fauci AS (2001) Infectious diseases: considerations for the 21st century. Clin Infect Dis 32(5):675–685
Fragoso YD, Adoni T, Gomes S, Goncalves MV, Parolin LF, Rosa G, Ruocco HH (2019) Severe exacerbation of multiple sclerosis following withdrawal of fingolimod. Clin Drug Investig 39:909–913
Fritz I, Wagner P, Bottai M, Eriksson H, Ingvar C, Krakowski I, Nielsen K, Olsson H (2020) Desloratadine and loratadine use associated with improved melanoma survival. Allergy 75(8):2096–2099
Gambhire VM, Salunkhe SM, Gambhire MS (2018) Atorvastatin-loaded lipid nanoparticles: antitumor activity studies on MCF-7 breast cancer cells. Drug Dev Ind Pharm 44(10):1685–1692
Gelosa P, Castiglioni L, Camera M, Sironi L (2020) Drug repurposing in cardiovascular diseases: opportunity or hopeless dream? Biochem Pharmacol 177:113894
Gennari L, Merlotti D, Paola VD, Nuti R (2008) Raloxifene in breast cancer prevention. Expert Opin Drug Saf 7(3):259–270
Goenka S, Simon SR (2020) Organogold drug Auranofin exhibits anti-melanogenic activity in B16F10 and MNT-1 melanoma cells. Arch Dermatol Res 312(3):213–221
Goodman A (2005) Tamoxifen, a cancer therapy, explored for als. Neurol Today 5(7):22–26
Goren A, Naccarato T (2018) Minoxidil in the treatment of androgenetic alopecia. Dermatol Ther 31(5):e12686
Grahovac J, Srdić-Rajić T, Santibañez JF, Pavlović M, Čavić M, Radulović S (2019) Telmisartan induces melanoma cell apoptosis and synergizes with vemurafenib in vitro by altering cell bioenergetics. Cancer Biol Med 16(2):247
Greene CS, Voight BF (2016) Pathway and network-based strategies to translate genetic discoveries into effective therapies. Hum Mol Genet 25(R2):R94–R98

Greene CS, Krishnan A, Wong AK, Ricciotti E, Zelaya RA, Himmelstein DS, Zhang R, Hartmann BM, Zaslavsky E, Sealfon SC, Chasman DI (2015) Understanding multicellular function and disease with human tissue-specific networks. Nat Genet 47(6):569–576
Guay DR (2005) Duloxetine for management of stress urinaryincontinence. Am J Geriatr Pharmacother 3(1):25–38
Günther R, Balck A, Koch JC, Nientiedt T, Sereda M, Bähr M, Lingor P, Tönges L (2017) Rho kinase inhibition with fasudil in the SOD1G93A mouse model of amyotrophic lateral sclerosis—symptomatic treatment potential after disease onset. Front Pharmacol 8:17
Hall-Jackson CA, Eyers PA, Cohen P, Goedert M, Boyle FT, Hewitt N, Plant H, Hedge P (1999) Paradoxical activation of Raf by a novel Raf inhibitor. Chem Biol 6(8):559–568
Hamsa TP, Kuttan G (2012) Antiangiogenic activity of berberine is mediated through the downregulation of hypoxia-inducible factor-1, VEGF, and proinflammatory mediators. Drug Chem Toxicol 35(1):57–70
Hanson K, Robinson SR, Al-Yousuf K, Hendry AE, Sexton DW, Sherwood V, Wheeler GN (2018) The anti-rheumatic drug, leflunomide, synergizes with MEK inhibition to suppress melanoma growth. Oncotarget 9(3):3815
Hartung HP, Gonsette R, Konig N, Kwiecinski H, Guseo A, Morrissey SP, Krapf H, Zwingers T (2002) Mitoxantrone in progressive multiple sclerosis: a placebo-controlled, double-blind, randomised, multicentre trial. Lancet 360(9350):2018–2025
Hu H, Dong Z, Tan P, Zhang Y, Liu L, Yang L, Liu Y, Cui H (2016) Antibiotic drug tigecycline inhibits melanoma progression and metastasis in a p21CIP1/Waf1-dependent manner. Oncotarget 7(3):3171
Huang H, Zhang G, Zhou Y, Lin C, Chen S, Lin Y, Mai S, Huang Z (2018) Reverse screening methods to search for the protein targets of chemopreventive compounds. Front Chem 6:138
Hurle MR, Yang L, Xie Q, Rajpal DK, Sanseau P, Agarwal P (2013) Computational drug repositioning: from data to therapeutics. Clin Pharmacol Ther 93(4):335–341
Iljin K, Ketola K, Vainio P, Halonen P, Kohonen P, Fey V, Grafström RC, Perälä M, Kallioniemi O (2009) High-throughput cell-based screening of 4910 known drugs and drug-like small molecules identifies disulfiram as an inhibitor of prostate cancer cell growth. Clin Cancer Res 15(19):6070–6078
Iorio F, Isacchi A, di Bernardo D, Brunetti-Pierri N (2010a) Identification of small molecules enhancing autophagic function from drug network analysis. Autophagy 6(8):1204–1205
Iorio F, Bosotti R, Scacheri E, Belcastro V, Mithbaokar P, Ferriero R, Murino L, Tagliaferri R, Brunetti-Pierri N, Isacchi A, di Bernardo D (2010b) Discovery of drug mode of action and drug repositioning from transcriptional responses. Proc Natl Acad Sci 107(33):14621–14626
Iorio F, Saez-Rodriguez J, Bernardo DD (2013) Network based elucidation of drug response: from modulators to targets. BMC Syst Biol 7(1):1–9
Iorio F, Knijnenburg TA, Vis DJ, Bignell GR, Menden MP, Schubert M, Aben N, Gonçalves E, Barthorpe S, Lightfoot H, Cokelaer T (2016) A landscape of pharmacogenomic interactions in cancer. Cell 166(3):740–754
Jensen PB, Jensen LJ, Brunak S (2012) Mining electronic health records: towards better research applications and clinical care. Nat Rev Genet 13(6):395–405
Jia H, Ren W, Feng Y, Wei T, Guo M, Guo J, Zhao J, Song X, Wang M, Zhao T, Wang H (2018) The enhanced antitumour response of pimozide combined with the IDO inhibitor L-MT in melanoma. Int J Oncol 53(3):949–960
Jin X, Wei Y, Liu Y, Lu X, Ding F, Wang J, Yang S (2019) Resveratrol promotes sensitization to doxorubicin by inhibiting epithelial-mesenchymal transition and modulating SIRT1/β-catenin signaling pathway in breast cancer. Cancer Med 8(3):1246–1257
Kanwal F, Ma M, Rehman MF, Khan FU, Elizur SE, Batool AI, Wang CC, Tabassum T, Lu C, Wang Y (2022) Aspirin repurposing in folate-decorated nanoparticles: another way to target breast cancer. Front Mol Biosci 8:788279
Karaman MW, Herrgard S, Treiber DK, Gallant P, Atteridge CE, Campbell BT, Chan KW, Ciceri P, Davis MI, Edeen PT, Faraoni R (2008) A quantitative analysis of kinase inhibitor selectivity. Nat Biotechnol 26(1):127–132

Katiyar SK, Meeran SM, Katiyar N, Akhtar S (2009) p53 cooperates berberine-induced growth inhibition and apoptosis of non-small cell human lung cancer cells in vitro and tumor xenograft growth in vivo. Mol. Carcinog 48(1):24–37

Katiyar SS, Muntimadugu E, Rafeeqi TA, Domb AJ, Khan W (2016) Co-delivery of rapamycin- and piperine-loaded polymeric nanoparticles for breast cancer treatment. Drug Deliv 23(7):2608–2616

Khan N, Syed DN, Ahmad N, Mukhtar H (2013) Fisetin: a dietary antioxidant for health promotion. Antioxid Redox Signal 19(2):151–162

Kharkar PS, Warrier S, Gaud RS (2014) Reverse docking: a powerful tool for drug repositioning and drug rescue. Future Med Chem 6(3):333–342

Khodaie F, Khazaei-Poul Y, Moini-Zanjani T (2017) Anti-proliferative effects of piroxicam and nimesulide on A431 human squamous carcinoma cell line. Int J Cancer Manag 10(4):e7565

Kitchen DB, Decornez H, Furr JR, Bajorath J (2004) Docking and scoring in virtual screening for drug discovery: methods and applications. Nat Rev Drug Discov 3(11):935–949

Klaeger S, Gohlke B, Perrin J, Gupta V, Heinzlmeir S, Helm D, Qiao H, Bergamini G, Handa H, Savitski MM, Bantscheff M (2016) Chemical proteomics reveals ferrochelatase as a common off-target of kinase inhibitors. ACS Chem Biol 11(5):1245–1254

Kremer S, Jones R (2014) Repurposed drugs: second time lucky. Life sciences intellectual property review.

Krishnappa P, Fernandez-Pascual E, Carballido J, Martinez-Salamanca JI (2019) Sildenafil/Viagra in the treatment of premature ejaculation. Int J Impot Res 31(2):65–70

Kudugunti SK, Vad NM, Ekogbo E, Moridani MY (2011) Efficacy of caffeic acid phenethyl ester (CAPE) in skin B16-F0 melanoma tumor bearing C57BL/6 mice. Investig New Drugs 29:52–62

Kumbhar P, Kole K, Khadake V, Marale P, Manjappa A, Nadaf S, Jadhav R, Patil A, Singh SK, Dua K, Jha NK (2022a) Nanoparticulate drugs and vaccines: breakthroughs and bottlenecks of repurposing in breast cancer. J Control Release 349:812–830

Kumbhar P, Manjappa A, Shah R, Jha NK, Singh SK, Dua K, Disouza J, Patravale V (2022b) Inhalation delivery of repurposed drugs for lung cancer: approaches, benefits and challenges. J Control Release 341:1–5

Kumbhar P, Kole K, Manjappa A, Jha NK, Disouza J, Patravale V (2023) Drug repurposing opportunities in cancer. In: Drug repurposing for emerging infectious diseases and cancer. Springer Nature Singapore, Singapore, pp 53–87

Lakhter AJ, Sahu RP, Sun Y, Kaufmann WK, Androphy EJ, Travers JB, Naidu SR (2013) Chloroquine promotes apoptosis in melanoma cells by inhibiting BH3 domain–mediated PUMA degradation. J Invest Dermatol 133(9):2247–2254

Lee JK, Shin JH, Lee JE, Choi EJ (2015) Role of autophagy in the pathogenesis of amyotrophic lateral sclerosis. Biochim Biophys Acta Mol Basis Dis 1852(11):2517–2524

Li W, Shi Y, Wang R, Pan L, Ma L, Jin F (2018a) Resveratrol promotes the sensitivity of small-cell lung cancer H446 cells to cisplatin by regulating intrinsic apoptosis. Int J Oncol 53(5):2123–2130

Li S, Zhao Q, Wang B, Yuan S, Wang X, Li K (2018b) Quercetin reversed MDR in breast cancer cells through down-regulating P-gp expression and eliminating cancer stem cells mediated by YB-1 nuclear translocation. Phytother Res 32(8):1530–1536

Liu Z, Ren L, Liu C, Xia T, Zha X, Wang S (2015) Phenformin induces cell cycle change, apoptosis, and mesenchymal-epithelial transition and regulates the AMPK/mTOR/p70s6k and MAPK/ERK pathways in breast cancer cells. PLoS One 10(6):e0131207

Liu W, Stachura P, Xu HC, Umesh Ganesh N, Cox F, Wang R, Lang KS, Gopalakrishnan J, Häussinger D, Homey B, Lang PA (2020) Repurposing the serotonin agonist Tegaserod as an anticancer agent in melanoma: molecular mechanisms and clinical implications. J Exp Clin Cancer Res 39(1):1–6

Lopez-Olivo MA, Urruela MA, McGahan L, Pollono EN, Suarez-Almazor ME (2015) Rituximab for rheumatoid arthritis. Cochrane Database Syst Rev 2:99–111

Lynch PM, Ayers GD, Hawk E, Richmond E, Eagle C, Woloj M, Church J, Hasson H, Patterson S, Half E, Burke CA (2010) The safety and efficacy of celecoxib in children with familial adenomatous polyposis. Am J Gastroenterol 105(6):1437–1443
Marcondes MC, Fernandes AC, Itabaiana JI, de Souza RO, Sola-Penna M, Zancan P (2015) Nanomicellar formulation of clotrimazole improves its antitumor action toward human breast cancer cells. PLoS One 10(6):e0130555
Martinez A, Palomo Ruiz MD, Perez DI, Gil C (2017) Drugs in clinical development for the treatment of amyotrophic lateral sclerosis. Expert Opin Investig Drugs 26(4):403–414
Miettinen TP, Björklund M (2014) NQO2 is a reactive oxygen species generating off-target for acetaminophen. Mol Pharm 11(12):4395–4404
Millwood IY, Bennett DA, Walters RG, Clarke R, Waterworth D, Johnson T, Chen Y, Yang L, Guo Y, Bian Z, Hacker A (2016) A phenome-wide association study of a lipoprotein-associated phospholipase A2 loss-of-function variant in 90 000 Chinese adults. Int J Epidemiol 45(5):1588–1599
Moffat JG, Vincent F, Lee JA, Eder J, Prunotto M (2017) Opportunities and challenges in phenotypic drug discovery: an industry perspective. Nat Rev Drug Discov 16(8):531–543
Molina DM, Jafari R, Ignatushchenko M, Seki T, Larsson EA, Dan C, Sreekumar L, Cao Y, Nordlund P (2013) Monitoring drug target engagement in cells and tissues using the cellular thermal shift assay. Science 341(6141):84–87
Pal D, Banerjee S, Mukherjee S, Roy A, Panda CK, Das S (2010) Eugenol restricts DMBA croton oil induced skin carcinogenesis in mice: downregulation of c-Myc and H-ras, and activation of p53 dependent apoptotic pathway. J Dermatol Sci 59:31
Munoz L (2017) Non-kinase targets of protein kinase inhibitors. Nat Rev Drug Discov 16(6):424–440
Nosengo N (2016) Can you teach old drugs new tricks? Nature 534(7607):314–316
O'Donoghue ML, Braunwald E, White HD, Steen DL, Lukas MA, Tarka E, Steg PG, Hochman JS, Bode C, Maggioni AP, Im K (2014) Effect of darapladib on major coronary events after an acute coronary syndrome: the SOLID-TIMI 52 randomized clinical trial. JAMA 312(10):1006–1015
Onódi Z, Koch S, Rubinstein J, Ferdinandy P, Varga ZV (2023) Drug repurposing for cardiovascular diseases: new targets and indications for probenecid. Br J Pharmacol 180(6):685–700
Pagadala NS, Syed K, Tuszynski J (2017) Software for molecular docking: a review. Biophys Rev 9:91–102
Paik H, Chung AY, Park HC, Park RW, Suk K, Kim J, Kim H, Lee K, Butte AJ (2015) Repurpose terbutaline sulfate for amyotrophic lateral sclerosis using electronic medical records. Sci Rep 5(1):8580
Paleacu D (2007) Tetrabenazine in the treatment of Huntington's disease. Neuropsychiatr Dis Treat 3(5):545–551
Pan HC, Jiang Q, Yu Y, Mei JP, Cui YK, Zhao WJ (2015) Quercetin promotes cell apoptosis and inhibits the expression of MMP-9 and fibronectin via the AKT and ERK signalling pathways in human glioma cells. Neurochem Int 80:60–71
Parvathaneni V, Chilamakuri R, Kulkarni NS, Baig NF, Agarwal S, Gupta V (2022) Exploring Amodiaquine's repurposing potential in breast cancer treatment—assessment of in-vitro efficacy & mechanism of action. Int J Mol Sci 23(19):11455
Paskas S, Mazzon E, Basile MS, Cavalli E, Al-Abed Y, He M, Rakocevic S, Nicoletti F, Mijatovic S, Maksimovic-Ivanic D (2019) Lopinavir-NO, a nitric oxide-releasing HIV protease inhibitor, suppresses the growth of melanoma cells in vitro and in vivo. Investig New Drugs 37:1014–1028
Paulmurugan R, Bhethanabotla R, Mishra K, Devulapally R, Foygel K, Sekar TV, Ananta JS, Massoud TF, Joy A (2016) Folate receptor–targeted polymeric micellar nanocarriers for delivery of orlistat as a repurposed drug against triple-negative breast cancer. Mol Cancer Ther 15(2):221–231
Peplow M (2016) The 100 000 genomes project. BMJ 353: i1757
Pushpakom S, Iorio F, Eyers PA, Escott KJ, Hopper S, Wells A, Doig A, Guilliams T, Latimer J, McNamee C, Norris A (2019) Drug repurposing: progress, challenges and recommendations. Nat Rev Drug Discov 18(1):41–58

Rico SR, Abbasi AZ, Ribeiro G, Ahmed T, Wu XY, de Oliveira Silva D (2017) Diruthenium (ii, iii) metallodrugs of ibuprofen and naproxen encapsulated in intravenously injectable polymer–lipid nanoparticles exhibit enhanced activity against breast and prostate cancer cells. Nanoscale 9(30):10701–10714

Riedel T, Demaria O, Zava O, Joncic A, Gilliet M, Dyson PJ (2018) Drug repurposing approach identifies a synergistic drug combination of an antifungal agent and an experimental organometallic drug for melanoma treatment. Mol Pharm 15(1):116–126

Sampaio EP, Hernandez MO, Carvalho DS, Sarno EN (2002) Management of erythema nodosum leprosum by thalidomide: thalidomide analogues inhibit M. Leprae-induced TNFα production in vitro. Biomed Pharmacother 56(1):13–19

Sangkum P, Badr R, Serefoglu EC, Hellstrom WJ (2013) Dapoxetine and the treatment of premature ejaculation. Transl Androl Urol 2(4):301

Sanseau P, Agarwal P, Barnes MR, Pastinen T, Richards JB, Cardon LR, Mooser V (2012) Use of genome-wide association studies for drug repositioning. Nat Biotechnol 30(4):317–320

Sarvi S, Crispin R, Lu Y, Zeng L, Hurley TD, Houston DR, von Kriegsheim A, Chen CH, Mochly-Rosen D, Ranzani M, Mathers ME (2018) ALDH1 bio-activates nifuroxazide to eradicate ALDHHigh melanoma-initiating cells. Cell Chem Biol 25(12):1456–1469

Sau S, Mondal SK, Kashaw SK, Iyer AK, Banerjee R (2017) Combination of cationic dexamethasone derivative and STAT3 inhibitor (WP1066) for aggressive melanoma: a strategy for repurposing a phase I clinical trial drug. Mol Cell Biochem 436:119–136

Scatena R, Martorana GE, Bottoni P, Botta G, Pastore P, Giardina B (2007) An update on pharmacological approaches to neurodegenerative diseases. Expert Opin Investig Drugs 16(1):59–72

Seashore-Ludlow B, Rees MG, Cheah JH, Cokol M, Price EV, Coletti ME, Jones V, Bodycombe NE, Soule CK, Gould J, Alexander B (2015) Harnessing connectivity in a large-scale small-molecule sensitivity dataset. Cancer Discov 5(11):1210–1223

Shin DH, Kim OH, Jun HS, Kang MK (2008) Inhibitory effect of capsaicin on B16-F10 melanoma cell migration via the phosphatidylinositol 3-kinase/Akt/Rac1 signal pathway. Exp Mol Med 40(5):486–494

Shukla SK, Kulkarni NS, Chan A, Parvathaneni V, Farrales P, Muth A, Gupta V (2019) Metformin-encapsulated liposome delivery system: an effective treatment approach against breast cancer. Pharmaceutics 11(11):559

Simbulan-Rosenthal CM, Dakshanamurthy S, Gaur A, Chen YS, Fang HB, Abdussamad M, Zhou H, Zapas J, Calvert V, Petricoin EF, Atkins MB (2017) The repurposed anthelmintic mebendazole in combination with trametinib suppresses refractory NRASQ61K melanoma. Oncotarget 8(8):12576

Singh R, Kesharwani P, Mehra NK, Singh S, Banerjee S, Jain NK (2015) Development and characterization of folate anchored Saquinavir entrapped PLGA nanoparticles for anti-tumor activity. Drug Dev Ind Pharm 41(11):1888–1901

Singh SK, Banerjee S, Acosta EP, Lillard JW, Singh R (2017) Resveratrol induces cell cycle arrest and apoptosis with docetaxel in prostate cancer cells via a p53/p21WAF1/CIP1 and p27KIP1 pathway. Oncotarget 8(10):17216

Sleire L, Førde HE, Netland IA, Leiss L, Skeie BS, Enger PØ (2017) Drug repurposing in cancer. Pharmacol Res 124:74–91

Sloane DA, Trikic MZ, Chu ML, Lamers MB, Mason CS, Mueller I, Savory WJ, Williams DH, Eyers PA (2010) Drug-resistant aurora A mutants for cellular target validation of the small molecule kinase inhibitors MLN8054 and MLN8237. ACS Chem Biol 5(6):563–576

Smith SB, Dampier W, Tozeren A, Brown JR, Magid-Slav M (2012) Identification of common biological pathways and drug targets across multiple respiratory viruses based on human host gene expression analysis. PLoS One 7(3):e33174

Smith SM, Meyer M, Trinkley KE (2013) Phentermine/topiramate for the treatment of obesity. Ann Pharmacother 47(3):340–349

Soprano M, Sorriento D, Rusciano MR, Maione AS, Limite G, Forestieri P, D'Angelo D, D'Alessio M, Campiglia P, Formisano P, Iaccarino G (2016) Oxidative stress mediates the antiproliferative effects of nelfinavir in breast cancer cells. PLoS One 11(6):e0155970

Srirangam A, Mitra R, Wang M, Gorski JC, Badve S, Baldridge L, Hamilton J, Kishimoto H, Hawes J, Li L, Orschell CM (2006) Effects of HIV protease inhibitor ritonavir on Akt-regulated cell proliferation in breast cancer. Clin Cancer Res 12(6):1883–1896

Stability Investigators (2014) Darapladib for preventing ischemic events in stable coronary heart disease. N Engl J Med 370(18):1702–1711

Su F, Viros A, Milagre C, Trunzer K, Bollag G, Spleiss O, Reis-Filho JS, Kong X, Koya RC, Flaherty KT, Chapman PB (2012) RAS mutations in cutaneous squamous-cell carcinomas in patients treated with BRAF inhibitors. N Engl J Med 366(3):207–215

Sun W, Weingarten RA, Xu M, Southall N, Dai S, Shinn P, Sanderson PE, Williamson PR, Frank KM, Zheng W (2016) Rapid antimicrobial susceptibility test for identification of new therapeutics and drug combinations against multidrug-resistant bacteria. Emerg Microbes Infect 5(1):1

Sun X, Song J, Li E, Geng H, Li Y, Yu D, Zhong C (2019) (−)-Epigallocatechin-3-gallate inhibits bladder cancer stem cells via suppression of sonic hedgehog pathway. Oncol Rep 42(1):425–435

Takata M, Tanaka H, Kimura M, Nagahara Y, Tanaka K, Kawasaki K, Seto M, Tsuruma K, Shimazawa M, Hara H (2013) Fasudil, a rho kinase inhibitor, limits motor neuron loss in experimental models of amyotrophic lateral sclerosis. Br J Pharmacol 170(2):341–351

Timilsina S, Rajamanickam S, Rao A, Subbarayalu P, Nirzhor S, Abdelfattah N, Viswanadhapalli S, Chen Y, Jatoi I, Brenner A, Rao MK (2022) The antidepressant imipramine inhibits breast cancer growth by targeting estrogen receptor signaling and DNA repair events. Cancer Lett 540:215717

Tomic T, Botton T, Cerezo M, Robert G, Luciano F, Puissant A, Gounon P, Allegra M, Bertolotto C, Bereder JM, Tartare-Deckert S (2011) Metformin inhibits melanoma development through autophagy and apoptosis mechanisms. Cell Death Dis 2(9):e199

Tönges L, Günther R, Suhr M, Jansen J, Balck A, Saal KA, Barski E, Nientied T, Götz AA, Koch JC, Mueller BK (2014) Rho kinase inhibition modulates microglia activation and improves survival in a model of amyotrophic lateral sclerosis. Glia 62(2):217–232

Troutman S, Moleirinho S, Kota S, Nettles K, Fallahi M, Johnson GL, Kissil JL (2016) Crizotinib inhibits NF2-associated schwannoma through inhibition of focal adhesion kinase 1. Oncotarget 7(34):54515

Curry SJ, Krist AH, Owens DK, Barry MJ, Caughey AB, Davidson KW, Doubeni CA, Epling JW, Kemper AR, Kubik M, Landefeld CS (2018) Screening and behavioral counseling interventions to reduce unhealthy alcohol use in adolescents and adults: US Preventive Services Task Force recommendation statement. Jama. 2018 Nov 13;320(18):1899–909

Viecceli C, Mattos AC, Costa MC, Melo RB, Rodrigues TD, Czepielewski MA (2022) Evaluation of ketoconazole as a treatment for Cushing's disease in a retrospective cohort. Front Endocrinol 13:1017331

Walker DJ, Mason O, Clemow DB, Day KA (2015) Atomoxetine treatment in adults with attention-deficit/hyperactivity disorder. Postgrad Med 127(7):686–701

Wang ZY, Zhang HY (2013) Rational drug repositioning by medical genetics. Nat Biotechnol 31(12):1080–1082

Warren GL, Andrews CW, Capelli AM, Clarke B, LaLonde J, Lambert MH, Lindvall M, Nevins N, Semus SF, Senger S, Tedesco G (2006) A critical assessment of docking programs and scoring functions. J Med Chem 49(20):5912–5931

Wei WQ, Denny JC (2015) Extracting research-quality phenotypes from electronic health records to support precision medicine. Genome Med 7:1–4

Wei G, Xu Y, Peng T, Yan J, Wang Z, Sun Z (2017) Sanguinarine exhibits antitumor activity via up-regulation of Fas-associated factor 1 in non-small cell lung cancer. J Biochem Mol Toxicol 31(8):e21914

Wicks P, Vaughan TE, Massagli MP, Heywood J (2011) Accelerated clinical discovery using self-reported patient data collected online and a patient-matching algorithm. Nat Biotechnol 29(5):411–414

Willyard C (2018) Expanded human gene tally reignites debate. Nature 558(7710):354–355

Wisniewski D, Lambek CL, Liu C, Strife A, Veach DR, Nagar B, Young MA, Schindler T, Bornmann WG, Bertino JR, Kuriyan J (2002) Characterization of potent inhibitors of the Bcr-Abl and the c-kit receptor tyrosine kinases. Cancer Res 62(15):4244–4255

Xu M, Lee EM, Wen Z, Cheng Y, Huang WK, Qian X, Tcw J, Kouznetsova J, Ogden SC, Hammack C, Jacob F (2016) Identification of small-molecule inhibitors of Zika virus infection and induced neural cell death via a drug repurposing screen. Nat Med 22(10):1101–1107

Xu K, Wang J, Yang L, Wan L, Wang Y (2019) Effect of lidocaine on the safety of postoperative skin reconstruction after malignant melanoma resection. Exp Ther Med 18(2):949–954

Yang L, Agarwal P (2011) Systematic drug repositioning based on clinical side-effects. PLoS One 6(12):e28025

Yang CE, Lee WY, Cheng HW, Chung CH, Mi FL, Lin CW (2019) The antipsychotic chlorpromazine suppresses YAP signaling, stemness properties, and drug resistance in breast cancer cells. Chem Biol Interact 302:28–35

Yuan P, Ito K, Perez-Lorenzo R, Del Guzzo C, Lee JH, Shen CH, Bosenberg MW, McMahon M, Cantley LC, Zheng B (2013) Phenformin enhances the therapeutic benefit of BRAFV600E inhibition in melanoma. Proc Natl Acad Sci 110(45):18226–18231

Zeng Y, Lv Y, Hu M, Guo F, Zhang C (2022) Curcumin-loaded hydroxypropyl-β-cyclodextrin inclusion complex with enhanced dissolution and oral bioavailability for epilepsy treatment. Xenobiotica 52(7):718–728

Zhang L, Yao HJ, Yu Y, Zhang Y, Li RJ, Ju RJ, Wang XX, Sun MG, Shi JF, Lu WL (2012) Mitochondrial targeting liposomes incorporating daunorubicin and quinacrine for treatment of relapsed breast cancer arising from cancer stem cells. Biomaterials 33(2):565–582

Zhang Y, Duan W, Owusu L, Wu D, Xin Y (2015) Epigallocatechin-3-gallate induces the apoptosis of hepatocellular carcinoma LM6 cells but not non-cancerous liver cells. Int J Mol Med 35(1):117–124

Zhao Y, Zhao W, Lim YC, Liu T (2019) Salinomycin-loaded gold nanoparticles for treating cancer stem cells by ferroptosis-induced cell death. Mol Pharm 16(6):2532–2539

Zheng W, Sun W, Simeonov A (2018a) Drug repurposing screens and synergistic drug-combinations for infectious diseases. Br J Pharmacol 175(2):181–191

Zheng ZY, Li J, Li F, Zhu Y, Cui K, Wong ST, Chang EC, Liao YH (2018b) Induction of N-Ras degradation by flunarizine-mediated autophagy. Sci Rep 8(1):16932

Zhou C, Chen X, Zeng W, Peng C, Huang G, Ouyang Z, Luo Y, Xu X, Xu B, Wang W, He R (2016) Propranolol induced G0/G1/S phase arrest and apoptosis in melanoma cells via AKT/MAPK pathway. Oncotarget 7(42):68314

Zhou X, Li Z, Wang X, Jiang G, Shan C, Liu S (2019) Metabolomics reveals the effect of valproic acid on MCF-7 and MDA-MB-231 cells. Xenobiotica 50:252–260

Zhu Y, Zuo W, Chen L, Bian S, Jing J, Gan C, Wu X, Liu H, Su X, Hu W, Guo Y (2019) Repurposing of the anti-helminthic drug niclosamide to treat melanoma and pulmonary metastasis via the STAT3 signaling pathway. Biochem Pharmacol 169:113610

Drug Repurposing: Harnessing the Therapeutics from Antidiabetic Molecules

3

Archana Bhatta, Priyanka Mahale, Harapriya Baral, D. K. Yuvaraj, and Ravinder K. Kaundal

Abstract

The exploration of drug repurposing, specifically the reapplication of antidiabetic drugs, offers a novel and promising approach in the field of drug discovery. This chapter delves into the potential of antidiabetic agents in treating a variety of nondiabetic conditions, leveraging their established safety profiles and mechanisms of action. By re-evaluating existing medications, such as metformin and sulfonylureas, significant strides can be made in accelerating the development of new therapeutic strategies. The discussion highlights the multifaceted benefits of antidiabetic drugs, including their roles in cardiovascular health, cancer prevention, and neuroprotection. The repurposing of these drugs not only expedites the drug development process but also enhances our understanding of their broader biological impacts. The chapter concludes with an outlook on the future of drug repurposing, emphasizing the importance of continued research and the potential for discovering novel applications of well-known medications.

Keywords

Drug repurposing · Antidiabetic drugs · Metformin · Sulfonylureas · Cardiovascular health · Cancer prevention · Neuroprotection · Drug discovery

A. Bhatta · P. Mahale · H. Baral · D. K. Yuvaraj · R. K. Kaundal (✉)
Department of Pharmacology and Toxicology, National Institute of Pharmaceutical Education and Research, Raebareli (NIPER-R), Transit Campus, Lucknow, Uttar Pradesh, India
e-mail: ravinder.kaundal@niperraebareli.edu.in

R. K. Kaundal et al. (eds.), *Drug Repurposing: Exploring Newer Therapeutic Potentials*,
https://doi.org/10.1007/978-981-95-4522-3_3

3.1 Introduction

The traditional path of drug discovery, involving the design, development, and approval of new pharmaceuticals, is often a lengthy and costly process. It typically takes over a decade and billions of dollars to bring a new drug from conception to market. In response to these challenges, drug repurposing, also referred to as drug repositioning or reprofiling, has emerged as a promising alternative (Jourdan et al. 2020). This strategy includes exploring medications already approved for other conditions, drugs discontinued due to adverse effects, and investigational compounds previously abandoned because of unfavorable clinical trial results (Ashburn and Thor 2004). Drug repurposing extends the utility of medications and breathes new life into those previously considered ineffective (Barratt and Frail 2012). By doing so, drug repurposing can significantly accelerate the drug development process. Since these drugs have already undergone substantial research and safety testing, repurposing allows us to bypass many of the lengthy and expensive phases of drug discovery. Drug repurposing holds particular promise for treating rare and orphan diseases, where traditional drug development often falls short due to limited economic incentives. As a result, patients can gain access to new treatments more quickly, which is particularly beneficial for those with diseases that currently lack effective therapies (Elder and Tindall 2020). A classic example of drug repurposing is sildenafil, which was initially developed to treat coronary artery disease but was proven effective for treating erectile dysfunction, leading to its successful repositioning (Novac 2013).

The expanding availability of public and open-source databases, along with advanced computational tools, has further propelled the field of drug repurposing (Ashburn and Thor 2004). This approach offers practical benefits, such as lowering the risk of unexpected side effects due to well-established safety profiles, and enhances our understanding of disease biology by uncovering new mechanisms of action for existing drugs. Antidiabetic drugs represent a particularly intriguing category of repurposing. Diabetes mellitus (DM), particularly type 2 diabetes (T2DM), is a chronic disease marked by high blood sugar levels and is associated with a range of comorbidities such as cardiovascular disease, renal dysfunction, and neurodegenerative disorders. These drugs, primarily used to manage blood glucose levels, have shown potential benefits beyond their original purpose. Research has demonstrated that antidiabetic medications may have therapeutic effects on conditions such as certain cancers, ischemic stroke, Alzheimer's disease (AD), and Parkinson's disease (PD), among others (Chaudhury et al. 2017). The extensive use of antidiabetic drugs, combined with their well-established safety profiles, makes them ideal candidates for repurposing (Gupta et al. 2016; Parvathaneni et al. 2019). This offers the possibility of finding new treatments for various diseases and deepens our understanding of the complex biological mechanisms it play. By exploring the potential of these medications in nondiabetic conditions, researchers can unlock new therapeutic options that could significantly affect patient care and improve health outcomes.

This chapter explores the exciting potential of antidiabetic drugs in managing various nondiabetic conditions. We will examine the mechanisms through which these drugs exert their effects, discuss key examples of successful repurposing, and consider the future directions of this emerging field. By harnessing the therapeutic potential of antidiabetic molecules, we open new avenues for improving patient care and expanding the horizons of modern medicine.

3.2 Classification of Antidiabetic Drugs

3.3 Sulfonylureas

The development of sulfonylureas for insulin stimulation began with their unexpected discovery in the 1940s when patients experienced hypoglycemia while taking these drugs (Levine 1984). By 1955, sulfonylureas had become the first pharmacological treatment for non-insulin-dependent DM, offering an alternative to insulin injections (Loubatières 1957). Early sulfonylureas included acetohexamide, chlorpropamide, tolbutamide, and tolazamide (Skillman and Feldman 1981). Second-generation sulfonylureas, such as gliclazide, glipizide, and glibenclamide, have largely replaced these first-generation drugs, with some researchers classifying glimepiride as a third-generation drug (DeFronzo 1999). Table 3.1 summarizes the classification of antidiabetic drugs, their mechanisms of action, and their therapeutic outcome.

Sulfonylureas show their insulinotropic effects by binding to sulfonylurea receptor 1 (SUR-1) on β-cells, leading to the closure of adenosine triphosphate (ATP)-sensitive potassium channels. This closure reduces potassium efflux and causes membrane depolarization. Depolarization then opens voltage-sensitive calcium channels, permitting calcium ions to enter the cell. The calcium influx initiates a cascade of events that moves insulin-containing granules to the cell membrane. These granules then fuse with the membrane and release insulin into the bloodstream, effectively reducing the blood glucose levels (DeFronzo 1999; Lang 1999). Due to potential side effects like hypoglycemia, weight gain, and hyperinsulinemia, it is essential to monitor patients regularly when using sulfonylureas (Ahrén 2011). Interestingly, sulfonylureas have shown benefits beyond diabetes management. A study of diabetic patients in the Hong Kong Chinese community found that sulfonylureas, glibenclamide, and gliclazide were linked to a significant decrease in cancer risk (Yang et al. 2010). SURs are part of the ATP-binding cassette transporter superfamily. Within this group, multidrug resistance protein 1 (MRP1) functions as an efflux pump, aiding cancer cell resistance by expelling anticancer drugs. In vitro studies with GLC4/Sb30 cells have shown that glibenclamide reduces MRP1 activity, thereby increasing the cells' sensitivity to vincristine. Additionally, glibenclamide has been reported to show antiproliferative properties (Abdul and Hoosein 2002). Beyond glucose regulation, these drugs influence various cellular processes,

Table 3.1 Overview of antidiabetic drug classes: mechanisms of action and pleiotropic effects

Classification	Drugs	Mechanism of action	Therapeutic outcome
Sulfonylureas	Gliclazide Glibenclamide Glimepiride Tolbutamide	Increase insulin secretion by activating the β-cell SUR-1	Reduces the chance of cancer
Meglitinides	Nateglinide Repaglinide Mitiglinde	Activate SUR-1 receptors on β-cells, this activation closes ATP-sensitive K+ channels and opens voltage-gated Ca2+ channels, enhancing insulin secretion	Antiproliferative effects
Biguanides	Metformin	Suppress hepatic gluconeogenesis, Enhance glucose uptake in skeletal muscle, Augment glucose absorption in the intestinal mucosa, and elevate circulating levels of GLP-1	Improved cognitive functions (e.g., learning and attention), reduced risk of AD and PD, antidepressant effects, benefits for skin disorders (e.g., acanthosis nigricans, acne), cardiovascular protection, anticancer and antibacterial properties
Thiazolidinediones	Pioglitazone Rosiglitazone	PPARγ activators enhance pre-adipocyte differentiation and improve insulin responsiveness in muscle, liver, and fat cells Increase the absorption of glucose in peripheral tissues	Anti-inflammatory effects, neuroprotection (decreases the risk of AD and stroke), anticancer properties, cardiovascular and renal protection
DPP-4 inhibitors	Saxagliptin Vildagliptin Alogliptin Sitagliptin	Inhibits DPP-4, resulting in higher levels of incretin hormones that increases insulin secretion and decreases glucagon release	Anticancer effects, neuroprotection, antioxidant properties, cardiovascular protection
SGLT-2 inhibitors	Dapagliflozin Canagliflozin Empagliflozin Tofogliflozin	Inhibit SGLT-2 in the renal proximal tubules, decrease glucose reabsorption, and enhance urinary glucose excretion	Cardiovascular protection, antineoplastic effects, anti-inflammatory, diuretic effects, reduced risk of heart failure

(continued)

Table 3.1 (continued)

Classification	Drugs	Mechanism of action	Therapeutic outcome
GLP-1 receptor agonists	Exenatide Dulaglutide Albiglutide Lixisenatide Liraglutide	Activate GLP-1 receptors in the pancreas, enhancing insulin secretion and decreasing glucagon release	Cardiovascular risk reduction, prevention of atherogenesis, neuroprotection (benefits for AD, PD, mood disorders), bone-building effects
α-Glucosidase inhibitors	Acarbose Voglibose Miglitol	Inhibit α-glucosidase in the intestinal brush border, slowing carbohydrate breakdown and glucose absorption	Antineoplastic effects, reduced cholesterol levels, antioxidant properties

including ion channel activity and signaling pathways, which can affect cell survival and proliferation.

Furthermore, sulfonylureas such as tolbutamide, chlorpropamide, and glyburide have exhibited significant inhibition of acetylcholinesterase (AChE), an enzyme involved in neurotransmitter degradation (Baruah et al. 2021). This suggests potential applications in treating neurological disorders where AChE inhibition is beneficial. Additionally, glibenclamide has demonstrated antiproliferative effects, including the induction of apoptosis, which could be valuable for cancer therapies. These diverse effects underscore the potential for sulfonylureas to be repurposed for various therapeutic applications, including cancer treatment and neurological conditions, highlighting their versatility beyond diabetes management.

3.4 Biguanides

Biguanides are antidiabetic agents that have been used for many decades, with metformin being the most prominent member. Initially, the antidiabetic potential of biguanides was discovered through the herb *Galega officinalis*, which contains guanidine compounds. These compounds were found to have hypoglycemic effects, leading to the development of guanidine derivatives in the 1920s. However, it wasn't until the 1950s that biguanides were fully recognized for their role in diabetes management (Bailey 2017). Phenformin, buformin, and metformin are the three main biguanides known for their ability to lower blood glucose levels. Among these, metformin is the most widely used due to its favorable safety profile compared to phenformin and buformin, which have been associated with a high risk of lactic acidosis and are now restricted in many countries (DeFronzo et al. 2016). Metformin improves insulin sensitivity and reduces blood glucose levels through multiple mechanisms. It primarily inhibits gluconeogenesis in the liver, decreasing glucose synthesis (Sardu et al. 2019).

Additionally, metformin improves glucose uptake by skeletal muscles, facilitating more efficient absorption of glucose from the bloodstream (Musi et al. 2002). This increase in muscle cell sensitivity to insulin helps lower blood sugar levels. Metformin also reduces glucose absorption from the gastrointestinal tract, particularly the small intestine, which helps reduce postprandial blood glucose levels (Ikeda et al. 2000). Furthermore, metformin can affect the release of gastrointestinal hormones such as peptide YY and glucagon-like peptide-1 (Bahne et al. 2018). These hormones regulate appetite, glucose uptake, and insulin secretion. Metformin also lowers circulating free fatty acids (FFAs), as elevated FFA levels can impair insulin action (Kappe et al. 2013). By reducing FFA levels, metformin enhances insulin sensitivity. Metformin activates the AMP-activated protein kinase (AMPK) enzyme to regulate cellular energy balance. This activation leads to increased glucose utilization in muscles, decreased glucose production in the liver, and improved lipid metabolism (Zhou et al. 2001). Beyond their traditional use in diabetes, biguanides-particularly metformin-have shown promise as candidates for drug repurposing. Metformin's ability to modulate various biological pathways makes it a strong candidate for treating other conditions. Metformin, primarily used to treat T2DM, has demonstrated significant efficacy in improving cardiovascular health, making it a valuable therapeutic agent for a range of cardiovascular conditions. Its cardiovascular benefits are evident in its ability to reduce complications related to elevated blood sugar levels, which often lead to heart issues (Smith and Gale 2010). Its cardiovascular benefits are attributed to its ability to activate AMPK. Metformin enhances nitric oxide (NO) pathways, improving endothelial function and reducing ventricular hypertrophy, a condition marked by the thickening of the heart's ventricular walls (Zhang et al. 2011). This reduction in ventricular hypertrophy is achieved through inhibiting angiotensin II-induced protein synthesis in cardiomyocytes and enhancing NO production, which is crucial for vasodilation and healthy cardiac function (Zhang et al. 2011). Metformin has shown potential in mitigating cardiac remodeling and hypertrophy by inhibiting myocardial fibrosis and collagen synthesis, partially through its effects on the transforming growth factor-beta 1 (TGF-β1) and Smad signaling pathways (Wang et al. 2011). Studies conducted in mice with ischemic myocardium have shown that metformin improves left ventricular structure and function, increasing survival rates (Lorell and Carabello 2000; Xiao et al. 2010).

These positive outcomes are linked to the activation of peroxisome proliferator-activated receptor gamma coactivator 1-alpha (PGC-1α) and endothelial nitric oxide synthase (eNOS), both of which contribute to improved cardiac efficiency (Vettor et al. 2014; Gundewar et al. 2009). In a clinical trial, participants with LVH, coronary artery disease, and insulin resistance who received metformin at a daily dose of 2 g for 1 year showed a significant reduction in left ventricular mass compared to those receiving a placebo (Mohan et al. 2019). Alongside this reduction, metformin was associated with other beneficial changes, including decreased blood pressure, reduced oxidative stress, and weight loss (Kamel et al. 2022). These findings highlight metformin's potential in reducing cardiac workload and improving overall heart health, as illustrated in Fig. 3.1.

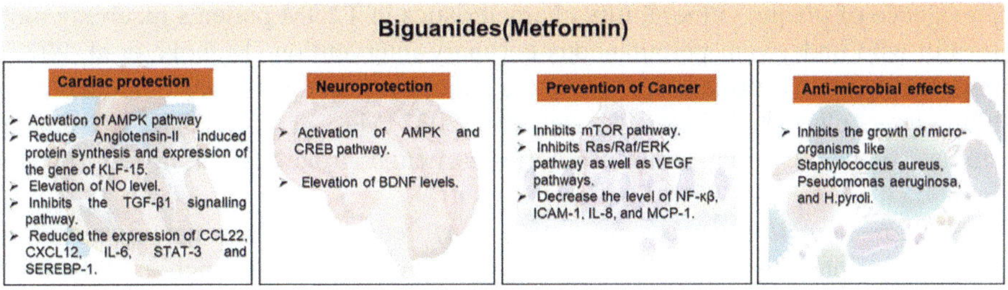

Fig. 3.1 Expanded therapeutic action of biguanides beyond glycemic control

Additionally, metformin has been shown to improve mitochondrial activity and cardiomyocyte metabolism in patients with heart failure, further underscoring its cardiovascular benefits. Metformin's cardiovascular effects extend to improved mitochondrial function and reduced atherogenesis. Metformin has also been effective in ameliorating atherosclerosis through multiple pharmacological pathways. It helps prevent insulin-induced weight gain and lowers traditional risk factors for atherosclerotic cardiovascular diseases, such as total cholesterol, low-density lipoprotein (LDL), and triglyceride levels, while promoting weight loss. It modulates the sterol regulatory element-binding protein 2-LDL receptor axis (SERBP2-LDLR axis), decreasing the levels of LDL and triglycerides while increasing high-density lipoprotein (HDL) cholesterol (Weng et al. 2020). In clinical trials, metformin has demonstrated a significant reduction in left ventricular mass and oxidative stress and improvements in endothelial dysfunction, which collectively lower the risk of heart failure and ischemic events (Patel et al. 2017).

Metformin has been reported to inhibit the tumor necrosis factor (TNF-α)-induced lipogenic response and reduce mRNA expression of inflammatory cytokines. Studies have demonstrated that metformin decreases the secretion of cytokines such as interleukin (IL)-12p40 and IL-6 by activated macrophages (Cameron et al. 2016). Compared to sulfonylureas, metformin is associated with lower levels of inflammatory cytokines like C-C motif chemokine ligand 22 (CCL22) and C-X-C motif chemokine ligand 12 (CXCL12), contributing to reduced systemic inflammation, as measured by the neutrophil-to-lymphocyte ratio (Cameron et al. 2016).

These anti-inflammatory effects holds potential for cardiovascular protection. Metformin suppresses the signal transducer and activator of transcription 3 (STAT3) pathway, reducing inflammation and the transformation of monocytes into macrophages, thereby decreasing monocyte infiltration and reducing atheromatous plaque development (Vasamsetti et al. 2015). Metformin also addresses endothelial dysfunction (Mone et al. 2021), a major factor in accelerated atherosclerosis (Gopoju et al. 2018). A clinical investigation involving prediabetic patients with non-obstructive coronary artery disorders showed that metformin use was associated with a reduced risk of adverse cardiovascular events and coronary artery endothelial dysfunction (Sardu et al. 2019). Recent studies have also indicated a decreased prevalence of plaque vulnerability characteristics in T2DM patients receiving metformin who underwent percutaneous coronary intervention (Kataoka et al. 2022). Peripheral vascular disease and stroke are major contributors to cardiovascular morbidity in people with diabetes (Kersten et al. 2022; Mary et al. 2017). Observational studies suggest that metformin use during ischemic stroke may be linked to better functional recovery in T2DM patients.

Moreover, metformin administration may reduce vascular calcification and increase peripheral vascular flow, benefiting the microvascular function of the heart (Mary et al. 2017; Sirtori et al. 1984; Khan et al. 2019). In a clinical trial involving 33 non-diabetic women with normal coronary angiography but positive exercise tolerance tests, metformin enhanced endothelial function and reduced ST changes during exercise, suggesting improved coronary artery function (Jadhav et al. 2006). Furthermore, metformin effectively reduces infarction and reperfusion injury by inhibiting mitochondrial complex I (Mohsin et al. 2019). These multifaceted cardiovascular benefits position metformin as a valuable therapeutic agent for improving cardiovascular health in both diabetic and non-diabetic populations.

Beyond its cardiovascular effects, metformin is recognized for its potential role in cancer prevention and therapy. The American Diabetes Association and the American Cancer Society have highlighted that individuals with T2DM face a high risk of several cancers, including those of the liver, pancreas, and endometrium (Giovannucci et al. 2010). The increased cancer risk in individuals with diabetes is linked to factors such as high insulin levels, insulin resistance, chronic high blood sugar, mild inflammation, and the accumulation of advanced glycation end products. These factors can enhance malignant transformation through oxidative stress and other pathways (Smith and Gale 2010; Azar and Lyons 2010). Epidemiological studies have demonstrated that metformin reduces the incidence of various cancers, including colorectal, lung, pancreatic, and breast cancers (Saraei et al., 2000; Anisimov et al. 2005). This effect is largely attributed to metformin's ability to lower circulating insulin and insulin-like growth factor 1 (IGF-1) levels (Memmott et al. 2010). Elevated insulin levels stimulate the liver to produce more IGF-1, which then binds to IGF-1 and insulin receptors, activating phosphoinositide 3-kinase (PI3K) and Akt/protein kinase B (PKB) (Fürstenberger and Senn 2002).

This activation promotes cellular proliferation by stimulating the mammalian targets of rapamycin complex-1 (mTORC-1) and Ras/Raf/ERK pathways, which are crucial for the metabolic alterations observed in tumor cells. Metformin counteracts these processes by inhibiting mTORC-1 through both AMPK-dependent and independent mechanisms. AMPK activates tuberous sclerosis complex protein 2 (TSC2), which inhibits mTORC-1, leading to reduced protein synthesis and cellular proliferation (Yoshida et al. 2011).

Metformin has demonstrated potential in cancer prevention and therapy, in part through its ability to inhibit angiogenesis. One proposed mechanism involves the phosphorylation of vascular endothelial growth factor receptor-2 (VEGFR-2), which interferes with the alternative splicing of VEGF-A mRNA. This disruption attenuates VEGFR activation and downstream VEGF signaling, thereby impairing neovascularization within the tumor microenvironment. Additionally, the anti-angiogenic effects of metformin may be partially mediated by its anti-inflammatory properties (Cao 2013; Yi et al. 2016). Metformin has been shown to suppress key pro-inflammatory mediators, including nuclear factor kappa B (NF-κB), IL-8, intracellular adhesion molecule-1 (ICAM-1), and monocyte chemoattractant protein-1 (MCP-1) (Han et al. 2018). In diabetic mouse models, treatment with metformin not only reduced these inflammatory markers but also decreased retinal leukostasis, further supporting its potential utility in cancer prevention and adjunctive therapy.

Furthermore, long-term use of metformin (for more than 4 years) was associated with a reduced incidence of both AD and PD (Shi et al. 2019). Clinical trials have demonstrated improvements in cognitive function, including learning, memory, and attention, in older adults taking metformin (Koenig et al. 2017). These effects were attributed to the activation of AMPK and cyclic AMP response element-binding protein (CREB) pathways, which increase brain-derived neurotrophic factor (BDNF) expression and increase synaptic plasticity. In animal models, metformin has shown antidepressant effects by preventing social defeat stress (SDS)-induced behavioral changes through histone acetylation and increased CREB phosphorylation (Fang et al. 2020).

Moreover, metformin displays antimicrobial properties against a range of pathogens, including *Pseudomonas aeruginosa, Mycobacterium tuberculosis, Staphylococcus aureus, and Helicobacter pylori* (Courtois et al. 2018; Garnett et al. 2013). Metformin exerts its antimicrobial activity by targeting bacterial nicotinamide adenine dinucleotide (NADH) dehydrogenase I (NADH-I), a homolog of mitochondrial complex I in eukaryotic cells, thereby impairing bacterial energy metabolism and inhibiting growth (Malik et al. 2018). Clinical studies have shown metformin's efficacy in reducing bacterial load and improving outcomes in patients with tuberculosis, particularly in those with diabetes, who are at a higher risk of infection (Lee et al. 2018). Metformin's modulation of pro-inflammatory and anti-inflammatory cytokines further supports its role in managing infections and inflammation-associated diseases (Kaufmann et al. 2018). The multifaceted mechanisms of metformin, including activation of AMPK pathways, modulation of lipid metabolism, and inhibition of inflammatory and proliferative signaling, underscore its broad therapeutic potential beyond glucose regulation.

3.5 Thiazolidinediones (TZDs)

TZDs, initially developed for the treatment of T2DM, are now being explored for their therapeutic potential in a range of nondiabetic conditions. TZDs, such as pioglitazone and rosiglitazone, function as agonists of peroxisome proliferator-activated receptor-γ (PPAR-γ), a nuclear receptor involved in glucose and lipid metabolism. Activation of PPAR-γ enhances insulin sensitivity by promoting glucose uptake in adipose tissue, skeletal muscle, and the liver. Additionally, TZDs reduce circulating FFAs, suppress pro-inflammatory cytokines, elevate adiponectin levels, and preserve pancreatic β-cell function. These multifaceted actions not only improve insulin resistance but also help mitigate β-cell exhaustion. The pleiotropic effects of TZDs have generated growing interest in their potential applications beyond glycemic control, including in the fields of oncology, neurodegeneration, and cardiovascular disease (Chaudhury et al. 2017).

Several studies have highlighted the neuroprotective properties of TZDs. In AD models and patients, TZDs have shown promise in halting disease progression (Cheng et al. 2016). Rosiglitazone and pioglitazone have been found to inhibit glycogen synthase kinase-3 (GSK-3), a key player in AD pathology (Toledo and Inestrosa 2010). TZDs also reduce infarct volume and improve neurological outcomes in models of cerebral ischemia (White and Murphy 2010). Additionally, TZDs exhibit anti-inflammatory effects by blocking microglial activation induced by amyloid β (Heneka et al. 2007). Oral administration of rosiglitazone has been observed to decrease the levels of phosphorylated tau protein in the hippocampus, reducing the formation of neurofibrillary tangles, a hallmark of AD (Yoon et al. 2010; LaFerla 2012). Pioglitazone has also been shown to enhance motor function in PD models, likely through the activation of the PPAR coactivator PGC-1 (Wang et al. 2017). Furthermore, pioglitazone may mitigate cerebral ischemia-reperfusion injury by decreasing oxidative stress and inhibiting NF-κB activation (Kaundal et al. 2009).

In the context of cardiovascular health, TZDs have demonstrated potential benefits. They can modulate serum lipid levels, enhance adiponectin levels, reduce inflammation, and improve cardiovascular outcomes (Della-Morte et al. 2014). Evidence suggests that pioglitazone can alleviate various cardiovascular risk factors, including lipid imbalance, endothelial dysfunction, and inflammation (Xue et al. 2022).

Randomized controlled trials have indicated that TZDs offer anti-atherosclerotic benefits, with meta-analyses showing improved adiponectin levels, reduced plaque volume in diabetic patients with coronary atherosclerosis, and enhanced endothelial function (Xue et al. 2022). Rosiglitazone, in particular, has been reported to reduce atherosclerotic lesions significantly (Pistrosch et al. 2004).

TZDs also exhibit renoprotective effects. Studies have reported reductions in urine albumin secretion in T2DM patients and diabetic rats (Fujii et al. 1997). Additionally, troglitazone has demonstrated renoprotective effects in nondiabetic animals with chronic renal impairment (Yoshida et al. 2001). TZDs positively affect renal matrix protein inhibition and reduce mesangial expansion, likely due to the predominance of PPAR-γ in renal cells, which enhances their responsiveness to TZDs (Guan et al. 2001).

Moreover, TZDs have been demonstrated to inhibit cancer cell growth and migration (Ferruzzi et al. 2005), induce G1 cell cycle arrest, and reduce tumor progression in various cancer models (Tsujie et al. 2003). They also lower the risk of liver carcinoma, as demonstrated in a study involving T2DM patients (Chang et al. 2012). While pioglitazone is associated with a higher risk of bladder cancer, clinical trials have overall indicated a lower risk of cancer among TZD users (Bosetti et al. 2013). Troglitazone exhibits species-specific effects, promoting proliferation in mouse cancer cell lines while inhibiting growth in human cancer cells

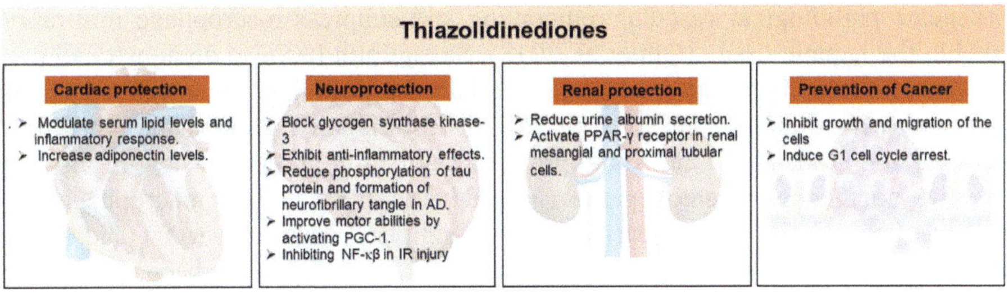

Fig. 3.2 Pleotropic effects of thiazolidinediones beyond glycemic control

(Kakiuchi-Kiyota et al. 2009). This multifaceted effect on cancer-related pathways underscores the potential of TZDs as adjunctive treatments or as preventive agents in oncology. The broad range of effects demonstrated by TZDs highlights their potential for repurposing across multiple therapeutic areas as illustrated in Fig. 3.2. Continued research into these diverse applications could lead to new and valuable treatment options.

3.6 Dipeptidyl-Dipeptidase-4 (DPP-4) Inhibitors

DPP-4 is an enzyme found throughout the body that regulates incretin hormones, namely glucagon-like peptide-1 (GLP-1) and glucose-dependent insulinotropic polypeptide (GIP). These hormones help control blood sugar levels by increasing insulin release and inhibiting glucagon production (Vella 2012). GLP-1, released by enteroendocrine L-cells, lowers blood glucose levels through insulin secretion, glucagon suppression, and delayed gastric emptying (Pathak and Bridgeman 2010). GIP, produced by neuroendocrine K-cells in the stomach and upper small intestine, has similar effects (Gupta and Kalra 2011). Because GLP-1 and GIP have short half-lives, they are quickly broken down by DPP-4 after taking food. DPP-4 inhibitors, such as saxagliptin, sitagliptin, linagliptin, vildagliptin, and alogliptin, block this breakdown, leading to higher levels of GLP-1 and GIP. This results in improved insulin release from beta cells and reduced blood sugar levels after meals and fasting (Capuano et al. 2013). Beyond their glucose-lowering properties, DPP-4 inhibitors exhibit a range of pleiotropic effects. These include antihypertensive, immunomodulatory, anti-inflammatory, and anti-apoptotic actions on the cardiovascular and renal systems, which appear to be independent of their incretin-mediated mechanisms (Lim et al. 2015).

Recent studies suggest that DPP-4 inhibitors might also help in reducing cancer risks. For example, high levels of DPP-4 are associated with worsening of hepatocellular carcinoma (Hsu et al. 2021; Kamada et al. 2021; Kaji et al. 2014). In preclinical studies, gliptins, such as vildagliptin, have demonstrated protective effects against diethylnitrosamine-induced hepatocarcinogenesis, particularly in high-fat models. This chemopreventive action is attributed to vildagliptin's ability to attenuate pathological vascular remodeling and suppress macrophage infiltration within the hepatic tissue (Qin et al. 2018). Saxagliptin has also demonstrated protective effects in liver injury models, potentially by inhibiting cancer-related signaling pathways (Abd Elhameed et al. 2021). Additionally, gliptins like anagliptin (Kawakubo et al. 2020) and sitagliptin (Kawaguchi et al. 2019) show promise in treating nonalcoholic steatohepatitis (NASH). Anagliptin inhibits macrophage activation through GLP-1 pathways, while sitagliptin reduces liver fat and tumor growth, possibly by affecting metabolic pathways. Tenegliptin has been noted for its potential to reduce liver fibrosis and precancerous lesions by targeting cell activation and oxidative stress (Ozutsumi et al. 2020).

DPP-4 inhibitors may also help prevent colorectal cancer by enhancing GLP-2 effects and reducing precancerous lesions (Masur et al. 2006). Sitagliptin has been associated with decreased hydrogen peroxide and intestinal inflammation, indicating its potential for cancer prevention (Femia et al. 2013). DPP-4 inhibitors also offer neuroprotective effects, particularly in stroke models. They reduce inflammation, oxidative stress, and neuronal damage, independent of their effects on blood glucose (Darsalia et al. 2018). Linagliptin, for instance, increases active stromal cell-derived factor-1α (SDF-1α) in the brain, which is associated with improved stroke outcomes through the SDF-1α/CXCL-4 pathway (Li et al. 2014; Chiazza et al. 2018). Additionally, DPP-4 inhibitors have shown beneficial effects on cardiovascular health by increasing lipid metabolism, enhancing endothelial function, and reducing oxidative stress and inflammation (Zhong et al. 2013; Wiciński et al. 2020). Anagliptin reduces stress-induced senescence in vascular endothelial cells by decreasing endoplasmic reticulum (ER) stress and reactive oxygen species (ROS) while activating protective signaling pathways (Kang et al. 2021). Gemigliptin has been shown to reduce vascular aging and calcification (Choi et al. 2017) and neointimal hyperplasia in various models, further emphasizing its role in reducing oxidative stress and inflammation (Choi et al. 2015). The wide-ranging applications of DPP-4 inhibitors in cancer prevention, neuroprotection, and cardiovascular health highlight the need for further research to thoroughly explore and validate these potential therapeutic benefits, illustrated in Fig. 3.3.

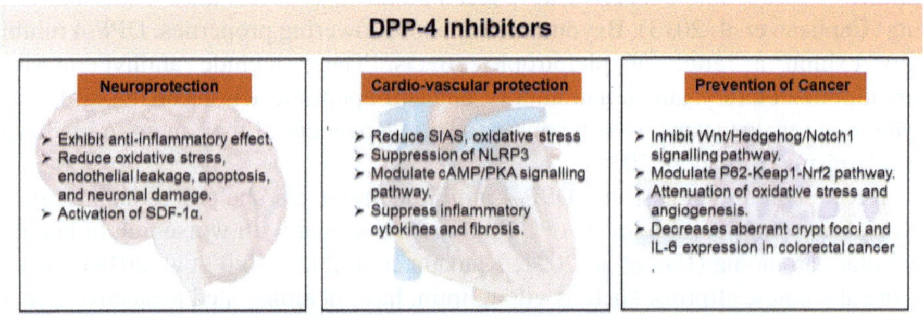

Fig. 3.3 Multifaceted roles of DPP-4 inhibitors extending beyond glucose regulation

3.7 Sodium-Glucose Transport Proteins (SGLT-2) Inhibitors

SGLT-2 inhibitors, initially developed for managing T2DM, have shown considerable promise beyond their primary use. First approved in the United States in 2012 and later in Japan with ipragliflozin in 2014, SGLT-2 inhibitors, such as dapagliflozin, ipragliflozin, canagliflozin, luseogliflozin, tofogliflozin, and empagliflozin, have received regulatory endorsements for their efficacy in glycemic control (Satoh 2018). These drugs work by preventing glucose reabsorption, thus enhancing glucose excretion in the urine and decreasing blood glucose levels independent of insulin secretion and pancreatic β-cell function, making them effective even in insulin-resistant states (Satoh 2018). Additionally, these drugs lead to increased urinary output and calorie loss, contributing to weight reduction and lower blood pressure (Kao and Huang 2022). The cardiovascular benefits of SGLT-2 inhibitors are well established. Clinical trials, such as EMPA-REG OUTCOME, have shown that empagliflozin significantly lowers systolic and diastolic blood pressure by 5/3 mmHg after 16 weeks (Zinman et al. 2015).

Similarly, the CANVAS trial reported reductions of 3.93/1.39 mmHg with canagliflozin (Neal et al. 2017), and the DECLARE-TIMI 58 trial with dapagliflozin demonstrated changes of 2.7/0.7 mmHg (Wiviott et al. 2019). A meta-analysis of 64 studies confirmed an average systolic and diastolic blood pressure decrease of 2.89/1.44 mmHg across diabetic populations (Tsapas et al. 2021). Canagliflozin also significantly lowers the risk of albuminuria progression by 23% and improves renal outcomes by 40%, including reductions in estimated glomerular filtration rate (eGFR) and the need for renal replacement therapy (Neal et al. 2017). These effects are partly attributed to the activation of the nitric oxide-cyclic guanosine monophosphate (NO-cGMP) pathway, which promotes vasodilation and blood pressure reduction (Zhang et al. 2019).

In addition to their cardiovascular benefits, SGLT-2 inhibitors may play a role in cancer treatment. Canagliflozin, for example, has shown potential in reducing the risk of gastrointestinal cancers. In a murine model mimicking human nonalcoholic steatohepatitis (NASH), canagliflozin demonstrated antineoplastic effects by reducing hepatic steatosis and liver tumor prevalence. It also showed the ability to decrease blood glucose levels and inhibit mitochondrial complex-1 function, thereby suppressing prostate and lung cancer cell proliferation. Another intriguing aspect of SGLT-2 inhibitors is their anti-inflammatory properties. SGLT-2 inhibitors have demonstrated notable anti-inflammatory properties in various models. For example, empagliflozin significantly inhibited the activation of NF-κB and MAPK in a murine model, leading to reduced macrophage infiltration and angiogenesis (Ortega et al. 2019). These effects suggest that SGLT-2 inhibitors may play a role in modulating inflammation.

Epicardial adipocytes regulate inflammatory responses through paracrine signaling, which is influenced by adipokines such as leptin and adiponectin. Conditions like obesity, insulin resistance, and diabetes alter adipokine levels, thereby affecting inflammation. SGLT-2 inhibitors are believed to counteract these inflammatory changes by lowering IL-6, leptin, and plasminogen activator inhibitor-1 levels and reducing ectopic epicardial fat. This modulation of inflammatory markers further supports the therapeutic versatility of SGLT-2 inhibitors. The unique mechanisms of action of SGLT-2 inhibitors and their anti-inflammatory and potential anticancer properties suggest that SGLT-2 inhibitors hold promise for drug repurposing across various medical fields beyond diabetes management. As research continues, these inhibitors may provide novel therapeutic options for cardiovascular diseases, kidney disease, and even certain types of cancer, highlighting their expanding role in modern medicine.

3.8 Glucagon-Like Peptide 1 (GLP-1) Receptor Agonist

GLP-1 receptor agonist, such as liraglutide, dulaglutide, albiglutide, semaglutide, exenatide, and lixisenatide, were initially developed to manage T2DM through mechanisms that increase insulin secretion and reduce glucagon release in a glucose-dependent manner (Drucker 2006). This action leads to a decrease in hepatic glucose production and improves glycemic control, as evidenced by reductions in glycosylated hemoglobin (HbA1c) and lower postprandial glucose levels (Inzucchi et al. 2012; Gerich 2013). Beyond their primary role in glycemic regulation, GLP-1 receptor agonist exhibit a broad spectrum of effects across various physiological systems, suggesting their potential for drug repurposing. These receptors are present in multiple organs, including the cardiac muscles, brain, gastrointestinal tract and kidneys, indicating their capacity to influence diverse biological processes (Vilsbøll and Garber 2012). Clinical trials have demonstrated that GLP-1 receptor agonist positively modify cardiovascular risk factors. They have been shown to improve lipid profiles by reducing cholesterol, triglycerides, and FFAs (Burgmaier et al. 2013; Yanovski and Yanovski 2014) and to lower both systolic and diastolic blood pressure (Wang et al. 2013). The antihypertensive effects of these drugs may involve diuretic actions in the kidneys (Filippatos and Elisaf 2013), vasodilation (Nyström et al. 2005), and potential central nervous system interactions (Sivertsen et al. 2012). Notably, liraglutide's effects on blood pressure have been linked to atrial natriuretic peptide (ANP) release in the cardiac atria (Kim et al. 2013).

GLP-1 receptor agonist have also been documented to prevent atherosclerosis. For instance, exendin-4 and liraglutide have effectively reduced vascular inflammation and improved endothelial function in preclinical models (Gaspari et al. 2013).

In apolipoprotein E-deficient mice, exendin-4 decreased monocyte and macrophage accumulation, while liraglutide inhibited early atherosclerosis, slowed plaque formation, and enhanced plaque stability (Gaspari et al. 2013). In myocardial infarction models, liraglutide improved survival rate, reduced infarct size, and enhanced cardiac function (Noyan-Ashraf et al. 2009). Similarly, derivatives such as AC3174 (Liu et al. 2010) and lixisenatide (Wohlfart et al. 2013) have shown benefits in cardiac function and infarct size reduction.

Neuroprotective effects of GLP-1 receptor agonist are also supported by emerging evidence. In murine models, liraglutide prevented memory impairment, preserved synaptic plasticity, and reduced cerebral plaque accumulation and inflammation (Hölscher 2014; McClean et al. 2011). Similarly, exenatide diminished amyloid-beta peptide levels, indicating its potential neuroprotective benefits (Li et al. 2010). The ability of GLP-1 receptor agonist to cross the blood-brain barrier and modulate cerebral glucose balance suggests therapeutic possibilities for neurodegenerative disorders (Lerche et al. 2008). Additionally, GLP-1 receptor agonists may have beneficial effects on mood and mental health. In animal models, liraglutide alleviated behavioral depression and metabolic disturbances induced by atypical antipsychotics (Sharma et al. 2015). Weekly administration of GLP-1 receptor agonists has been associated with improved health-related quality of life, with semaglutide showing particularly high treatment satisfaction (Billings et al. 2018). Furthermore, exenatide's antinociceptive effects through spinal microglial stimulation highlight its potential for pain management (Gong et al. 2014).

Research also indicates that GLP-1 receptor agonist have positive effects on bone health. These drugs, acting on receptors in osteoclasts, osteocytes, and osteoblasts, enhance calcitonin levels and reduce bone resorption and bone formation (Yamada et al. 2008; Mabilleau et al. 2018; Ma et al. 2013; Gier et al. 2012; Bjerre Knudsen et al. 2010). They promote osteoblast activity, improve bone morphology, and lower fracture risk (Ma et al. 2013). Meta-analyses have confirmed a significant reduction in fracture risk with liraglutide use (Su et al. 2015), with exenatide identified as the safest option for this purpose (Zhang et al. 2018). Preliminary studies suggested that GLP-1 receptor agonists may also benefit skin diseases through their antioxidant, anti-inflammatory, and neuromodulatory properties. For example, liraglutide has been reported to resolve psoriasis in a patient with T2DM (Faurschou et al. 2014). Overall, the diverse actions of GLP-1 receptor agonist beyond glycemic regulation underscore their potential for drug repurposing across a range of therapeutic areas, as supported by a growing body of scientific evidence, illustrated in Fig. 3.4.

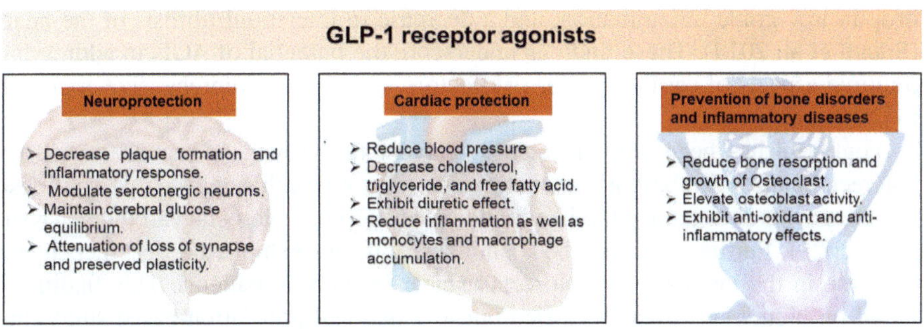

Fig. 3.4 GLP-1R agonist as multifunctional therapeutics beyond glucose control

3.9 α-Glucosidase Inhibitors

Several antidiabetic agents with α-glucosidase-inhibitory properties, including acarbose, voglibose, emiglitate, and miglitol, are currently available to manage postprandial hyperglycemia. All α-glucosidase inhibitors (AGIs) share structural similarities with disaccharides or oligosaccharides, forming complexes with a higher affinity than the carbohydrate-α-glucosidase complex. This interaction leads to competitive inhibition of α-glucosidase activity, reducing carbohydrate hydrolysis and impeding glucose absorption within the brush border of the small intestine. Most unhydrolyzed carbohydrates undergo subsequent breakdown in the lower part of the small intestine, resulting in delayed glucose absorption after food intake. This unique mechanism of action exhibited by AGIs effectively decreases postprandial hyperglycemia, representing an additional facet of their effectiveness. AGIs delay the digestion of polysaccharides and elevate local carbohydrate concentration in the lower gut. Given the ample quantity of GLP1-secreting cells in the lower gut, the delayed carbohydrate absorption stimulates GLP1 secretion. Consequently, AGIs facilitate GLP1 secretion from intestinal L cells, subsequently promoting insulin release (Hossain et al. 2020). Additionally, AGIs have demonstrated effectiveness and safety, both as monotherapy and adjuvant therapy, and they exhibit additional effects such as the stabilization of carotid plaques and the reduction of inflammation (Derosa and Maffioli 2012).

Carotid intima-media thickening (CIMT), a significant risk factor for cardiovascular diseases (CVD), and microalbuminuria, a CVD surrogate highly predictive of mortality and morbidity, contribute to the cardiovascular disease risk profile. Numerous European and Asian studies have affirmed that acarbose plays a preventive role in endothelial dysfunction and restores blood flow to normal conditions in individuals with T2DM. In acute coronary syndrome individuals and postprandial hyperglycemic patients, another AGI, miglitol, shows promise in ameliorating postprandial endothelial dysfunction, suggesting a possible beneficial effect. Moreover, acarbose exhibits cardiovascular benefits by influencing surrogate outcomes associated with CVD. These include a reduction in nephrology and microalbuminuria, a drop in the thickening of the carotid intima-media, decreases in oxidative stress, a drop in low-grade inflammation, and a decrease in interstitial fibrosis of the heart (Standl et al. 2014). These findings underscore the potential of AGIs in addressing multiple aspects of cardiovascular risk factors, offering a comprehensive approach to CVD management.

Voglibose has been demonstrated to prevent the growth of lesions caused by colorectal cancer in obese and diabetic mice. This research suggests that voglibose has the potential to reduce the likelihood of developing colorectal carcinoma (Kato et al. 2020). Furthermore, acarbose, another AGI, has exhibited a dose-dependent decrease in the incidence of colorectal cancer in diabetic patients. This finding is based on biomarker research carried out in a national population-based study that used a sizable diabetes cohort from the Taiwan National Health Insurance Research Database (Tseng et al. 2015). The combined evidence highlights the potential of AGIs, including voglibose and acarbose, in mitigating colorectal cancer risk, offering valuable insights into their role in diabetes management. Acarbose, an AGI, has continuously shown efficacy in slowing the evolution of the thickness of the carotid intima-media in randomized controlled studies carried out in Europe and Asia (CIMT). Furthermore, a prearranged post hoc analysis of specified secondary results showed that acarbose significantly decreased myocardial infarction and other cardiovascular events. The collective evidence underscores the potential cardiovascular benefits associated with acarbose use, prompting further exploration of its role in managing people with reduced glucose tolerance and coronary heart disease.

3.10 Conclusion and Future Prospects

Based on the comprehensive analysis presented in this chapter, it is evident that antidiabetic drugs exhibit multifaceted therapeutic potential beyond their primary function of glycemic control. The mechanisms through which these drugs exert their effects are diverse, encompassing not only metabolic regulation but also anti-inflammatory, anti-atherosclerotic, and anticancer properties. This underscores the significance of repurposing antidiabetic agents for a broader spectrum of therapeutic applications. Future research should continue to explore the multifaceted benefits of these medications beyond glucose control. Antidiabetic drugs have shown significant promise in reducing cardiovascular risks, improving cognitive function in neurodegenerative diseases, providing antimicrobial effects in infectious diseases, preventing bone loss, and reducing the risk of certain cancers. Continued investigation into the mechanisms through which these drugs produce their beneficial effects is crucial. This could potentially lead to new therapeutic strategies for managing cardiovascular diseases, neurodegenerative diseases, cancer, and various infectious diseases in diabetic and nondiabetic populations. By continuing to investigate the diverse therapeutic potentials of antidiabetic drugs, we can harness their full benefits to address a broad spectrum of diseases. This multifaceted approach holds great promise for improving patient outcomes and expanding the therapeutic toolkit available to healthcare providers.

References

Abd Elhameed AG, Helal MG, Said E, Salem HA (2021) Saxagliptin defers thioacetamide-induced hepatocarcinogenesis in rats: a novel suppressive impact on Wnt/Hedgehog/Notch1 signaling. Environ Toxicol Pharmacol 86:103668

Abdul M, Hoosein N (2002) Expression and activity of potassium ion channels in human prostate cancer. Cancer Lett 186(1):99–105

Ahrén B (2011) Are sulfonylureas less desirable than DPP-4 inhibitors as add-on to metformin in the treatment of type 2 diabetes? Curr Diab Rep 11:83–90

Anisimov VN, Berstein LM, Egormin PA, Piskunova TS, Popovich IG, Zabezhinski MA et al (2005) Effect of metformin on life span and on the development of spontaneous mammary tumors in HER-2/neu transgenic mice. Exp Gerontol 40(8–9):685–693

Ashburn TT, Thor KB (2004) Drug repositioning: identifying and developing new uses for existing drugs. Nat Rev Drug Discov 3(8):673–683

Azar M, Lyons TJ (2010) Diabetes, insulin treatment, and cancer risk: what is the evidence? F1000 Med Rep 2:4

Bahne E, Sun EW, Young RL, Hansen M, Sonne DP, Hansen JS et al (2018) Metformin-induced glucagon-like peptide-1 secretion contributes to the actions of metformin in type 2 diabetes. JCI Insight 3(23):e93936

Bailey CJ (2017) Metformin: historical overview. Diabetologia 60(9):1566–1576

Barratt MJ, Frail DE (2012) Drug repositioning: bringing new life to shelved assets and existing drugs. John Wiley & Sons

Baruah P, Das A, Paul D, Chakrabarty S, Aguan K, Mitra S (2021) Sulfonylurea class of antidiabetic drugs inhibit acetylcholinesterase activity: unexplored auxiliary pharmacological benefit toward Alzheimer's disease. ACS Pharmacol Transl Sci 4(1):193–205

Billings LK, Handelsman Y, Heile M, Schneider D, Wyne K (2018) Health-related quality of life assessments with once-weekly glucagon-like peptide-1 receptor agonists in type 2 diabetes mellitus. J Manag Care Spec Pharm 24(9-a Suppl):S30–S41

Bjerre Knudsen L, Madsen LW, Andersen S, Almholt K, de Boer AS, Drucker DJ, Gotfredsen C, Egerod FL, Hegelund AC, Jacobsen H, Jacobsen SD (2010) Glucagon-like peptide-1 receptor agonists activate rodent thyroid C-cells causing calcitonin release and C-cell proliferation. Endocrinology 151(4):1473

Bosetti C, Rosato V, Buniato D, Zambon A, La Vecchia C, Corrao G (2013) Cancer risk for patients using thiazolidinediones for type 2 diabetes: a meta-analysis. Oncologist 18(2):148–156

Burgmaier M, Heinrich C, Marx N (2013) Cardiovascular effects of GLP-1 and GLP-1-based therapies: implications for the cardiovascular continuum in diabetes? Diabet Med 30(3):289–299

Cameron AR, Morrison VL, Levin D, Mohan M, Forteath C, Beall C et al (2016) Anti-inflammatory effects of metformin irrespective of diabetes status. Circ Res 119(5):652–665

Cao Y (2013) Angiogenesis and vascular functions in modulation of obesity, adipose metabolism, and insulin sensitivity. Cell Metab 18(4):478–489

Capuano A, Sportiello L, Maiorino MI, Rossi F, Giugliano D, Esposito K (2013) Dipeptidyl peptidase-4 inhibitors in type 2 diabetes therapy–focus on alogliptin. Drug Des Devel Ther 7:989–1001

Chang CH, Lin JW, Wu LC, Lai MS, Chuang LM, Arnold Chan K (2012) Association of thiazolidinediones with liver cancer and colorectal cancer in type 2 diabetes mellitus. Hepatology 55(5):1462–1472

Chaudhury A, Duvoor C, Reddy Dendi VS, Kraleti S, Chada A, Ravilla R et al (2017) Clinical review of antidiabetic drugs: implications for type 2 diabetes mellitus management. Front Endocrinol 8:6

Cheng H, Shang Y, Jiang L, Shi T-l, Wang L (2016) The peroxisome proliferators activated receptor-gamma agonists as therapeutics for the treatment of Alzheimer's disease and mild-to-moderate Alzheimer's disease: a meta-analysis. Int J Neurosci 126(4):299–307

Chiazza F, Tammen H, Pintana H, Lietzau G, Collino M, Nyström T et al (2018) The effect of DPP-4 inhibition to improve functional outcome after stroke is mediated by the SDF-1α/CXCR4 pathway. Cardiovasc Diabetol 17:1–13

Choi SH, Park S, Oh CJ, Leem J, Park K-G, Lee I-K (2015) Dipeptidyl peptidase-4 inhibition by gemigliptin prevents abnormal vascular remodeling via NF-E2-related factor 2 activation. Vasc Pharmacol 73:11–19

Choi S-Y, Ryu H-M, Oh E-J, Choi J-Y, Cho J-H, Kim C-D et al (2017) Dipeptidyl peptidase-4 inhibitor gemigliptin protects against vascular calcification in an experimental chronic kidney disease and vascular smooth muscle cells. PLoS One 12(7):e0180393

Courtois S, Bénéjat L, Izotte J, Mégraud F, Varon C, Lehours P et al (2018) Metformin can inhibit Helicobacter pylori growth. Future Microbiol 13(14):1575–1583

Darsalia V, Klein T, Nyström T, Patrone C (2018) Glucagon-like receptor 1 agonists and DPP-4 inhibitors: antidiabetic drugs with anti-stroke potential. Neuropharmacology 136:280–286

DeFronzo RA (1999) Pharmacologic therapy for type 2 diabetes mellitus. Ann Intern Med 131(4):281–303

DeFronzo R, Fleming GA, Chen K, Bicsak TA (2016) Metformin-associated lactic acidosis: current perspectives on causes and risk. Metabolism 65(2):20–29
Della-Morte D, Palmirotta R, Rehni AK, Pastore D, Capuani B, Pacifici F et al (2014) Pharmacogenomics and pharmacogenetics of thiazolidinediones: role in diabetes and cardiovascular risk factors. Pharmacogenomics 15(16):2063–2082
Derosa G, Maffioli P (2012) α-Glucosidase inhibitors and their use in clinical practice. Arch Med Sci 8(5):899
Drucker DJ (2006) The biology of incretin hormones. Cell Metab 3(3):153–165
Elder D, Tindall S (2020) The many advantages of repurposing existing drugs. Eur Pharm Rev 25(3):34–37
Fang W, Zhang J, Hong L, Huang W, Dai X, Ye Q et al (2020) Metformin ameliorates stress-induced depression-like behaviors via enhancing the expression of BDNF by activating AMPK/CREB-mediated histone acetylation. J Affect Disord 260:302–313
Faurschou A, Knop F, Thyssen J, Zachariae C, Skov L, Vilsbøll T (2014) Improvement in psoriasis after treatment with the glucagon-like peptide-1 receptor agonist liraglutide. Acta Diabetol 51:147–150
Femia AP, Raimondi L, Maglieri G, Lodovici M, Mannucci E, Caderni G (2013) Long-term treatment with Sitagliptin, a dipeptidyl peptidase-4 inhibitor, reduces colon carcinogenesis and reactive oxygen species in 1, 2-dimethylhydrazine-induced rats. Int J Cancer 133(10):2498–2503
Ferruzzi P, Ceni E, Tarocchi M, Grappone C, Milani S, Galli A et al (2005) Thiazolidinediones inhibit growth and invasiveness of the human adrenocortical cancer cell line H295R. J Clin Endocrinol Metab 90(3):1332–1339
Filippatos TD, Elisaf MS (2013) Effects of glucagon-like peptide-1 receptor agonists on renal function. World J Diabetes 4(5):190
Fujii M, Takemura R, Yamaguchi M, Hasegawa G, Shigeta H, Nakano K et al (1997) Troglitazone (CS-045) ameliorates albuminuria in streptozotocin-induced diabetic rats. Metabolism 46(9):981–983
Fürstenberger G, Senn H-J (2002) Insulin-like growth factors and cancer. Lancet Oncol 3(5):298–302
Garnett JP, Baker EH, Naik S, Lindsay JA, Knight GM, Gill S et al (2013) Metformin reduces airway glucose permeability and hyperglycaemia-induced Staphylococcus aureus load independently of effects on blood glucose. Thorax 68(9):835–845
Gaspari T, Welungoda I, Widdop RE, Simpson RW, Dear AE (2013) The GLP-1 receptor agonist liraglutide inhibits progression of vascular disease via effects on atherogenesis, plaque stability and endothelial function in an ApoE−/− mouse model. Diab Vasc Dis Res 10(4):353–360
Gerich J (2013) Pathogenesis and management of postprandial hyperglycemia: role of incretin-based therapies. Int J Gen Med 6:877–895
Gier B, Butler PC, Lai CK, Kirakossian D, DeNicola MM, Yeh MW (2012) Glucagon like peptide-1 receptor expression in the human thyroid gland. J Clin Endocrinol Metabol 97(1):121–131
Giovannucci E, Harlan DM, Archer MC, Bergenstal RM, Gapstur SM, Habel LA et al (2010) Diabetes and cancer: a consensus report. CA Cancer J Clin 60(4):207–221
Gong N, Xiao Q, Zhu B, Zhang C-Y, Wang Y-C, Fan H et al (2014) Activation of spinal glucagon-like peptide-1 receptors specifically suppresses pain hypersensitivity. J Neurosci 34(15):5322–5334
Gopoju R, Panangipalli S, Kotamraju S (2018) Metformin treatment prevents SREBP2-mediated cholesterol uptake and improves lipid homeostasis during oxidative stress-induced atherosclerosis. Free Radic Biol Med 118:85–97
Guan Y, Zhang Y, Schneider A, Davis L, Breyer RM, Breyer MD (2001) Peroxisome proliferator-activated receptor-γ activity is associated with renal microvasculature. Am J Physiol Renal Physiol 281(6):F1036–F1F46
Gundewar S, Calvert JW, Jha S, Toedt-Pingel I, Yong Ji S, Nunez D et al (2009) Activation of AMP-activated protein kinase by metformin improves left ventricular function and survival in heart failure. Circ Res 104(3):403–411
Gupta V, Kalra S (2011) Choosing a gliptin. Indian J Endocrinol Metab 15(4):298

Gupta P, Bala M, Gupta S, Dua A, Dabur R, Injeti E et al (2016) Efficacy and risk profile of anti-diabetic therapies: conventional vs traditional drugs—a mechanistic revisit to understand their mode of action. Pharmacol Res 113:636–674

Han J, Li Y, Liu X, Zhou T, Sun H, Edwards P et al (2018) Metformin suppresses retinal angiogenesis and inflammation in vitro and in vivo. PLoS One 13(3):e0193031

Heneka MT, Landreth GE, Hüll M (2007) Drug insight: effects mediated by peroxisome proliferator-activated receptor-γ in CNS disorders. Nat Clin Pract Neurol 3(9):496–504

Hölscher C (2014) Central effects of GLP-1: new opportunities for treatments of neurodegenerative diseases. J Endocrinol 221(1):T31–T41

Hossain U, Das AK, Ghosh S, Sil PC (2020) An overview on the role of bioactive α-glucosidase inhibitors in ameliorating diabetic complications. Food Chem Toxicol 145:111738

Hsu W-H, Sue S-P, Liang H-L, Tseng C-W, Lin H-C, Wen W-L et al (2021) Dipeptidyl peptidase 4 inhibitors decrease the risk of hepatocellular carcinoma in patients with chronic hepatitis C infection and type 2 diabetes mellitus: a nationwide study in Taiwan. Front Public Health 9:711723

Ikeda T, Iwata K, Murakami H (2000) Inhibitory effect of metformin on intestinal glucose absorption in the perfused rat intestine. Biochem Pharmacol 59(7):887–890

Inzucchi SE, Bergenstal RM, Buse JB, Diamant M, Ferrannini E, Nauck M et al (2012) Management of hyperglycaemia in type 2 diabetes: a patient-centered approach. Position statement of the American Diabetes Association (ADA) and the European Association for the Study of Diabetes (EASD). Diabetologia 55:1577–1596

Jadhav S, Ferrell W, Greer IA, Petrie JR, Cobbe SM, Sattar N (2006) Effects of metformin on microvascular function and exercise tolerance in women with angina and normal coronary arteries: a randomized, double-blind, placebo-controlled study. J Am Coll Cardiol 48(5):956–963

Jourdan J-P, Bureau R, Rochais C, Dallemagne P (2020) Drug repositioning: a brief overview. J Pharm Pharmacol 72(9):1145–1151

Kaji K, Yoshiji H, Ikenaka Y, Noguchi R, Aihara Y, Douhara A et al (2014) Dipeptidyl peptidase-4 inhibitor attenuates hepatic fibrosis via suppression of activated hepatic stellate cell in rats. J Gastroenterol 49:481–491

Kakiuchi-Kiyota S, Vetro JA, Suzuki S, Varney ML, Han H-Y, Nascimento M et al (2009) Effects of the PPARγ agonist troglitazone on endothelial cells in vivo and in vitro: differences between human and mouse. Toxicol Appl Pharmacol 237(1):83–90

Kamada S, Namekawa T, Ikeda K, Suzuki T, Kagawa M, Takeshita H et al (2021) Functional inhibition of cancer stemness-related protein DPP4 rescues tyrosine kinase inhibitor resistance in renal cell carcinoma. Oncogene 40(22):3899–3913

Kamel AM, Sabry N, Farid S (2022) Effect of metformin on left ventricular mass and functional parameters in nondiabetic patients: a meta-analysis of randomized clinical trials. BMC Cardiovasc Disord 22(1):405

Kang SM, Jung HS, Kwon MJ, Lee SH, Park JH (2021) Effects of anagliptin on the stress induced accelerated senescence of human umbilical vein endothelial cells. Ann Transl Med 9(9):750

Kao T-W, Huang C-C (2022) Pleiotropic effect of sodium-glucose cotransporter 2 inhibitors on blood pressure. Front Cardiovasc Med 9:1086672

Kappe C, Patrone C, Holst JJ, Zhang Q, Sjöholm Å (2013) Metformin protects against lipoapoptosis and enhances GLP-1 secretion from GLP-1-producing cells. J Gastroenterol 48:322–332

Kataoka Y, Nicholls SJ, Andrews J, Uno K, Kapadia SR, Tuzcu EM et al (2022) Plaque microstructures during metformin therapy in type 2 diabetic subjects with coronary artery disease: optical coherence tomography analysis. Cardiovasc Diagn Ther 12(1):77

Kato J, Shirakami Y, Mizutani T, Kubota M, Sakai H, Ibuka T et al (2020) Alpha-glucosidase inhibitor voglibose suppresses azoxymethane-induced colonic preneoplastic lesions in diabetic and obese mice. Int J Mol Sci 21(6):2226

Kaufmann SH, Dorhoi A, Hotchkiss RS, Bartenschlager R (2018) Host-directed therapies for bacterial and viral infections. Nat Rev Drug Discov 17(1):35–56

Kaundal RK, Iyer S, Kumar A, Sharma SS (2009) Protective effects of pioglitazone against global cerebral ischemic-reperfusion injury in gerbils. J Pharmacol Sci 109(3):361–367

Kawaguchi T, Nakano D, Koga H, Torimura T (2019) Effects of a DPP4 inhibitor on progression of NASH-related HCC and the p62/Keap1/Nrf2-pentose phosphate pathway in a mouse model. Liver Cancer 8(5):359–372

Kawakubo M, Tanaka M, Ochi K, Watanabe A, Saka-Tanaka M, Kanamori Y et al (2020) Dipeptidyl peptidase-4 inhibition prevents non-alcoholic steatohepatitis–associated liver fibrosis and tumor development in mice independently of its antidiabetic effects. Sci Rep 10(1):983

Kersten C, Knottnerus I, Heijmans E, Haalboom M, Zandbergen A, den Hertog H (2022) Effect of metformin on outcome after acute ischemic stroke in patients with type 2 diabetes mellitus. J Stroke Cerebrovasc Dis 31(9):106648

Khan SZ, Rivero M, Nader ND, Cherr GS, Harris LM, Dryjski ML et al (2019) Metformin is associated with improved survival and decreased cardiac events with no impact on patency and limb salvage after revascularization for peripheral arterial disease. Ann Vasc Surg 55:63–77

Kim M, Platt MJ, Shibasaki T, Quaggin SE, Backx PH, Seino S et al (2013) GLP-1 receptor activation and Epac2 link atrial natriuretic peptide secretion to control of blood pressure. Nat Med 19(5):567–575

Koenig AM, Mechanic-Hamilton D, Xie SX, Combs MF, Cappola AR, Xie L et al (2017) Effects of the insulin sensitizer metformin in Alzheimer's disease: pilot data from a randomized placebo-controlled crossover study. Alzheimer Dis Assoc Disord 31(2):107

LaFerla FM (2012) Pre-clinical success against Alzheimer's disease with an old drug. N Engl J Med 367(6):570–572

Lang J (1999) Molecular mechanisms and regulation of insulin exocytosis as a paradigm of endocrine secretion. Eur J Biochem 259(1–2):3–17

Lee M-C, Chiang C-Y, Lee C-H, Ho C-M, Chang C-H, Wang J-Y et al (2018) Metformin use is associated with a low risk of tuberculosis among newly diagnosed diabetes mellitus patients with normal renal function: a nationwide cohort study with validated diagnostic criteria. PLoS One 13(10):e0205807

Lerche S, Brock B, Rungby J, Bøtker HE, Møller N, Rodell A et al (2008) Glucagon-like peptide-1 inhibits blood-brain glucose transfer in humans. Diabetes 57(2):325–331

Levine R (1984) Sulfonylureas: background and development of the field. Diabetes Care 7:3–7

Li Y, Duffy KB, Ottinger MA, Ray B, Bailey JA, Holloway HW et al (2010) GLP-1 receptor stimulation reduces amyloid-β peptide accumulation and cytotoxicity in cellular and animal models of Alzheimer's disease. J Alzheimer's Dis 19(4):1205–1219

Li Y, Huang J, He X, Tang G, Tang Y-H, Liu Y et al (2014) Postacute stromal cell–derived factor-1α expression promotes neurovascular recovery in ischemic mice. Stroke 45(6):1822–1829

Lim SW, Jin JZ, Jin L, Jin J, Li C (2015) Role of dipeptidyl peptidase-4 inhibitors in new-onset diabetes after transplantation. Korean J Intern Med 30(6):759

Liu Q, Anderson C, Broyde A, Polizzi C, Fernandez R, Baron A et al (2010) Glucagon-like peptide-1 and the exenatide analogue AC3174 improve cardiac function, cardiac remodeling, and survival in rats with chronic heart failure. Cardiovasc Diabetol 9:1–14

Lorell BH, Carabello BA (2000) Left ventricular hypertrophy: pathogenesis, detection, and prognosis. Circulation 102(4):470–479

Loubatières A (1957) The hypoglycemic sulfonamides: history and development of the problem from 1942 to 1955. Ann N Y Acad Sci 71(1):4–11

Ma X, Meng J, Jia M, Bi L, Zhou Y, Wang Y et al (2013) Exendin-4, a glucagon-like peptide-1 receptor agonist, prevents osteopenia by promoting bone formation and suppressing bone resorption in aged ovariectomized rats. J Bone Miner Res 28(7):1641–1652

Mabilleau G, Pereira M, Chenu C (2018) Novel skeletal effects of glucagon-like peptide-1 (GLP-1) receptor agonists. J Endocrinol 236(1):R29–R42

Malik F, Mehdi SF, Ali H, Patel P, Basharat A, Kumar A et al (2018) Is metformin poised for a second career as an antimicrobial? Diabetes Metab Res Rev 34(4):e2975

Mary A, Hartemann A, Liabeuf S, Aubert CE, Kemel S, Salem JE et al (2017) Association between metformin use and below-the-knee arterial calcification score in type 2 diabetic patients. Cardiovasc Diabetol 16:1–9

Masur K, Schwartz F, Entschladen F, Niggemann B, Zaenker K (2006) DPPIV inhibitors extend GLP-2 mediated tumour promoting effects on intestinal cancer cells. Regul Pept 137(3):147–155
McClean PL, Parthsarathy V, Faivre E, Hölscher C (2011) The diabetes drug liraglutide prevents degenerative processes in a mouse model of Alzheimer's disease. J Neurosci 31(17):6587–6594
Memmott RM, Mercado JR, Maier CR, Kawabata S, Fox SD, Dennis PA (2010) Metformin prevents tobacco carcinogen–induced lung tumorigenesis. Cancer Prev Res 3(9):1066–1076
Mohan M, Al-Talabany S, McKinnie A, Mordi IR, Singh JS, Gandy SJ et al (2019) A randomized controlled trial of metformin on left ventricular hypertrophy in patients with coronary artery disease without diabetes: the MET-REMODEL trial. Eur Heart J 40(41):3409–3417
Mohsin AA, Chen Q, Quan N, Rousselle T, Maceyka MW, Samidurai A et al (2019) Mitochondrial complex I inhibition by metformin limits reperfusion injury. J Pharmacol Exp Ther 369(2):282–290
Mone P, Gambardella J, Pansini A, de Donato A, Martinelli G, Boccalone E et al (2021) Cognitive impairment in frail hypertensive elderly patients: role of hyperglycemia. Cells 10(8):2115
Musi N, Hirshman MF, Nygren J, Svanfeldt M, Bavenholm P, Rooyackers O et al (2002) Metformin increases AMP-activated protein kinase activity in skeletal muscle of subjects with type 2 diabetes. Diabetes 51(7):2074–2081
Neal B, Perkovic V, Mahaffey KW, De Zeeuw D, Fulcher G, Erondu N et al (2017) Canagliflozin and cardiovascular and renal events in type 2 diabetes. N Engl J Med 377(7):644–657
Novac N (2013) Challenges and opportunities of drug repositioning. Trends Pharmacol Sci 34(5):267–272
Noyan-Ashraf MH, Momen MA, Ban K, Sadi AM, Zhou YQ, Riazi AM, Baggio LL, Henkelman RM, Husain M, Drucker DJ (2009) GLP-1R agonist liraglutide activates cytoprotective pathways and improves outcomes after experimental myocardial infarction in mice. Diabetes 58:975–983
Nyström T, Gonon AT, Sjöholm Å, Pernow J (2005) Glucagon-like peptide-1 relaxes rat conduit arteries via an endothelium-independent mechanism. Regul Pept 125(1–3):173–177
Ortega R, Collado A, Selles F, Gonzalez-Navarro H, Sanz M-J, Real JT et al (2019) SGLT-2 (sodium-glucose cotransporter 2) inhibition reduces Ang II (angiotensin II)–induced dissecting abdominal aortic aneurysm in ApoE (apolipoprotein E) knockout mice. Arterioscler Thromb Vasc Biol 39(8):1614–1628
Ozutsumi T, Namisaki T, Shimozato N, Kaji K, Tsuji Y, Kaya D et al (2020) Combined treatment with sodium-glucose cotransporter-2 inhibitor (Canagliflozin) and dipeptidyl peptidase-4 inhibitor (teneligliptin) alleviates NASH progression in a nondiabetic rat model of steatohepatitis. Int J Mol Sci 21(6):2164
Parvathaneni V, Kulkarni NS, Muth A, Gupta V (2019) Drug repurposing: a promising tool to accelerate the drug discovery process. Drug Discov Today 24(10):2076–2085
Patel SK, Wai B, Lang CC, Levin D, Palmer CN, Parry HM et al (2017) Genetic variation in Kruppel like factor 15 is associated with left ventricular hypertrophy in patients with type 2 diabetes: discovery and replication cohorts. EBioMedicine 18:171–178
Pathak R, Bridgeman MB (2010) Dipeptidyl peptidase-4 (DPP-4) inhibitors in the management of diabetes. Pharm Ther 35(9):509
Pistrosch F, Passauer J, Fischer S, Fuecker K, Hanefeld M, Gross P (2004) In type 2 diabetes, rosiglitazone therapy for insulin resistance ameliorates endothelial dysfunction independent of glucose control. Diabetes Care 27(2):484–490
Qin C-J, Zhao L-H, Zhou X, Zhang H-L, Wen W, Tang L et al (2018) Inhibition of dipeptidyl peptidase IV prevents high fat diet-induced liver cancer angiogenesis by downregulating chemokine ligand 2. Cancer Lett 420:26–37
Saraei P, Asadi I, Kakar M, Moradi-Kor N (2019) The beneficial effects of metformin on cancer prevention and therapy: a comprehensive review of recent advances. Cancer Manag Res 11:3295–3313. 2000
Sardu C, Paolisso P, Sacra C, Mauro C, Minicucci F, Portoghese M et al (2019) Effects of metformin therapy on coronary endothelial dysfunction in patients with prediabetes with stable

angina and nonobstructive coronary artery stenosis: the CODYCE multicenter prospective study. Diabetes Care 42(10):1946–1955
Satoh H (2018) Pleiotropic effects of SGLT2 inhibitors beyond the effect on glycemic control. Diabetol Int 9(4):212–214
Sharma AN, Ligade SS, Sharma JN, Shukla P, Elased KM, Lucot JB (2015) GLP-1 receptor agonist liraglutide reverses long-term atypical antipsychotic treatment associated behavioral depression and metabolic abnormalities in rats. Metab Brain Dis 30:519–527
Shi Q, Liu S, Fonseca VA, Thethi TK, Shi L (2019) Effect of metformin on neurodegenerative disease among elderly adult US veterans with type 2 diabetes mellitus. BMJ Open 9(7):e024954
Sirtori CR, Franceschini G, Gianfranceschi G, Sirtori M, Montanari G, Bosisio E et al (1984) Metformin improves peripheral vascular flow in nonhyperlipidemic patients with arterial disease. J Cardiovasc Pharmacol 6(5):914–923
Sivertsen J, Rosenmeier J, Holst JJ, Vilsbøll T (2012) The effect of glucagon-like peptide 1 on cardiovascular risk. Nat Rev Cardiol 9(4):209–222
Skillman TG, Feldman JM (1981) The pharmacology of sulfonylureas. Am J Med 70:361–372
Smith U, Gale E (2010) Cancer and diabetes: are we ready for prime time? Diabetologia 53:1541–1544
Standl E, Theodorakis MJ, Erbach M, Schnell O, Tuomilehto J (2014) On the potential of acarbose to reduce cardiovascular disease. Cardiovasc Diabetol 13:1–12
Su B, Sheng H, Zhang M, Bu L, Yang P, Li L et al (2015) Risk of bone fractures associated with glucagon-like peptide-1 receptor agonists' treatment: a meta-analysis of randomized controlled trials. Endocrine 48:107–115
Toledo E, Inestrosa N (2010) Activation of Wnt signaling by lithium and rosiglitazone reduced spatial memory impairment and neurodegeneration in brains of an APPswe/PSEN1ΔE9 mouse model of Alzheimer's disease. Mol Psychiatry 15(3):272–285
Tsapas A, Karagiannis T, Kakotrichi P, Avgerinos I, Mantsiou C, Tousinas G et al (2021) Comparative efficacy of glucose-lowering medications on body weight and blood pressure in patients with type 2 diabetes: a systematic review and network meta-analysis. Diabetes Obes Metab 23(9):2116–2124
Tseng Y-H, Tsan Y-T, Chan W-C, Sheu WH-H, Chen P-C (2015) Use of an α-glucosidase inhibitor and the risk of colorectal cancer in patients with diabetes: a nationwide, population-based cohort study. Diabetes Care 38(11):2068–2074
Tsujie M, Nakamori S, Okami J, Hayashi N, Hiraoka N, Nagano H et al (2003) Thiazolidinediones inhibit growth of gastrointestinal, biliary, and pancreatic adenocarcinoma cells through activation of the peroxisome proliferator-activated receptor γ/retinoid X receptor α pathway. Exp Cell Res 289(1):143–151
Vasamsetti SB, Karnewar S, Kanugula AK, Thatipalli AR, Kumar JM, Kotamraju S (2015) Metformin inhibits monocyte-to-macrophage differentiation via AMPK-mediated inhibition of STAT3 activation: potential role in atherosclerosis. Diabetes 64(6):2028–2041
Vella A (2012) Mechanism of action of DPP-4 inhibitors—new insights. J Clin Endocrinol Metab, Oxford University Press 97:2626–2628
Vettor R, Valerio A, Ragni M, Trevellin E, Granzotto M, Olivieri M et al (2014) Exercise training boosts eNOS-dependent mitochondrial biogenesis in mouse heart: role in adaptation of glucose metabolism. Am J Physiol Endocrinol Metab 306(5):E519–EE28
Vilsbøll T, Garber A (2012) Non-glycaemic effects mediated via GLP-1 receptor agonists and the potential for exploiting these for therapeutic benefit: focus on liraglutide. Diabetes Obes Metab 14:41–49
Wang XF, Zhang JY, Li L, Zhao XY, Tao HL, Zhang L (2011) Metformin improves cardiac function in rats via activation of AMP-activated protein kinase. Clin Exp Pharmacol Physiol 38(2):94–101
Wang B, Zhong J, Lin H, Zhao Z, Yan Z, He H et al (2013) Blood pressure-lowering effects of GLP-1 receptor agonists exenatide and liraglutide: a meta-analysis of clinical trials. Diabetes Obes Metab 15(8):737–749

Wang Y, Zhao W, Li G, Chen J, Guan X, Chen X et al (2017) Neuroprotective effect and mechanism of thiazolidinedione on dopaminergic neurons in vivo and in vitro in Parkinson's disease. PPAR Res 2017:4089214
Weng S, Luo Y, Zhang Z, Su X, Peng D (2020) Effects of metformin on blood lipid profiles in nondiabetic adults: a meta-analysis of randomized controlled trials. Endocrine 67:305–317
White AT, Murphy AN (2010) Administration of thiazolidinediones for neuroprotection in ischemic stroke: a pre-clinical systematic review. J Neurochem 115(4):845–853
Wiciński M, Górski K, Wódkiewicz E, Walczak M, Nowaczewska M, Malinowski B (2020) Vasculoprotective effects of vildagliptin. Focus on atherogenesis. Int J Mol Sci 21(7):2275
Wiviott SD, Raz I, Bonaca MP, Mosenzon O, Kato ET, Cahn A et al (2019) Dapagliflozin and cardiovascular outcomes in type 2 diabetes. N Engl J Med 380(4):347–357
Wohlfart P, Linz W, Hübschle T, Linz D, Huber J, Hess S et al (2013) Cardioprotective effects of lixisenatide in rat myocardial ischemia-reperfusion injury studies. J Transl Med 11:1–12
Xiao H, Ma X, Feng W, Fu Y, Lu Z, Xu M et al (2010) Metformin attenuates cardiac fibrosis by inhibiting the TGFβ1–Smad3 signalling pathway. Cardiovasc Res 87(3):504–513
Xue CY, Zhou MQ, Zheng QY, Zhang JH, Cheng WT, Bai XH et al (2022) Thiazolidinediones play a positive role in the vascular endothelium and inhibit plaque progression in diabetic patients with coronary atherosclerosis: a systematic review and meta-analysis. Front Cardiovasc Med 9:1043406
Yamada C, Yamada Y, Tsukiyama K, Yamada K, Udagawa N, Takahashi N et al (2008) The murine glucagon-like peptide-1 receptor is essential for control of bone resorption. Endocrinology 149(2):574–579
Yang X, So WY, Ma RC, Linda W, Ko GT, Kong AP et al (2010) Use of sulphonylurea and cancer in type 2 diabetes—the Hong Kong Diabetes Registry. Diabetes Res Clin Pract 90(3):343–351
Yanovski SZ, Yanovski JA (2014) Long-term drug treatment for obesity: a systematic and clinical review. JAMA 311(1):74–86
Yi Q-Y, Deng G, Chen N, Bai Z-S, Yuan J-S, Wu G-H et al (2016) Metformin inhibits development of diabetic retinopathy through inducing alternative splicing of VEGF-A. Am J Transl Res 8(9):3947
Yoon S-Y, Park J-S, Choi J-E, Choi J-M, Lee W-J, Kim S-W et al (2010) Rosiglitazone reduces tau phosphorylation via JNK inhibition in the hippocampus of rats with type 2 diabetes and tau transfected SH-SY5Y cells. Neurobiol Dis 40(2):449–455
Yoshida K, Kohzuki M, Xu H-L, Wu X-M, Kamimoto M, Sato T (2001) Effects of troglitazone and temocapril in spontaneously hypertensive rats with chronic renal failure. J Hypertens 19(3):503–510
Yoshida S, Hong S, Suzuki T, Nada S, Mannan AM, Wang J et al (2011) Redox regulates mammalian target of rapamycin complex 1 (mTORC1) activity by modulating the TSC1/TSC2-Rheb GTPase pathway. J Biol Chem 286(37):32651–32660
Zhang CX, Pan SN, Meng RS, Peng CQ, Xiong ZJ, Chen BL et al (2011) Metformin attenuates ventricular hypertrophy by activating the AMP-activated protein kinase–endothelial nitric oxide synthase pathway in rats. Clin Exp Pharmacol Physiol 38(1):55–62
Zhang Y, Weng W, Xie B, Meng Y, Hao Y, Liang Y et al (2018) Glucagon-like peptide-1 receptor agonists and fracture risk: a network meta-analysis of randomized clinical trials. Osteoporos Int 29:2639–2644
Zhang N, Feng B, Ma X, Sun K, Xu G, Zhou Y (2019) Dapagliflozin improves left ventricular remodeling and aorta sympathetic tone in a pig model of heart failure with preserved ejection fraction. Cardiovasc Diabetol 18(1):1–15
Zhong J, Rao X, Rajagopalan S (2013) An emerging role of dipeptidyl peptidase 4 (DPP4) beyond glucose control: potential implications in cardiovascular disease. Atherosclerosis 226(2):305–314
Zhou G, Myers R, Li Y, Chen Y, Shen X, Fenyk-Melody J et al (2001) Role of AMP-activated protein kinase in mechanism of metformin action. J Clin Invest 108(8):1167–1174
Zinman B, Wanner C, Lachin JM, Fitchett D, Bluhmki E, Hantel S et al (2015) Empagliflozin, cardiovascular outcomes, and mortality in type 2 diabetes. N Engl J Med 373(22):2117–2128

Drug Repurposing for Cardioprotective Therapy

4

Swarnima Negi, Krishna Jadhav, Raghuraj Singh,
Agrim Jhilta, and Rahul K. Verma

Abstract

Drug repurposing is a promising strategy that involves identifying new therapeutic uses for existing or experimental drugs. It offers significant advantages over developing new drugs, such as reduced risk, faster development timelines, and lower costs. In the context of cardiovascular diseases (CVDs), which have the highest mortality rate across the globe, drug repurposing presents an opportunity to address the growing need for effective therapies. Despite the importance of CVDs, there has been a noticeable decline in the development of novel cardiovascular drugs. Drug repurposing is the strategy of taking an existing, approved, or investigational drug and using it for a new. Drug repurposing can bridge this gap by identifying existing drugs that could be effective for cardiovascular conditions. The process involves computational approaches, such as molecular docking and signature matching, along with experimental techniques like phenotypic screening and clinical data analysis, for the purpose of discovering new uses for drugs. However, despite its potential benefits, drug repurposing faces regulatory challenges. Regulatory authorities, such as the FDA and EMA, may require more conclusive data to ensure the safety and efficacy of repurposed drugs for new indications. The lack of clarity regarding the regulatory pathway for repurposed drugs can result in delays and increased costs. Additionally, the repurposing of drugs for cardiovascular conditions requires rigorous data from

S. Negi · K. Jadhav · A. Jhilta · R. K. Verma (✉)
Chemical Biology Unit, Institute of Nano Science and Technology (INST),
Knowledge City, Mohali, Punjab, India
e-mail: rahulverma@inst.ac.in

R. Singh
Chemical Biology Unit, Institute of Nano Science and Technology (INST),
Knowledge City, Mohali, Punjab, India

Academy of Scientific and Innovative Research (AcSIR), Ghaziabad, UP, India

R. K. Kaundal et al. (eds.), *Drug Repurposing: Exploring Newer Therapeutic Potentials*,
https://doi.org/10.1007/978-981-95-4522-3_4

preclinical and clinical studies to demonstrate their effectiveness and manage any unforeseen side effects. Despite these hurdles, drug repurposing remains a promising and efficient strategy for developing new cardiovascular therapies to address the growing burden of cardiovascular diseases worldwide.

Keywords

Drug repurposing · Cardiovascular diseases · Phenotype screening · Molecular docking · Signature matching

4.1 Introduction

Drug repurposing, also known as drug repositioning, reprofiling, or retasking, is the process of discovering new therapeutic indications for existing or experimental drugs that were originally developed for different medical conditions (Pushpakom et al. 2018; March-Vila et al. 2017). This strategy provides multiple advantages over developing an entirely new drug. One significant benefit is the lower risk of failure, as the repurposed drug has already undergone safety assessments in previous clinical trials. Another advantage is a shorter development timeline, as the drugs have already passed the preclinical and safety evaluations checkpoints. Additionally, repurposing a drug generally requires less financial investment. Although the costs can vary depending on the stage of development, a substantial savings can be achieved in preclinical and initial phases of the clinical trials. However, the regulatory expenses and phase 3 clinical trial costs may be comparable to those of a new drug (Abdelsayed et al. 2022).

Traditionally, drug repurposing has been largely driven by serendipity. Many successful instances have arisen where a medication was found to produce an unforeseen off-target effect or a newly identified on-target effect, which was later developed for commercial purposes (Pushpakom et al. 2018; Ashburn and Thor 2004). Interestingly, some of the most notable examples of drug repurposing did not result from structured strategies. For example, sildenafil citrate was originally indicated for hypertension and later repurposed for erectile dysfunction based on retrospective clinical observations (Novac 2013). When Pfizer rebranded it as Viagra, it dominated the erectile dysfunction drug market, holding a 47% market share in 2012, where its global sales were around $2.05 billion (Pushpakom et al. 2018; Ashburn and Thor 2004). Similarly, thalidomide, originally introduced as a sedative in 1957, which was later withdrawn from the market within a few years due to birth defects, such as phocomelia, when taken during pregnancy. However, in 1964, it was unexpectedly found to be effective in treating erythema nodosum leprosum (ENL), and later, in 1999, it was approved to treat multiple myeloma. Its effectiveness in treating multiple myeloma also led to the creation of improved derivatives of thalidomide, such as lenalidomide, which generated worldwide sales of $8.2 billion in 2017 (Pushpakom et al. 2018; Ashburn and Thor 2004; Novac 2013).

In the field of cardioprotective therapy, many repurposed drugs have been identified based on their known pharmacological properties or through retrospective analyses of clinical outcomes when used for their original indication. These approaches continue to play a crucial role in identifying new therapeutic applications for existing drugs. New drug repurposing strategies leverage extensive phenotyping from electronic health records and analyze drug effects on the intercellular mechanism and gene expression of proteins (Abdelsayed et al. 2022; Gelosa et al. 2020). By identifying drugs that counteract disease-related molecular changes, these approaches offer a systematic, mechanism-based method for repurposing. Moreover, high-throughput screening using realistic models can identify compounds that enhance disease outcomes without the need for prior knowledge of their targets (March-Vila et al. 2017). Emphasizing therapeutic mechanisms has increased accuracy in drug discovery, similar to how mechanism-based diagnoses have improved clinical care. Aligning drugs with molecular mechanisms can broaden treatment possibilities and refine clinical approaches. This is particularly crucial for rare cardiovascular diseases, which affect up to 3% of the population but often lack dedicated drug development due to limited financial incentives. Many of these conditions have monogenic origins, making them ideal candidates for modern computational and experimental drug discovery techniques (Lacaze et al. 2021).

This chapter highlights the significance of drug repurposing in cardioprotective therapies, along with different approaches used in drug repurposing. Since cardiovascular disease (CVD) incidences are rising at an alarming rate in older as well as in younger generations across the globe, many new approaches have made the whole process of drug repurposing much easier, by combining different experimental approaches with computational approaches that can yield beneficial outcomes.

4.2 Need for Novel Cardioprotective Therapies

The World Health Organization (WHO) states that CVDs are the leading cause of death worldwide, responsible for approximately 17.9 million fatalities annually. This category includes various conditions that impact the heart and blood vessels, such as coronary heart disease, cerebrovascular disease, and rheumatic heart disease. As of 2024, CVDs continue to lead the chart with the highest mortality rates among noncommunicable disease. In the WHO European region alone, CVDs account for over 42.5% of all deaths annually, equating to approximately 10,000 deaths every day (World Health Organization n.d.). The American Heart Association's 2024 report reveals that around 6.7 million Americans over the age of 20 are living with heart failure, with this figure expected to grow to 8.7 million by 2030 (Martin et al. 2024). Notably, there has been a rise in heart failure cases among individuals aged 35 to 64, highlighting a concerning trend of this condition affecting younger populations. Comparing the development of novel drugs for cardiovascular diseases to cancer, there has been a notable decline over the past few decades. Sixteen percent of phase 1 clinical trials were conducted for cardiovascular disease in the year 1990–1995, where this figure has seen a massive decline in the following decade to 5% between 2005–2012, compared to clinical trials for cancer therapeutics. This has also been highlighted by the U.S. Food and Administration's (FDA), where they have approved only four new medical interventions for cardiovascular interventions compared to 49 new drugs for cancer management (Hwang et al. 2016). Over the last few decades there have been very few cardiovascular drugs entered in clinical trials, out of which most of them fail in phase 3 clinical trials due to safety concerns and inadequate efficacy. CVDs remain the leading cause of death globally, yet the development of new cardiovascular therapies has not kept pace with the growing burden of these conditions. The diminishing pipeline of novel cardiovascular interventions is inadequate for patient needs, creating a huge gap in treatment options. In this scenario, drug repurposing emerges as a vital solution by identifying new cardiovascular applications for existing drugs. Repurposing offers a faster, cost-effective, and safer approach to developing effective therapies (Fordyce et al. 2015). Since repurposed drugs already have established safety profiles, they can be rapidly advanced through clinical trials, ensuring quicker availability for patients. This approach not only speeds up the development of treatment options but also holds the potential to substantially lower CVD-related mortality, tackling one of the most pressing challenges in global healthcare (Fordyce et al. 2015; Kort and Jovinge 2021).

4.3 Approaches Used in Drug Repurposing

The drug repurposing strategy involves three steps. The first consists of identifying a potential active drug for a disease based on scientific research along with background data analysis. Further, the selected drug undergoes preclinical testing, which includes lab experiments and animal studies, to assess its efficacy on the new proposed indication. The drug is then directly moved to the phase 2 clinical trial upon receiving all the safety clearance from the previously held clinical trials. Therefore, selecting the right drug candidate is crucial for the rigorous studies ahead. The initial selection of the drug has been made easy by combining computational studies with experimental studies. The computational approach gives a boost to data mining by effectively helping to analyze large databases and predicting drug-disease interactions through computer models, while experimental approaches include laboratory testing to validate findings. These systematic methods help to accelerate and refine drug repurposing efforts, making it a promising strategy for developing new treatments efficiently. With the advent of new genomics, transcriptomics, and functional screening technologies, the identification of novel applications of existing drugs has become much easier (Fig. 4.1). Previously, unintended outcomes of many drugs went unnoticed or regarded as off-target effects or side effects, but intense profiling of these drugs gives leverages these side effects in different pathological contexts, where they can play a significant role. Therefore, the amalgamation of newer modern approaches with observational strategies and wet lab experiments can enhance the productivity and precision of repurposed drugs.

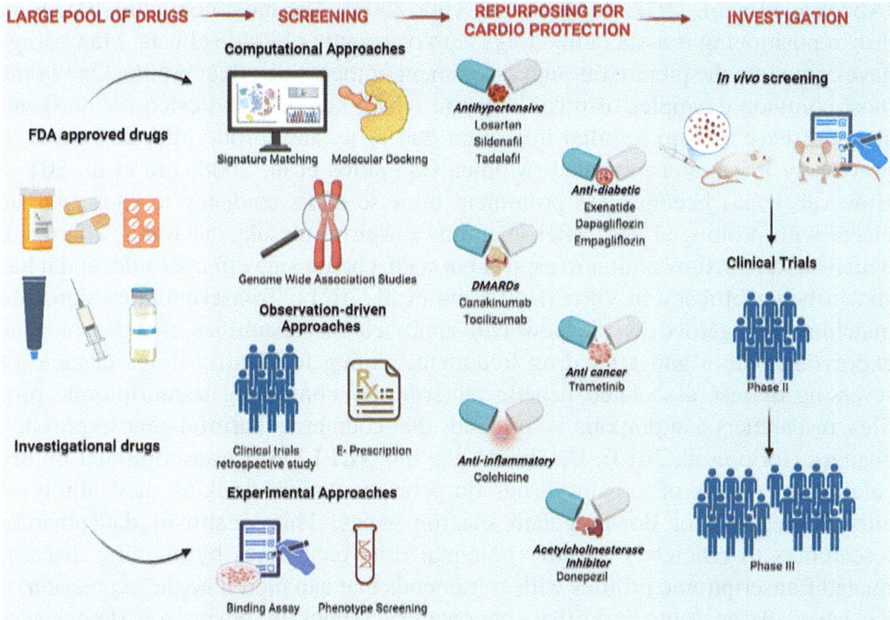

Fig. 4.1 Illustration of the steps involved in the drug repurposing process. (Created with BioRender.com)

4.3.1 Computational Approaches

4.3.1.1 Molecular Docking

Molecular docking is the structure-based method used in drug repurposing to predict a drug receptor interaction. It consists of two types: conventional docking and inverse docking. Conventional docking involves multiple drugs being tested for a known target or receptor, whereas inverse docking is when one single drug is being screened for multiple targets to find new interactions (Shamim et al. 2024). These two approaches can not only provide an array of new molecules for known receptors, but conversely can give information about a drug that can target multiple receptors, therefore having many pathological indications. This computational approach led to the discovery of mebendazole's potential to inhibit VEGFR2, a key protein in angiogenesis (Pushpakom et al. 2018). The docking process can predict ligand posture at the binding location of a receptor and any possible physical and chemical interactions.

4.3.1.2 Signature Matching

Signature matching is one of the most useful and powerful tools in drug repurposing, comparing various drugs' unique characteristics such as transcriptomic, proteomic, metabolomic profiles, chemical structures, and adverse events to other drugs, diseases, and clinical phenotypes to unveil new therapeutic applications

(Abdelsayed et al. 2022; Ashburn and Thor 2004). The most common strategy in drug repositioning is associating drugs with one another by side effects. Many drugs have come into the picture through assessment of their side effect profile. One of the most common examples of this is aspirin, which has been indicated for coronary artery disease and myocardial infarction due to its antifibrotic effects, for which previously it was not indicated (Monica Campillos et al. 2008; Liu et al. 2017). However, it has become less prominent because of its tendency to cause gastric ulcers with prolonged use. After examining aspirin, mesalazine was synthesized, which has a structure similar to aspirin but with a better side effect profile, and it has shown better efficacy in vitro (Hoffmann et al. 2021). Transcriptomics signature matching is regarded as the powerful approach that examines changes in gene expression before and after drug treatment, aiming to identify drugs capable of reversing disease-associated genetic patterns. By comparing transcriptomic profiles, researchers can pinpoint compounds that counteract harmful gene expression changes (Iorio et al. 2013). Databases like the NIH LINCS transcriptional library catalog the effects of existing drugs on gene regulation, tracking their ability to either upregulate or downregulate specific genes. This wealth of data enables researchers to efficiently identify potential drug candidates by aligning disease-related transcriptomic profiles with compounds that can modulate the expression of key genes. By applying drugs that either activate (knock-in) or suppress (knock-out) certain genes, this approach offers a systematic way to discover new therapeutic interventions (Pushpakom et al. 2018). Another such repurposing strategy is to find similarity between the chemical structures of existing drugs. Chemical structure matching actively compares the molecular framework of drugs to predict proposed biological activity, whereas adverse effect matching links drugs with similar side effects to common pathways. This approach involves the screening of chemical compounds for their ability to bind to a target structure, followed by evaluating their effectiveness (Dudley et al. 2011). The researchers have predicted 184 new targets, with 70% success rate, which have a higher dissociation constant than the drugs previously approved for the target site. Drug-to-drug similarity on the other hand works to detect a similar mechanism of action between structurally dissimilar drugs, which gives the information about new and off-target effects of a drug. The principle is called "guilt by association," where any coinciding transcriptomic hit between two drugs could be a signal where they share the same therapeutic application. Researchers have identified 1302 drugs and 41,047 similarities between drug pairs based on transcriptomic profiles, suggesting that drugs with similar profiles may share mechanisms of action. Utilizing this information, they predicted the previously unrecognized mechanism of fasudil, a Rho-kinase (ROCK) inhibitor, by analyzing its similarity to known autophagy enhancers such as 2-deoxy-D-glucose and tamoxifen (Iorio et al. 2010). This approach underscores the potential of transcriptomic data in uncovering novel drug mechanisms and repurposing existing medications for new therapeutic applications. Similarly, tamoxifen, primarily known as a selective estrogen receptor modulator, has been identified as an autophagy enhancer, indicating its broader therapeutic potential beyond its conventional use (Ashburn and Thor 2004).

4.3.1.3 Genome-Wide Association Studies (GWAS)

With the completion of the Human Genome Project, numerous genome-wide association studies (GWAS) have been conducted using advanced genotyping technologies. GWAS play a crucial role in identifying genetic variants linked to diseases, offering valuable insights into potential drug targets. By analyzing the genomes of large populations, GWAS detects single-nucleotide polymorphisms (SNPs) that occur more frequently in individuals with a particular condition compared to those without it (Sanseau et al. 2012). The U.S. National Human Genome Research Institute (NHGRI) has conducted comprehensive studies showing that genes closely linked to disease traits are more likely to encode proteins that can be targeted by drugs, compared to other areas of the genome. GWAS identifies common genetic variations with smaller individual effects but significant contributions to complex diseases like diabetes, cardiovascular diseases, and cancer. These studies provide valuable insights into disease mechanisms, potential therapeutic targets, and opportunities for drug repurposing. A study found 92 genes with traits different from their original drug indications, suggesting potential for drug repositioning (Sanseau et al. 2012). For example, statins, highly effective despite small effect sizes in the HMGCR gene. Although GWAS has shown promise in identifying new drug targets, clinical trials based on this strategy, particularly in cardiovascular disease, are still lacking (Pushpakom et al. 2018; Abdelsayed et al. 2022). Denosumab, an antibody targeting RANKL (TNFSF11), is approved for osteoporosis treatment but has also been implicated in Crohn's disease, where TNFSF11 activation is observed. Here, GWAS identified a TNFSF11 variant (rs2062305) linked to Crohn's disease, prompting research into denosumab's potential role (Franke et al. 2010). Studies in human cells showed that this variant affects TNFSF11 expression, suggesting a causal link to Crohn's disease. In a mouse colitis model, denosumab reduced inflammation and altered gut microbiota, supporting its potential use for inflammatory bowel disease (Franke et al. 2010).

4.3.2 Experimental Approaches

4.3.2.1 Binding Assays

To identify drug interactions with a target protein, techniques such as affinity chromatography and mass spectroscopy are widely exploited. One such method, the Cellular ThermoStability Assay (CETSA), measures target engagement by assessing the thermal stabilization of proteins upon binding to drug-like molecules, offering a powerful tool for understanding drug interactions within cells. High-throughput binding assays have identified thousands of kinase-drug interactions, leading to repurposing successes, such as imatinib for gastrointestinal stromal tumors.

4.3.2.2 Phenotypic Screening

Phenotypic screening identifies compounds that produce disease-relevant effects in model systems, often without prior knowledge of the drug's target. This approach can uncover potential new uses for approved or experimental drugs. For example, a high-throughput screen of 4910 small-molecule compounds in prostate cancer cells identified disulfiram (an alcohol abuse treatment) as a potential cancer therapy (Iljin et al. 2009). Additionally, whole-organism assays like zebrafish models have been used to evaluate FDA-approved drugs for conditions like tobacco dependence, finding that apomorphine and topiramate modified the behavior of nicotine and ethanol (Cousin et al. 2014).

4.3.2.3 iPSC Models of Heart Disease

Induced pluripotent stem cell (iPSC) models are now gaining more interest in screening the drugs for various cardiac pathologies. They effectively replicate the characteristics of different myopathic, electrophysiological, and metabolic disorders. iPSC-derived cardiomyocytes (iPSC-CMs) respond to known drugs in ways that mimic physiological cardiac functions, making them valuable tools for testing treatments (del Álamo et al. 1863). These cells are used for screening of drugs for inherited cardiovascular abnormalities. Because iPSCs retain the genetic makeup of their donor, they are particularly useful for studying congenital heart diseases. For instance, iPSC-CMs from patients with hypertrophic cardiomyopathy have shown that the TRPC3 inhibitor Pyr3 can reduce elevated calcium levels and hypertrophy (Hnatiuk et al. 2021). Similarly, CRISPR-edited iPSC-CMs have been used to identify treatments for dilated cardiomyopathy by targeting defects in PLN and RBM20 genes. Although iPSC-based drug screening has not yet led to new cardiovascular drugs in clinical use, it has played a role in testing repurposed drugs for other conditions, such as amyotrophic lateral sclerosis. Some large-scale studies have used iPSC-CM models to explore diabetic cardiomyopathy, oxidative stress, and arrhythmias, revealing novel therapeutic targets and improving drug efficacy (Fiedler et al. 2019).

4.3.3 Observation-Driven Approach

4.3.3.1 Retrospective Clinical Data Analysis

Observation-driven drug repurposing refers to the identification of new therapeutic indications for existing drugs based on clinical observations in the clinical trials. This approach has resulted in the identification of various drug repurposing opportunities. Recently, the recognition of inflammation's role in heart disease has prompted the investigation of anti-inflammatory drugs for cardiovascular treatments (Abdelsayed et al. 2022). Colchicine, traditionally used for gout, has been investigated for coronary artery disease and myocardial infarction, although clinical trials have had mixed results (Tardif et al. 2019). Similarly, hydralazine, known for its antihypertensive properties, has been studied for its antioxidant and anti-apoptotic effects in myocardial injury following infarction. Another observation-driven strategy involves discovering synergistic drug combinations (Kalkhoran et al. 2022). By cross-referencing drug targets with known signaling pathways involved in diseases, researchers identified a potential combination therapy of allethrin and menthol for heart failure, which has been validated in vitro (Pushpakom et al. 2018). The drugs indicated for cardioprotective therapies based on the retrospective analysis have been tabulated in Table 4.1. Data from clinical trials and post-marketing surveillance provide great insight into potential new interventions, and combining them with a computational approach can facilitate drug repurposing. Electronic health records (EHR) aid the observation-driven approach as healthcare has made it easier to conduct in silico clinical trials, where computational analysis of existing data can suggest new drug-disease pairings. This approach uses large-scale clinical data, including treated and control patient cohorts, to analyze whether repurposed drugs show beneficial effects. However, challenges like incomplete data, biases, and difficulties in analyzing free-text fields in EHRs persist (Abdelsayed et al. 2022).

4.4 Repurposed Drugs Indicated for the Cardio Protection

4.4.1 Canakinumab (IL-1β Antibody)

Acute myocardial infarction (AMI) and heart failure (HF) involve a significant inflammatory response that exacerbates tissue damage and dysfunction. The inflammation is characterized by increased levels of cytokines like IL-1β, IL-6, and C-reactive protein, correlating with disease progression, mortality, and exercise capacity. Specifically, the process activates the inflammasome, driving the production of interleukin (IL)-1β, a central cytokine in both local and systemic inflammation commonly known as "the fever molecule" (Toldo et al. 2017). Experimental studies have shown that these cytokines affect nitric oxide production, myocyte contractility, and left ventricular dysfunction, which has sparked interest in anti-inflammatory therapies for heart failure (HF) prevention and treatment. Consequently, inhibiting IL-1β with canakinumab could potentially enhance HF outcomes (Everett et al. 2019). Canakinumab is a human IgGκ monoclonal antibody developed by Novartis, targetting IL-1β to treat immune-related disorders. It has high affinity for IL-1β and forms the canakinumab/IL-1β complex, which eventually prevents IL-1β from interacting with its cell surface receptor, thereby blocking down the line IL-1β-dependent signaling. Its antigenic epitope, including Glu 64, is crucial for IL-1β recognition but lies outside the IL-1β/IL-1R binding interface. Two clinical trials investigated the effects of canakinumab, on inflammation and cardiovascular outcomes. The Canakinumab Anti-inflammatory Thrombosis Outcomes Study (CANTOS), sponsored by Novartis Pharma, involved 10,061 patients with a history of myocardial infarction and elevated high-sensitivity C-reactive protein (hsCRP). It evaluated the effects of canakinumab (50, 150, or 300 mg) administered every three months over a median follow-up of 3.7 years. The results showed a dose-dependent reduction in heart failure hospitalization (HHF) and associated mortality, although none of the doses were significantly different from placebo (Everett et al.

Table 4.1 Repurposed Drugs for Cardiovascular Indications

Drugs	Original indication	Repurposed indication	Approved for original/ repurposed indication	Results	References
Canakinumab (IL-1β antibody)	Rheumatoid arthritis; approved for cryopyrin-associated periodic syndrome by the FDA	Acute coronary syndrome, Heart Failure	2009/–	RCT indicated a reduction in hospitalizations for HF.	Everett et al. (2019), Ridker et al. (2012)
Donepezil (Acetylcholinesterase inhibitor)	Alzheimer's Disease	HF and coronary outcomes	2010/–	Still under consideration	Sato et al. (2010)
Dapagliflozin, Empagliflozin (SGLT2 inhibitors)	Diabetes Mellitus—Type II	Heart Failure	2014/2020	Reduction in HF or related events	Nassif et al. (2021), Zannad et al. (2020)
Colchicine (Anti-inflammatory)	Gout	Acute coronary syndrome	2009/–	Reduction in MACE after myocardial infarction	Tardif et al. (2019), Crittenden et al. (2012)
Exenatide (GLP1R agonist)	Diabetes Mellitus—Type II	Acute coronary syndrome	2005/–	Enhances the myocardial salvage index in patients with STEMI.	Ratner et al. (2011)
Allopurinol (Xanthine oxidase inhibitor)	Gout	Heart Failure	2000/–		Neogi et al. (2012a)
Losartan (AT1 antagonist)	Hypertension	Marfan Syndrome	1995/–	Long-term clinical benefit of losartan in combination with β- blocker.	Habashi et al. (2006)

Sildenafil, tadalafil (PDE5 inhibitors)	Pulmonary Arterial hypertension	Congestive Heart Failure	2005/2017	Improvement in the rate survival after acute myocardial infarction	Andersson et al. (2017), Hwang et al. (2017)
Trametinib (MEK inhibitor)	Malignant melanoma with *BRAF*V600E variant	Arteriovenous Malformations	2018/–	Phase—II clinical trial (ongoing)	
Tocilizumab (IL-6 receptor antibody)	Rheumatoid arthritis	Acute Coronary Syndrome	2009–2010/–	Enhances the myocardial salvage index in patients with STEMI	Kim et al. (2017b)

2019). The second trial, a double-bind, phase IIb clinical trial with 556 adults at high cardiovascular risk and well-controlled diabetes, administered canakinumab (at doses of 5, 15, 50, or 150 mg) monthly for 4 months. This study showed that canakinumab reduced inflammation markers like C-reactive protein, interleukin-6, and fibrinogen, but had minimal impact on glucose, insulin, or cholesterol levels. These findings support canakinumab's potential for reducing inflammation without significantly affecting lipid profiles (Ridker et al. 2012).

4.4.2 Donepezil

Donepezil, an acetylcholinesterase inhibitor (AChI), is indicated for treating Alzheimer's disease (AD). Upon chronic intake of AChIs, they are often linked to side effects, particularly cardiovascular issues like QT interval prolongation, bradycardia, and ventricular tachycardia, and they may also offer cardiovascular benefits (Sato et al. 2010). Research has shown that donepezil can inhibit atherogenesis and enhance hemodynamic. Additionally, it has shown potential to improve cardiac remodeling in chronic heart failure (Howes 2014). A retrospective cohort study aimed to evaluate the impact of donepezil. Given that donepezil increases acetylcholine availability, it could potentially influence heart function through enhanced vagal activity, which might affect sinus node and conduction system function, raising concerns about bradycardia. The study analyzed 1004 patients, with 76 receiving donepezil and 80 untreated, matched for age, sex, and race (Sato et al. 2010). Over a mean follow-up of approximately 29 months, patients given donepezil demonstrated significantly lower total mortality and cardiovascular mortality compared to untreated controls. These findings suggest a potential survival benefit, possibly related to an enhanced vagal tone or other cardiovascular protective effects of acetylcholine, such as antiarrhythmic properties or modulation of endothelial and platelet activation. However, due to limitations like small sample size, selection bias, and unknown medication adherence, the results should be interpreted with caution, and further prospective studies are required to validate these findings and explore the underlying cause (Sato et al. 2010).

4.4.3 Dapagliflozin, Empagliflozin (SGLT2 Inhibitors)

Sodium-glucose cotransporter 2 (SGLT2) protein is a symporter present on the proximal tubules of the nephron, in the kidney. It is responsible for the reabsorption of glucose and its return to the bloodstream. Normally, it reabsorbs around 90% of the glucose back to the bloodstream (Lopaschuk and Verma 2020). Therefore, it is one of the target proteins in the management of type 2 diabetes mellitus. Dapagliflozin and empagliflozin act as SGLT2 inhibitors, preventing glucose reabsorption into the bloodstream (Preda et al. 2024). Both have shown beneficial effects on the kidney and the heart, in cases of heart failure with reduced ejection fraction (HFrEF) (Preda et al. 2024). They help reduce fluid retention by promoting sodium and water excretion, relieving symptoms like edema and shortness of breath. In addition, they offer kidney-protective benefits and may slow the progression of chronic kidney disease, a condition that frequently occurs alongside heart failure. Their anti-inflammatory and metabolic benefits further contribute to improved outcomes in heart failure patients (Preda et al. 2024). A multicenter, double-blind, randomized trial investigated the impact of dapagliflozin in patients with heart failure with preserved ejection fraction (HFpEF). The study found that dapagliflozin significantly enhanced symptoms and physical function, leading to meaningful improvements in the Kansas City Cardiomyopathy Questionnaire Clinical Summary Score (KCCQ-CS) and a 20-meter increase in the 6-min walk test (6MWT) after 12 weeks. These benefits were consistent in cohorts presenting with or without diabetes. Potential mechanisms for these improvements included reduced pulmonary artery pressure, increased myocardial energy, and modest weight loss. Despite the short duration and U.S.-focused population, dapagliflozin was well tolerated and emerged as a promising treatment for HFpEF (Nassif et al. 2021). Further meta-analysis by Zannad et al. reviewed two clinical trials that assessed the effects of SGLT2 inhibitors (dapagliflozin and empagliflozin) in patients with heart failure with reduced ejection fraction (HFrEF), regardless of diabetes status (Zannad et al. 2020). The analysis found that both SGLT2 inhibitors significantly reduced the risks of cardiovascular death, hospitalizations for heart failure, and severe renal outcomes, with a 30% reduction in hospitalizations for heart failure and a 14% reduction in cardiovascular death. The benefits were consistent across various patient subgroups, particularly those with impaired renal function, and they appeared to result from cardioprotective and nephroprotective effects rather than glucose control (Zannad et al. 2020).

4.4.4 Colchicine

Colchicine is an FDA-approved medication for prophylaxis and the treatment of acute gout. It has a broad anti-inflammatory effect, which made it an attractive molecule to treat various other conditions, including cardiovascular diseases. It acts by disrupting microtubules (Bhattacharyya et al. 2008) and cellular functions of white blood cells. It inhibits neutrophil from attaching to endothelium, reduces the release of leukotriene B4, a key chemotactic factor, and influences neutrophil deformability (Leung et al. 2015). Additionally, colchicine inhibits NALP3 inflammasome-mediated caspase-1 activation, IL-1β processing and release, and the production of superoxide and various cytokines and pyrogens (Martinon et al. 2006). Over the last few decades, it has been indicated for the management of acute coronary syndrome (ACS), primarily due to its anti-inflammatory effects (Imazio 2015). In ACS, inflammation plays a central role in the rupture of atherosclerotic plaques and the subsequent formation of blood clots, which can obstruct coronary arteries and lead to a heart attack. By targeting inflammation, colchicine may help reduce the risk of plaque rupture, prevent further clot formation, and improve outcomes (Imazio 2015). Two studies have investigated the impact of colchicine on cardiovascular disease management, especially in patients with a history of myocardial infarction (MI) and gout. The first study was a retrospective, cross-sectional analysis involving 1288 gout patients, 576 of whom were using colchicine. It found that colchicine users had a significantly lower prevalence of MI (1.2%) compared to non-users (2.6%). While people treated with colchicine also showed trends toward reduced all-cause mortality and lower levels of C-reactive protein (CRP), these differences did not reach statistical significance. These findings suggest that colchicine might reduce the risk of MI and inflammation in gout patients, but the results are hypothesis generating and warrant further investigation (Crittenden et al. 2012). The second study was a large, randomized, double-blind trial that included 4745 patients who had experienced a recent MI (Tardif et al. 2019). These patients were assigned to either colchicine (0.5 mg daily) or placebo and followed for a median of 22.6 months. The primary composite endpoint, encompassing cardiovascular death, resuscitated cardiac arrest, myocardial infarction (MI), stroke, and urgent hospitalization for angina, occurred in 5.5% of the colchicine group, compared to 7.1% in the placebo group (Tardif et al. 2019). It has also shown significant reduction in the risk of angina and stroke. However, while colchicine was generally well-tolerated, it was associated with a slightly higher incidence of diarrhea and a higher rate of pneumonia as a serious adverse event. When combined, both studies suggest that colchicine may have beneficial effects in reducing the risk of cardiovascular events, including MI, stroke, and urgent hospitalization for angina. In gout patients, colchicine use was associated with a lower prevalence of MI, and in post-MI patients, colchicine led to a significant reduction in ischemic events. However, the safety profile should be carefully considered, as there were some adverse events like diarrhea and pneumonia (Tardif et al. 2019). Overall, these findings indicate the need for further research to confirm colchicine's potential cardiovascular benefits, mechanisms, and long-term safety in different patient populations.

4.4.5 Exenatide

Exenatide, a GLP-1 receptor agonist, mimics the action of the natural GLP-1 hormone and plays a major role in glucose metabolism and appetite regulation. As a receptor agonist, the drug binds to GLP-1 receptors on pancreatic beta cells, stimulates the secretion of the insulin in response to the elevated glucose levels, and suppresses glucagon release, thereby managing blood glucose levels. Therefore, the drug plays a significant role in the management of type -II diabetes. Additionally, exenatide has shown an ability to lower blood pressure, significantly improve lipid profiles, and promote weight loss, all of which can contribute to better cardiovascular outcomes (Lorber 2013). Studies indicate that exenatide does not elevate cardiovascular risk and may even reduce the incidence of cardiovascular events, such as heart attack and stroke, offering benefits for patients with type 2 diabetes who are at increased cardiovascular risk (Best et al. 2011). In animal models of atherosclerosis and myocardial infarction, GLP-1 receptor agonists have shown various nonglycemic benefits that may contribute to lowered cardiovascular risk (Ussher and Drucker 2023). These agents have been shown to increase the activity of endothelial nitric oxide synthase (eNOS), improve endothelial function, enhance cardiac output, and boost exercise capacity and respiratory efficiency. They also reduce the expression of endothelial adhesion molecules, lower monocyte/macrophage accumulation in arterial walls, decrease systolic blood pressure, reduce the size of myocardial infarction (MI), and activate prosurvival kinases and cytoprotective genes, leading to a decrease in cardiomyocyte apoptosis (Gelosa et al. 2020). One study examined the cardiovascular safety of twice-daily exenatide (exenatide BID) in patients with type 2 diabetes by analyzing data from 12 randomized controlled trials. The primary outcome of major adverse cardiovascular (MACE) showed no increased risk with exenatide compared to placebo or insulin groups. Exenatide-treated patients had a lower exposure-adjusted incidence rate of MACE (18.73 vs. 23.17 per 1000 patient-years) and fewer serious CV events. Kaplan-Meier survival analysis showed a higher percentage of exenatide patients free of MACE over a one-year period. The findings suggest exenatide does not increase cardiovascular risk and may offer cardiovascular benefits in type-2 diabetes patients (Ratner et al. 2011).

4.4.6 Losartan

Losartan is an angiotensin II receptor blocker (ARB) used primarily to treat hypertension, heart failure, and diabetic nephropathy. It works by blocking the effects of angiotensin II, powerful vasoconstrictor involved in blood pressure regulation and fluid balance. Losartan reduces aldosterone levels, leading to less sodium and water retention, which further lowers blood pressure and reduces fluid overload in conditions like heart failure (Paul et al. 2006). Apart from its effect in blocking the angiotensin receptor, it has shown positive effect in inhibiting TGF-β signaling. In addition to its known role in lowering blood pressure, losartan has been observed to reduce TGF-β activity in several animal models of cardiovascular diseases, including chronic renal insufficiency and cardiomyopathy (Xu et al. 2009). In Marfan syndrome (MFS), excessive TGF-β signaling is a key contributor to the progressive dilation of the aorta, a major life-threatening feature of the disease (Habashi et al. 2006). The researchers hypothesized that losartan, by blocking the AT1 receptor, could reduce TGF-β signaling and thereby prevent or slow aortic enlargement in MFS. This was supported by the observation that losartan treatment in *Fbn1C1039G/+* mice normalized aortic wall architecture, reduced elastic fiber fragmentation, and slowed the progression of aortic root enlargement, while also reducing pSmad2 nuclear accumulation, a marker of TGF-β signaling (van Andel et al. 2020). Losartan has the ability to antagonize TGF-β signaling through mechanisms beyond its effects on blood pressure. Specifically, it may prevent the upregulation of TGF-β ligands and receptors induced by AT1 receptor activation, thus contributing to the observed therapeutic benefits in this model of MFS (Habashi et al. 2006).

4.4.7 Sildenafil and Tadalafil

Sildenafil and tadalafil are potent phosphodiesterase 5 inhibitors (PDE5), they work by enhancing nitric oxide mediated vasodilation, leading to increased blood flow. These agents were initially developed for the treatment of coronary artery disease and pulmonary arterial hypertension (Anderson et al. 2015), but they are now commonly used to treat erectile dysfunction (Boolell et al. 1996). Many experimental studies have supported the use of PDE5 inhibitors (PDE5) in heart failure (HF) in vivo (Salloum et al. 2014) and in patients as well (De Santo et al. 2008; Schofield 2003). These inhibitors safeguard cardiomyocytes from apoptosis by activating protein kinase G (PKG), which subsequently upregulates nitric oxide synthase (NOS) proteins and increases the Bcl-2 to Bax ratio. Moreover, they display anti-inflammatory properties by decreasing the expression of IL-6, IL-18, and cyclin-dependent kinase inhibitor 2a (Korkmaz-Icöz et al. 2018). Moreover, PDE5 inhibitors help prevent cardiac hypertrophy at both the chamber and cellular levels, contributing to improved heart function. A Swedish Study examined the association between treatment for erectile dysfunction and subsequent risk of death and cardiovascular outcomes in men who experienced their first myocardial infarction (MI) between 2007 and 2013. The study population included 43,145 men aged 18–80 years, identified from the Swedish Patient Register, and excluded those with prior MI, revascularization, or certain pelvic surgeries, ensuring that the sample was free from conditions that could confound ED status (Andersson et al. 2017). Exposure was defined as having at least one dispensed prescription for ED medication, with a focus on phosphodiesterase-5 (PDE5) inhibitors and alprostadil, as recorded in the comprehensive Prescribed Drug Register. Men were followed for an average of 3.3 years for outcomes including all-cause mortality, myocardial infarction, heart failure hospitalizations, revascularization, and a composite of major adverse cardiac events (MACEs). The results indicated that men treated for ED had a significantly lower risk of death and heart failure compared to those who were not treated, with PDE5 inhibitor users demonstrating a dose-dependent reduction in risk, whereas alprostadil users did not show a similar benefit. However, these findings suggest a potential cardioprotective effect from PDE5 inhibitors (Andersson et al. 2017). Hwang et al. (2017) have done meta-analysis of the effects of phosphodiesterase 5 inhibitors (PDE5i) on patients with heart failure, focusing on both HFrEF (heart failure with reduced ejection fraction) and HFpEF (heart failure with preserved ejection fraction). The study included 13 randomized controlled trials (RCTs) and two sub-analysis studies, encompassing a total of 898 patients. Data were extracted on exercise capacity (e.g., peak VO_2, 6-min walking distance, and VE/VCO_2 slope), cardiac performance (e.g., diastolic function, left ventricular ejection fraction, cardiac output, and pulmonary hemodynamics), as well as clinical outcomes like hospitalization and mortality. In patients with HFrEF, PDE5i significantly improved exercise capacity, enhanced cardiac performance, and reduced pulmonary pressures and hospitalization events while maintaining a favorable safety profile. Overall, the study indicates that PDE5 inhibitors have a promising therapeutic role in heart failure, particularly in HFrEF and in select HFpEF patients characterized by advanced pulmonary hypertension, underscoring the importance of considering individual pulmonary hemodynamic profiles in treatment strategies (Hwang et al. 2017).

4.4.8 Allopurinol

Allopurinol, primarily indicated for acute gout and related conditions, functions by inhibiting xanthine oxidase, an enzyme involved in converting hypoxanthine to xanthine and xanthine to uric acid in the purine degradation pathway. By blocking this enzyme, allopurinol decreases the production of uric acid, which in turn lowers serum and urinary uric acid levels. This reduction helps prevent the formation of urate crystals (Pacher et al. 2006). Additionally, allopurinol is metabolized into alloxanthin, which also acts as a xanthine oxidase inhibitor, further contributing to its therapeutic effect. Recent extensive epidemiological studies have consistently indicated that elevated serum uric acid levels are an independent risk factor for a range of cardiovascular diseases, including heart failure and coronary artery disease (Neogi et al. 2012a). The elevated uric acid levels can initiate a cascade of inflammatory responses; here xanthine oxidase contributes to oxidative stress in endothelial cells and impairs their function independently of uric acid production. In vivo studies have shown that increased uric acid levels cause cardiomyocyte hypertrophy and impaired diastolic relaxation (Neogi et al. 2012b). Treatment with allopurinol significantly reduced left ventricular dilation, hypertrophy, fibrosis, and dysfunction by inhibiting pathways involving S6 kinase-1 (S6K1), TGFβ1, and MMP-9, suggesting that allopurinol may serve as a potential therapeutic option for managing cardiovascular disease (Neogi et al. 2012b; Salim et al. 2025). Two RCTs indicated that these inhibitors might enhance myocardial efficiency in heart failure patients. In one study with 50 patients, allopurinol (300 mg/day for 3 months) reduced BNP levels but did not improve exercise capacity. In another trial involving 60 chronic heart failure patients, oxypurinol (600 mg/day for 4 weeks) improved left ventricular ejection fraction (LVEF) by an average of 6.8% from baseline in patients with LVEF $\leq$ 40% (Cingolani et al. 2006; Gavin 2005). In summary, while several studies, often with small sample sizes, suggest that xanthine oxidase inhibitors like allopurinol may confer beneficial effects in the context of coronary angioplasty, bypass surgery, and left ventricular function in heart failure, especially in patients with elevated serum uric acid levels, large-scale prospective studies are needed to definitively establish their clinical utility (Hare et al. 2008).

4.4.9 Tocilizumab

Tocilizumab is a monoclonal antibody that targets the interleukin-6 (IL-6) receptor. IL-6 is a cytokine that plays a central role in immune responses and is a key mediator of inflammation. By blocking both membrane-bound and soluble IL-6 receptors, tocilizumab inhibits IL-6 signaling. This suppression reduces the downstream effects of IL-6, which include the production of proinflammatory cytokines, the activation of T cells, and the stimulation of B cell differentiation and antibody production (Nishimoto et al. 2005). Several ongoing studies are currently exploring the effects of tocilizumab in the cardiovascular field, including conditions such as cardiac arrest and myocardial infarction (Castagné et al. 2019). These studies specifically focus on evaluating tocilizumab's impact on markers of cardiac injury and inflammation, as well as its effects on cardiac function and survival in patients who have experienced out-of-hospital cardiac arrest (Castagné et al. 2019). A Multi-Database Cohort Study compared the cardiovascular safety of tocilizumab (TCZ) and tumor necrosis factor inhibitors (TNFi) in patients with rheumatoid arthritis (RA), particularly focusing on cardiovascular events like MI and stroke. The cohort included 28,028 RA patients who had previously used biologic or targeted synthetic DMARDs, and were initiated on either TCZ (9218 patients) or TNFi (18,810 patients) after propensity score matching (Kim et al. 2017a). The analysis, based on three large U.S. insurance databases, revealed no significant difference in the risk of composite cardiovascular events (MI or stroke) between TCZ and TNFi groups. The incidence rate of cardiovascular events was slightly lower in the TCZ group (0.52 per 100 person-years) compared to the TNFi group (0.59 per 100 person-years) (Kleveland et al. 2016). Subgroup analyses showed no increased cardiovascular risk with TCZ, even in patients with preexisting cardiovascular disease or diabetes or those taking statins, steroids, or methotrexate. The study's findings suggest that while TCZ may raise LDL cholesterol levels, this does not translate into an increased risk of clinical cardiovascular events when compared to TNFi (Kleveland et al. 2016). These results are significant for clinical practice, as they offer reassurance on the cardiovascular safety of TCZ in RA patients, despite concerns about its lipid-increasing effects. However, the study acknowledges some limitations including potential residual confounding and a short mean follow-up <1 year, necessitating longer-term studies to confirm the findings (Kim et al. 2017b).

4.5 Regulatory Considerations in Drug Repurposing

Regulatory frameworks are essential in shaping the approval and commercialization process for repurposed drugs, providing specific pathways to facilitate market entry while offering varying levels of exclusivity. In the European Union (EU), the centralized procedure serves as the primary route for obtaining approval, enabling a single application to be reviewed and authorized across all member states. Research conducted by Murteira and colleagues showed that this pathway was the most commonly utilized in the UK, France, and Germany (Murteira et al. 2014a). In the United States, repurposed drugs are typically submitted under different classifications within the New Drug Application (NDA) system. These include NDA type 1 for new molecular entities, NDA type 6 for new indications, and supplemental NDAs (sNDAs) for expanded uses of existing drugs. Additionally, NDA type 3 for new dosage forms and NDA type 4 for new combinations can be utilized for drug repurposing or reformulation.

The study also emphasized that the majority of repurposed drugs were approved prior to the expiration of the original product's patent, with approval rates surpassing 80% in the EU and reaching 69.6% in the U.S. For drugs granted orphan designation in the EU, a 10-year market exclusivity period is provided, preventing competition from similar treatments targeting the same condition. If the drug meets the requirements of a Pediatric Investigation Plan (PIP), an additional 2 years of exclusivity may be awarded. These applications must go through the centralized approval process (Murteira et al. 2014b).

Repurposed drugs without orphan status are eligible for a 10-year data exclusivity period under Article 8(3) for complete dossier submissions. Moreover, applications filed under Article 10(5) for new indications of well-established substances may qualify for a 1-year data exclusivity period. However, modifications to existing marketing authorizations are not eligible for data exclusivity protections. The USFDA grants a 3-year data exclusivity period for new indications of previously approved drugs. However, this time frame is often insufficient for companies to recover the costs associated with repurposing efforts. Another significant challenge for repurposed drugs, particularly generics, is the potential for off-label prescription, where healthcare professionals use the drug for the new indication without the company formally securing approval for that use. This practice can further diminish the commercial incentives for drug repurposing, making it financially less viable, despite potential therapeutic benefits.

4.6 Conclusion

Drug repurposing offers a promising and cost-effective approach to addressing the urgent need for new cardiovascular therapies. By leveraging existing drugs, this strategy reduces development time, risk, and costs while accelerating the availability of effective treatments for cardiovascular diseases (CVDs). Computational and experimental techniques play a crucial role in identifying potential candidates for repurposing, thus providing innovative solutions where novel drug development has slowed. However, regulatory challenges and the need for robust preclinical and clinical validation remain significant hurdles. Overcoming these barriers through well-defined regulatory frameworks and comprehensive research efforts will be essential for full realization of the potential of drug repurposing in cardiovascular medicine. Despite these challenges, drug repurposing remains a viable and efficient strategy to address the growing burden of cardiovascular diseases, offering a pathway to faster and more affordable therapeutic solutions for patients worldwide.

References

Abdelsayed M, Kort EJ, Jovinge S, Mercola M (2022) Repurposing drugs to treat cardiovascular disease in the era of precision medicine. Nat Rev Cardiol 19:751–764. https://doi.org/10.1038/s41569-022-00717-6

Anderson J, Nawarskas J, Henrie A (2015) Clinical utility of tadalafil in the treatment of pulmonary arterial hypertension: an evidence-based review. Core Evid 99. https://doi.org/10.2147/CE.S58457

Andersson DP, Trolle Lagerros Y, Grotta A, Bellocco R, Lehtihet M, Holzmann MJ (2017) Association between treatment for erectile dysfunction and death or cardiovascular outcomes after myocardial infarction. Heart 103:1264–1270. https://doi.org/10.1136/heartjnl-2016-310746

Ashburn TT, Thor KB (2004) Drug repositioning: identifying and developing new uses for existing drugs. Nat Rev Drug Discov 3:673–683. https://doi.org/10.1038/nrd1468

Best JH, Hoogwerf BJ, Herman WH, Pelletier EM, Smith DB, Wenten M, Hussein MA (2011) Risk of cardiovascular disease events in patients with type 2 diabetes prescribed the Glucagon-Like Peptide 1 (GLP-1) Receptor agonist exenatide twice daily or other glucose-lowering therapies. Diabetes Care 34:90–95. https://doi.org/10.2337/dc10-1393

Bhattacharyya B, Panda D, Gupta S, Banerjee M (2008) Anti-mitotic activity of colchicine and the structural basis for its interaction with tubulin. Med Res Rev 28:155–183. https://doi.org/10.1002/med.20097

Boolell M, Allen MJ, Ballard SA, Gepi-Attee S, Muirhead GJ, Naylor AM, Osterloh IH, Gingell C (1996) Sildenafil: an orally active type 5 cyclic GMP-specific phosphodiesterase inhibitor for the treatment of penile erectile dysfunction. Int J Impot Res 8:47–52

Castagné B, Viprey M, Martin J, Schott A-M, Cucherat M, Soubrier M (2019) Cardiovascular safety of tocilizumab: a systematic review and network meta-analysis. PLoS One 14:e0220178. https://doi.org/10.1371/journal.pone.0220178

Cingolani HE, Plastino JA, Escudero EM, Mangal B, Brown J, Pérez NG (2006) The effect of xanthine oxidase inhibition upon ejection fraction in heart failure patients: La Plata study. J Card Fail 12:491–498. https://doi.org/10.1016/j.cardfail.2006.05.005

Cousin MA, Ebbert JO, Wiinamaki AR, Urban MD, Argue DP, Ekker SC, Klee EW (2014) Larval Zebrafish model for FDA-approved drug repositioning for tobacco dependence treatment. PLoS One 9:e90467. https://doi.org/10.1371/journal.pone.0090467

Crittenden DB, Lehmann RA, Schneck L, Keenan RT, Shah B, Greenberg JD, Cronstein BN, Sedlis SP, Pillinger MH (2012) Colchicine use is associated with decreased prevalence of myocardial infarction in patients with gout. J Rheumatol 39:1458–1464. https://doi.org/10.3899/jrheum.111533

De Santo LS, Mastroianni C, Romano G, Amarelli C, Marra C, Maiello C, Galdieri N, Della Corte A, Cotrufo M, Caianiello G (2008) Role of Sildenafil in acute posttransplant right ventricular dysfunction: successful experience in 13 consecutive patients. Transplant Proc 40:2015–2018. https://doi.org/10.1016/j.transproceed.2008.05.055

del Álamo JC, Lemons D, Serrano R, Savchenko A, Cerignoli F, Bodmer R, Mercola M (1863) High throughput physiological screening of iPSC-derived cardiomyocytes for drug development. Biochim Biophys Acta, Mol Cell Res 2016:1717–1727. https://doi.org/10.1016/j.bbamcr.2016.03.003

Dudley JT, Deshpande T, Butte AJ (2011) Exploiting drug-disease relationships for computational drug repositioning. Brief Bioinform 12:303–311. https://doi.org/10.1093/bib/bbr013

Everett BM, Cornel JH, Lainscak M, Anker SD, Abbate A, Thuren T, Libby P, Glynn RJ, Ridker PM (2019) Anti-inflammatory therapy with canakinumab for the prevention of hospitalization for heart failure. Circulation 139:1289–1299. https://doi.org/10.1161/CIRCULATIONAHA.118.038010

Fiedler LR, Chapman K, Xie M, Maifoshie E, Jenkins M, Golforoush PA, Bellahcene M, Noseda M, Faust D, Jarvis A, Newton G, Paiva MA, Harada M, Stuckey DJ, Song W, Habib J, Narasimhan P, Aqil R, Sanmugalingam D, Yan R, Pavanello L, Sano M, Wang SC, Sampson RD, Kanayaganam S, Taffet GE, Michael LH, Entman ML, Tan T-H, Harding SE, Low CMR, Tralau-Stewart C, Perrior T, Schneider MD (2019) MAP4K4 inhibition promotes survival of human stem cell-derived cardiomyocytes and reduces infarct size in vivo. Cell Stem Cell 24:579–591.e12. https://doi.org/10.1016/j.stem.2019.01.013

Fordyce CB, Roe MT, Ahmad T, Libby P, Borer JS, Hiatt WR, Bristow MR, Packer M, Wasserman SM, Braunstein N, Pitt B, Demets DL, Cooper-Arnold K, Armstrong PW, Berkowitz SD, Scott R, Prats J, Galis ZS, Stockbridge N, Peterson ED, Califf RM (2015) Cardiovascular drug development: is it dead or just hibernating? J Am Coll Cardiol 65:1567–1582. https://doi.org/10.1016/j.jacc.2015.03.016

Franke A, McGovern DPB, Barrett JC, Wang K, Radford-Smith GL, Ahmad T, Lees CW, Balschun T, Lee J, Roberts R, Anderson CA, Bis JC, Bumpstead S, Ellinghaus D, Festen EM, Georges M, Green T, Haritunians T, Jostins L, Latiano A, Mathew CG, Montgomery GW, Prescott NJ, Raychaudhuri S, Rotter JI, Schumm P, Sharma Y, Simms LA, Taylor KD, Whiteman D, Wijmenga C, Baldassano RN, Barclay M, Bayless TM, Brand S, Büning C, Cohen A, Colombel J-F, Cottone M, Stronati L, Denson T, De Vos M, D'Inca R, Dubinsky M, Edwards C, Florin T, Franchimont D, Gearry R, Glas J, Van Gossum A, Guthery SL, Halfvarson J, Verspaget HW, Hugot J-P, Karban A, Laukens D, Lawrance I, Lemann M, Levine A, Libioulle C, Louis E, Mowat C, Newman W, Panés J, Phillips A, Proctor DD, Regueiro M, Russell R, Rutgeerts P, Sanderson J, Sans M, Seibold F, Steinhart AH, Stokkers PCF, Torkvist L, Kullak-Ublick G, Wilson D, Walters T, Targan SR, Brant SR, Rioux JD, D'Amato M, Weersma RK, Kugathasan S, Griffiths AM, Mansfield JC, Vermeire S, Duerr RH, Silverberg MS, Satsangi J, Schreiber S, Cho JH, Annese V, Hakonarson H, Daly MJ, Parkes M (2010) Genome-wide meta-analysis increases to 71 the number of confirmed Crohn's disease susceptibility loci. Nat Genet 42:1118–1125. https://doi.org/10.1038/ng.717

Gavin AD (2005) Allopurinol reduces B-type natriuretic peptide concentrations and haemoglobin but does not alter exercise capacity in chronic heart failure. Heart 91:749–753. https://doi.org/10.1136/hrt.2004.040477

Gelosa P, Castiglioni L, Camera M, Sironi L (2020) Drug repurposing in cardiovascular diseases: opportunity or hopeless dream? Biochem Pharmacol 177:113894. https://doi.org/10.1016/j.bcp.2020.113894

Habashi JP, Judge DP, Holm TM, Cohn RD, Loeys BL, Cooper TK, Myers L, Klein EC, Liu G, Calvi C, Podowski M, Neptune ER, Halushka MK, Bedja D, Gabrielson K, Rifkin DB, Carta L, Ramirez F, Huso DL, Dietz HC (2006) Losartan, an AT1 antagonist, prevents aortic aneurysm in a mouse model of Marfan syndrome. Science (80-.) 312:117–121. https://doi.org/10.1126/science.1124287

Hare JM, Mangal B, Brown J, Fisher C, Freudenberger R, Colucci WS, Mann DL, Liu P, Givertz MM, Schwarz RP (2008) Impact of Oxypurinol in patients with symptomatic heart failure. J Am Coll Cardiol 51:2301–2309. https://doi.org/10.1016/j.jacc.2008.01.068

Hnatiuk AP, Briganti F, Staudt DW, Mercola M (2021) Human iPSC modeling of heart disease for drug development. Cell Chem Biol 28:271–282. https://doi.org/10.1016/j.chembiol.2021.02.016

Hoffmann M et al (2021) Repurposing mesalazine against cardiac fibrosis in vitro. Naunyn Schmiedeberg's Arch Pharmacol 394:533–543

Howes LG (2014) Cardiovascular effects of drugs used to treat Alzheimer's disease. Drug Saf 37:391–395. https://doi.org/10.1007/s40264-014-0161-z

Hwang TJ, Lauffenburger JC, Franklin JM, Kesselheim AS (2016) Temporal trends and factors associated with cardiovascular drug development, 1990 to 2012. JACC Basic Transl Sci 1:301–308. https://doi.org/10.1016/j.jacbts.2016.03.012

Hwang I, Kim Y, Park J, Yoon YE, Lee S, Kim H, Cho G, Sohn D (2017) Retraction: therapeutic benefits of phosphodiesterase-5 inhibition in chronic heart failure: a meta-analysis. Interv Med Appl Sci 9:3 (123–135). https://doi.org/10.1556/1646.9.2017.26. Interv. Med. Appl. Sci. 9 (2017) 235. 10.1186/s12872-017-0576-4

Iljin K, Ketola K, Vainio P, Halonen P, Kohonen P, Fey V, Grafström RC, Perälä M, Kallioniemi O (2009) High-throughput cell-based screening of 4910 known drugs and drug-like small molecules identifies Disulfiram as an inhibitor of prostate cancer cell growth. Clin Cancer Res 15:6070–6078. https://doi.org/10.1158/1078-0432.CCR-09-1035

Imazio M (2015) Colchicine for pericarditis. Trends Cardiovasc Med 25:129–136. https://doi.org/10.1016/j.tcm.2014.09.011

Iorio F, Bosotti R, Scacheri E, Belcastro V, Mithbaokar P, Ferriero R, Murino L, Tagliaferri R, Brunetti-Pierri N, Isacchi A, di Bernardo D (2010) Discovery of drug mode of action and drug repositioning from transcriptional responses. Proc Natl Acad Sci 107:14621–14626. https://doi.org/10.1073/pnas.1000138107

Iorio F, Rittman T, Ge H, Menden M, Saez-Rodriguez J (2013) Transcriptional data: a new gateway to drug repositioning? Drug Discov Today 18:350–357. https://doi.org/10.1016/j.drudis.2012.07.014

Kalkhoran SB, Kriston-Vizi J, Hernandez-Resendiz S, Crespo-Avilan GE, Rosdah AA, Lees JG, Da Costa JRS, Ling NXY, Holien JK, Samangouei P, Chinda K, Yap EP, Riquelme JA, Ketteler R, Yellon DM, Lim SY, Hausenloy DJ (2022) Hydralazine protects the heart against acute ischaemia/reperfusion injury by inhibiting Drp1-mediated mitochondrial fission. Cardiovasc Res 118:282–294. https://doi.org/10.1093/cvr/cvaa343

Kim SC, Solomon DH, Rogers JR, Gale S, Klearman M, Sarsour K, Schneeweiss S (2017a) Cardiovascular safety of Tocilizumab versus tumor necrosis factor inhibitors in patients with rheumatoid arthritis: a multi-database cohort study. Arthritis Rheumatol 69:1154–1164. https://doi.org/10.1002/art.40084

Kim SC, Solomon DH, Rogers JR, Gale S, Klearman M, Sarsour K, Schneeweiss S (2017b) Cardiovascular safety of Tocilizumab versus tumor necrosis factor inhibitors in patients with rheumatoid arthritis: a multi-database Cohort study. Arthritis Rheumatol 69:1154–1164. https://doi.org/10.1002/art.40084

Kleveland O, Kunszt G, Bratlie M, Ueland T, Broch K, Holte E, Michelsen AE, Bendz B, Amundsen BH, Espevik T, Aakhus S, Damås JK, Aukrust P, Wiseth R, Gullestad L (2016) Effect of a single dose of the interleukin-6 receptor antagonist tocilizumab on inflammation and troponin T release in patients with non-ST-elevation myocardial infarction: a double-blind, randomized, placebo-controlled phase 2 trial. Eur Heart J 37:2406–2413. https://doi.org/10.1093/eurheartj/ehw171

Korkmaz-Icöz S, Radovits T, Szabó G (2018) Targeting phosphodiesterase 5 as a therapeutic option against myocardial ischaemia/reperfusion injury and for treating heart failure. Br J Pharmacol 175:223–231. https://doi.org/10.1111/bph.13749

Kort E, Jovinge S (2021) Drug repurposing: claiming the full benefit from drug development. Curr Cardiol Rep 23:1–7. https://doi.org/10.1007/s11886-021-01484-5

Lacaze P, Sebra R, Riaz M, Ingles J, Tiller J, Thompson BA, James PA, Fatkin D, Semsarian C, Reid CM, Tonkin AM, Winship I, Schadt E, McNeil JJ (2021) Genetic variants associated with inherited cardiovascular disorders among 13,131 asymptomatic older adults of European descent. Npj Genom Med 6. https://doi.org/10.1038/s41525-021-00211-x

Leung YY, Yao Hui LL, Kraus VB (2015) Colchicine—update on mechanisms of action and therapeutic uses. Semin Arthritis Rheum 45:341–350. https://doi.org/10.1016/j.semarthrit.2015.06.013

Liu P-P, Liu H-H, Sun S-H, Shi X-X, Yang W-C (2017) Aspirin alleviates cardiac fibrosis in mice by inhibiting autophagy. Acta Pharmacol Sin 38:488–497

Lopaschuk GD, Verma S (2020) Mechanisms of cardiovascular benefits of Sodium Glucose Co-Transporter 2 (SGLT2) inhibitors. JACC Basic Transl Sci 5:632–644. https://doi.org/10.1016/j.jacbts.2020.02.004

Lorber D (2013) GLP -1 receptor agonists: effects on cardiovascular risk reduction. Cardiovasc Ther 31:238–249. https://doi.org/10.1111/1755-5922.12000

March-Vila E, Pinzi L, Sturm N, Tinivella A, Engkvist O, Chen H, Rastelli G (2017) On the integration of in silico drug design methods for drug repurposing. Front Pharmacol 8:1–7. https://doi.org/10.3389/fphar.2017.00298

Martin SS, Aday AW, Almarzooq ZI, Anderson CAM, Arora P, Avery CL, Baker-Smith CM, Barone Gibbs B, Beaton AZ, Boehme AK, Commodore-Mensah Y, Currie ME, Elkind MSV, Evenson KR, Generoso G, Heard DG, Hiremath S, Johansen MC, Kalani R, Kazi DS, Ko D, Liu J, Magnani JW, Michos ED, Mussolino ME, Navaneethan SD, Parikh NI, Perman SM, Poudel R, Rezk-Hanna M, Roth GA, Shah NS, St-Onge MP, Thacker EL, Tsao CW, Urbut SM, Van Spall HGC, Voeks JH, Wang NY, Wong ND, Wong SS, Yaffe K, Palaniappan LP (2024) 2024 heart disease and stroke statistics: a report of US and Global Data from the American Heart Association. Circulation. https://doi.org/10.1161/CIR.0000000000001209

Martinon F, Pétrilli V, Mayor A, Tardivel A, Tschopp J (2006) Gout-associated uric acid crystals activate the NALP3 inflammasome. Nature 440:237–241. https://doi.org/10.1038/nature04516

Monica Campillos PB, Kuhn M, Gavin A-C, Jensen LJ (2008) Drug target identification using side-effect similarity. Science (80-) 321:263–266

Murteira S, Millier A, Ghezaiel Z, Lamure M (2014a) Drug reformulations and repositioning in the pharmaceutical industry and their impact on market access: regulatory implications. J Mark Access Heal Policy 2:22813. https://doi.org/10.3402/jmahp.v2.22813

Murteira S, Millier A, Ghezaiel Z, Lamure M (2014b) Drug reformulations and repositioning in the pharmaceutical industry and their impact on market access: regulatory implications. J. Mark Access Heal Policy. 2:22813. https://doi.org/10.3402/jmahp.v2.22813

Nassif ME, Windsor SL, Borlaug BA, Kitzman DW, Shah SJ, Tang F, Khariton Y, Malik AO, Khumri T, Umpierrez G, Lamba S, Sharma K, Khan SS, Chandra L, Gordon RA, Ryan JJ, Chaudhry SP, Joseph SM, Chow CH, Kanwar MK, Pursley M, Siraj ES, Lewis GD, Clemson BS, Fong M, Kosiborod MN (2021) The SGLT2 inhibitor dapagliflozin in heart failure with preserved ejection fraction: a multicenter randomized trial. Nat Med 27:1954–1960. https://doi.org/10.1038/s41591-021-01536-x

Neogi T, George J, Rekhraj S, Struthers AD, Choi H, Terkeltaub RA (2012a) Are either or both hyperuricemia and xanthine oxidase directly toxic to the vasculature?: a critical appraisal. Arthritis Rheum 64:327–338. https://doi.org/10.1002/art.33369

Neogi T, George J, Rekhraj S, Struthers AD, Choi H, Terkeltaub RA (2012b) Are either or both hyperuricemia and xanthine oxidase directly toxic to the vasculature? A critical appraisal. Arthritis Rheum 64:327–338. https://doi.org/10.1002/art.33369

Nishimoto N, Kanakura Y, Aozasa K, Johkoh T, Nakamura M, Nakano S, Nakano N, Ikeda Y, Sasaki T, Nishioka K, Hara M, Taguchi H, Kimura Y, Kato Y, Asaoku H, Kumagai S, Kodama F, Nakahara H, Hagihara K, Yoshizaki K, Kishimoto T (2005) Humanized anti–interleukin-6 receptor antibody treatment of multicentric Castleman disease. Blood 106:2627–2632. https://doi.org/10.1182/blood-2004-12-4602

Novac N (2013) Challenges and opportunities of drug repositioning. Trends Pharmacol Sci 34:267–272. https://doi.org/10.1016/j.tips.2013.03.004

Pacher P, Nivorozhkin A, Szabó C (2006) Therapeutic effects of Xanthine oxidase inhibitors: renaissance half a century after the discovery of Allopurinol. Pharmacol Rev 58:87–114. https://doi.org/10.1124/pr.58.1.6

Paul M, Poyan Mehr A, Kreutz R (2006) Physiology of local renin-angiotensin systems. Physiol Rev 86:747–803. https://doi.org/10.1152/physrev.00036.2005

Preda A, Montecucco F, Carbone F, Camici GG, Lüscher TF, Kraler S, Liberale L (2024) SGLT2 inhibitors: from glucose-lowering to cardiovascular benefits. Cardiovasc Res 120:443–460. https://doi.org/10.1093/cvr/cvae047

Pushpakom S, Iorio F, Eyers PA, Escott KJ, Hopper S, Wells A, Doig A, Guilliams T, Latimer J, McNamee C, Norris A, Sanseau P, Cavalla D, Pirmohamed M (2018) Drug repurposing: Progress, challenges and recommendations. Nat Rev Drug Discov 18:41–58. https://doi.org/10.1038/nrd.2018.168

Ratner R, Han J, Nicewarner D, Yushmanova I, Hoogwerf BJ, Shen L (2011) Cardiovascular safety of exenatide BID: an integrated analysis from controlled clinical trials in participants with type 2 diabetes. Cardiovasc Diabetol 10:1–10. https://doi.org/10.1186/1475-2840-10-22

Ridker PM, Howard CP, Walter V, Everett B, Libby P, Hensen J, Thuren T (2012) Effects of interleukin-1β inhibition with canakinumab on hemoglobin A1c, lipids, C-reactive protein, interleukin-6, and fibrinogen a phase IIb randomized, placebo-controlled trial. Circulation 126:2739–2748. https://doi.org/10.1161/CIRCULATIONAHA.112.122556

Salim AA, Kawasoe S, Kubozono T, Ojima S, Yamaguchi S, Higuchi K, Ikeda Y, Miyahara H, Tokushige K, Ohishi M (2025) Association between serum uric acid levels and left ventricular hypertrophy based on electrocardiographic findings: a sex-specific analysis across cardiometabolic diseases. Sci Rep 15:2319. https://doi.org/10.1038/s41598-025-86497-8

Salloum FN, Chau VQ, Hoke NN, Kukreja RC (2014) Tadalafil prevents acute heart failure with reduced ejection fraction in mice. Cardiovasc Drugs Ther 28:493–500. https://doi.org/10.1007/s10557-014-6559-0

Sanseau P, Agarwal P, Barnes MR, Pastinen T, Richards JB, Cardon LR, Mooser V (2012) Use of genome-wide association studies for drug repositioning. Nat Biotechnol 30:317–320. https://doi.org/10.1038/nbt.2151

Schofield R (2003) Vascular effects of sildenafil in hypertensive cardiac transplant recipients. Am J Hypertens 16:874–877. https://doi.org/10.1016/S0895-7061(03)01006-9

Shamim S, Munawar R, Rashid Y, Muhammad S, Qadar Z, Bushra R, Begum I, Quds T. (2024) Molecular docking: An insight from drug discovery to drug repurposing approach. IntechOpen. https://doi.org/10.5772/intechopen.1005526

Tardif J-C, Kouz S, Waters DD, Bertrand OF, Diaz R, Maggioni AP, Pinto FJ, Ibrahim R, Gamra H, Kiwan GS, Berry C, López-Sendón J, Ostadal P, Koenig W, Angoulvant D, Grégoire JC, Lavoie M-A, Dubé M-P, Rhainds D, Provencher M, Blondeau L, Orfanos A, L'Allier PL, Guertin M-C, Roubille F (2019) Efficacy and safety of low-dose Colchicine after myocardial infarction. N Engl J Med 381:2497–2505. https://doi.org/10.1056/nejmoa1912388

Toldo S, Van Tassell BW, Abbate A (2017) Interleukin-1 blockade in acute myocardial infarction and heart failure: getting closer and closer. JACC Basic Transl Sci 2:431–433. https://doi.org/10.1016/j.jacbts.2017.07.006

Sato K, Urbano RR, Yu C, Yamasaki F, Sato T, Jordan J, Robertson D, Diedrich A (2010) The effect of donepezil treatment on cardiovascular mortality. Clin Pharmacol Ther 88:335–338. https://doi.org/10.1038/clpt.2010.98

Ussher JR, Drucker DJ (2023) Glucagon-like peptide 1 receptor agonists: cardiovascular benefits and mechanisms of action. Nat Rev Cardiol 20:463–474. https://doi.org/10.1038/s41569-023-00849-3

van Andel MM, Indrakusuma R, Jalalzadeh H, Balm R, Timmermans J, Scholte AJ, van den Berg MP, Zwinderman AH, Mulder BJM, de Waard V, Groenink M (2020) Long-term clinical outcomes of losartan in patients with Marfan syndrome: follow-up of the multicentre randomized controlled COMPARE trial. Eur Heart J 41:4181–4187. https://doi.org/10.1093/eurheartj/ehaa377

World Health Organization (n.d.) Cardiovascular diseases kill 10 000 people in the WHO European Region every day, with men dying more frequently than women

Xu F, Mao C, Hu Y, Rui C, Xu Z, Zhang L (2009) Cardiovascular effects of Losartan and its relevant clinical application. Curr Med Chem 16:3841–3857. https://doi.org/10.2174/092986709789178046

Zannad F, Ferreira JP, Pocock SJ, Anker SD, Butler J, Filippatos G, Brueckmann M, Ofstad AP, Pfarr E, Jamal W, Packer M (2020) SGLT2 inhibitors in patients with heart failure with reduced ejection fraction: a meta-analysis of the EMPEROR-reduced and DAPA-HF trials. Lancet 396:819–829. https://doi.org/10.1016/S0140-6736(20)31824-9

Zheng C, Guo Z, Huang C, Wu Z, Li Y, Chen X, Fu Y, Ru J, Ali Shar P, Wang Y, Wang Y (2015) Large-scale direct targeting for drug repositioning and discovery. Sci Rep 5:1–10. https://doi.org/10.1038/srep11970

Drug Repurposing for Autoimmune Disorders

5

Neelu Singh, Kunal Agam Kanaujia, Abhishek Verma, Alka, and Nidhi Mishra

Abstract

Drug repurposing, identifying new therapeutic uses for existing drugs, has emerged as a crucial and promising strategy in the urgent fight against autoimmune disorders. This chapter aims to instill hope and optimism by exploring the potential of drug repurposing to offer rapid and cost-effective solutions to the complex and multifaceted nature of autoimmune diseases. Leveraging established pharmacological profiles and clinical data, repurposed drugs can bypass many early stage developmental hurdles, expediting the path to clinical application.

We delve into the mechanisms by which repurposed drugs can modulate immune responses, highlighting several inspiring success stories. These stories demonstrate how drugs originally developed for other indications have shown efficacy in treating autoimmune conditions, such as rheumatoid arthritis, multiple sclerosis, and systemic lupus erythematosus, and they are intended to inspire and motivate. Key examples include the use of antimalarial drugs like hydroxychloroquine in lupus and the employment of JAK inhibitors, initially designed for cancer therapy, to manage rheumatoid arthritis.

Furthermore, this chapter underscores the pivotal role of computational tools and bioinformatics in identifying candidate drugs for repurposing. These tools empower researchers to uncover novel drug-disease associations and are intended to inform and enlighten. Through comprehensive case studies and an overview of the current landscape, this chapter aims to provide a detailed understanding of the potential and limitations of drug repurposing in the fight against autoimmune diseases, ultimately contributing to more effective and personalized treatment options for patients.

N. Singh · K. A. Kanaujia · A. Verma · Alka · N. Mishra (✉)
Department of Pharmaceutical Sciences, Babasaheb Bhimrao Ambedkar University Lucknow (A Central University), Uttar Pradesh, Lucknow, Uttar Pradesh, India

R. K. Kaundal et al. (eds.), *Drug Repurposing: Exploring Newer Therapeutic Potentials*,
https://doi.org/10.1007/978-981-95-4522-3_5

Keywords

Autoimmune disorders · Personalized medication · Nanotechnology · Computational tools · Multi-omics approach

5.1 Introduction

Drug repurposing (DR)—also known as drug reprofiling, drug repositioning, drug re-tasking, therapeutic switching, and drug rescuing—encompasses identifying new pharmacological applications for drugs that are old, existing, failed, under investigation, already in the market, or approved by the Food and Drug Administration (FDA) (Ashburn and Thor 2004). These drugs are then applied to treat diseases different from their intended purposes (Dey 2019). This entails discovering fresh therapeutic applications for already familiar drugs, such as those approved, discontinued, abandoned, or in experimental stages (Mohammad Jafari et al. 2018). Developing novel drugs is a lengthy and intricate process characterized by high investment and low success rates (Bhagat and Butle 2021). Therefore, it is imperative to explore alternative strategies for drug discovery. Drug repositioning, also referred to as repurposing existing drugs for novel applications, stands out as a practical approach for identifying fresh therapeutic indications for established medications. It is characterized by its high efficiency, affordability, and reduced risk compared to traditional drug development processes. Despite being an established strategy, drug repurposing has gained significant traction in the past decade. Approximately one-third of recent drug approvals are attributed to repurposed drugs, contributing to about 25% of the pharmaceutical industry's annual revenue (Talevi and Bellera 2020). Throughout the history of medicine, there have been instances of drugs finding new uses, some of which are pretty old and often discovered by chance. Modern techniques, especially data mining, have been developed to identify potential candidates for drug repurposing (Jourdan et al. 2020a). Historically, drug repurposing was often accidental; when a drug showed unexpected effects, it was explored for commercial purposes (Pushpakom et al. 2019). Approximately 30% of FDA-approved drugs and vaccines are repositioned, and the market for repurposed drugs was valued at $24.4 billion in 2016 and is projected to reach $35.1 billion in 2025 (Rudrapal et al. 2020). Over the past decade, drug discovery projects have shifted from large pharmaceutical companies to academic, nonprofit, and small business research units, focusing on orphan and neglected diseases (Oprea et al. 2011). Development of a novel drug is expensive and time-consuming, taking 10–15 years. This lengthy process has increased interest in drug repurposing, where existing drugs approved by regulatory agencies like the FDA, EMA, and MHRA are used for new indications (Parvathaneni et al. 2019). Various pharmaceutical companies are embracing drug repurposing because it offers a faster, cheaper, and less risky alternative. Repurposing allows researchers to bypass earlier safety stages, saving time and costs. This approach speeds up therapy availability and reduces development expenses, making therapies more affordable for patients. Knowledge of safety, efficacy, and administration routes from prior use significantly cuts development time and costs, making it easier to bring repositioned drugs to the market (Padhy and Gupta 2011). However, it's crucial to ensure that finding a new disease target doesn't hinder the drug's original use potential (Sleigh and Barton 2010). During the COVID-19 pandemic, drug repurposing gained attention when the FDA authorized several drugs, like remdesivir, developed initially for RNA-based viruses and evaluated for Ebola, as COVID-19 treatments (Krishnamurthy et al. 2022). Drug repurposing involves three concepts: extending the use of approved drugs, repurposing molecules for new indications based on existing data, and rescuing abandoned molecules for new uses (Begley et al. 2021). Despite its drawbacks and challenges, this approach holds several advantages, particularly in addressing the attrition rates in new drug discovery (6). Contrary to developing an entirely new drug for a specific purpose, this strategy offers numerous benefits. Firstly, and most crucially, the risk of unsuccessfulness is minimized since the repurposed drug has previously proven its safety in preclinical evaluations and early human trials; it's less likely to fail, particularly regarding safety, during subsequent efficacy trials. Secondly, the timeline for drug development can be significantly shortened, as most of the preclinical testing, safety evaluations, and sometimes even formulation development have already been completed. Thirdly, it requires less investment, although the extent varies based on the development stage and process of the repurposed candidate (Breckenridge and Jacob 2019). However, drug repurposing also has its limitations and challenges, such as the potential for off-target effects and the need for careful consideration of intellectual property rights when repurposing existing drugs.

5.1.1 Advantages of Drug Repurposing

Drug repurposing offers several distinct advantages:

5.1.1.1 Reduced Development Time

Repurposing existing drugs significantly reduces the time required to bring a new treatment to patients. The average development time of a new drug is 10–15 years. However, the average development time for a repurposed drug is 3–5 years (Pushpakom et al. 2019).

5.1.1.2 Cost-Efficiency

Developing new drugs can be prohibitively expensive, while repurposing leverages existing research and development investments, reducing costs. Developing a new drug costs $2.6 billion; however, repurposing a drug costs $318 million (DiMasi et al. 2015).

5.1.1.3 Safety Profiles

Repurposed drugs often have well-documented safety profiles, which can expedite regulatory approval. A study published by Cha et al. found that repurposed drugs have well-documented safety profiles, which can expedite regulatory approval (Cha et al. 2018).

5.2 Approaches to Drug Repurposing

Typically, the development of a novel drug requires a span of 10 to 15 years. Yet, the likelihood of successfully bringing a new molecular entity to the market averages only around 2.01%, as revealed in a study conducted by the FDA (Luo et al. 2016). Furthermore, the count of FDA-approved medications has declined since 1995, according to the same FDA report. Additionally, investment in drug development has been steadily rising, as indicated in research from the Pharmaceutical Research and Manufacturers of America (PhRMA) (Luo et al. 2016). This underscores the continuous escalation in the expenses associated with new drug development. Traditional drug development typically involves five sequential stages: discovery and preclinical research, safety evaluation, clinical trials, FDA regulatory review, and post-market safety monitoring, as outlined in references (Lee et al. 2021; Sun et al. 2016). In contrast, drug repositioning streamlines this process in four key steps: compound identification, compound acquisition, development, and FDA post-market safety monitoring. Scientists now require 1–2 years to pinpoint new drug targets and approximately 8 years to develop a repurposed drug, on average (Lipinski et al. 2019). Moreover, the research and development costs associated with drug repositioning are notably lower compared to traditional strategies. This cost-effective approach mitigates financial burdens for many nations, with developing a new drug through drug repositioning averaging a mere $1.6 billion, starkly contrasting to the $12 billion required by traditional methods (Bhagat and Butle 2021; Jourdan et al. 2020b; Xue et al. 2018). The various approaches to drug repurposing are as follows:

5.2.1 Phenotypic Screening

This approach involves testing existing drugs in relevant biological or disease models to identify new therapeutic applications. It doesn't rely on prior knowledge of the drug's mechanism of action. This method often uses high-throughput screening of compound libraries, which involves testing existing drugs in relevant biological or disease models to identify new therapeutic applications. Researchers use cellular or animal models that mimic the disease features to identify potential drug candidates when applied to autoimmune diseases. This method can uncover drugs that modulate immune responses or reduce inflammation, both crucial aspects of autoimmune disorders (Reaume 2011; Ilg et al. 2022).

5.2.2 Target-Based Repurposing

In this approach, existing drugs are screened against specific molecular targets or pathways known to be associated with a disease. If a drug has a known mechanism of action relevant to the target, it may be considered for repurposing. In autoimmune disorders, the immune system often targets specific molecules or pathways (Ni et al. 2023). Target-based drug repurposing involves screening existing drugs against these known targets or pathways. For example, medications developed initially for cancer (targeting immune checkpoints) have been repurposed for autoimmune diseases like rheumatoid arthritis (Rajan et al. 2023; Unal et al. 2022).

5.2.3 Network Pharmacology

This method analyzes the interactions between drugs, proteins, and diseases in a network or systems biology context. Researchers can identify potential repurposing candidates by understanding how drugs affect different parts of a biological network. Autoimmune disorders involve complex interactions among various immune cells and signaling pathways. Network pharmacology examines the broader context of drug interactions within biological networks. Researchers use this approach to identify existing drugs that modulate multiple immune system components simultaneously (Shi et al. 2022).

5.2.4 Structural Bioinformatics

Computational methods, such as molecular docking and virtual screening, can predict how existing drugs may interact with new targets at a molecular level. This approach relies on the 3D structures of drugs and target proteins (Grammer and Lipsky 2017; Luo et al. 2016).

5.2.5 Clinical Observations

Clinical observations and epidemiological data can sometimes suggest new uses for existing drugs. For example, if a drug is found to have unexpected benefits in patients with a specific condition, further research may be conducted to validate and understand this effect. Through clinical observations, existing drugs may have demonstrated unexpected benefits in patients with autoimmune disorders. For example, medications prescribed for other conditions may reduce autoimmune symptoms when given to patients. This observation can lead to further clinical studies to validate the repurposing potential (Lee et al. 2021).

5.2.6 Drug Combination Therapies

Repurposing can also involve combining existing drugs to create a synergistic effect in treating a disease. This approach is particularly relevant in diseases with complex mechanisms (Sun et al. 2016).

5.2.7 Data Mining and Artificial Intelligence

Machine learning and data mining techniques can analyze large datasets of drug compounds, genomic information, clinical records, and scientific literature to identify potential drug repurposing opportunities. Advanced computational methods, including machine learning and AI, can analyze large datasets of genomic information, clinical records, and drug databases to predict potential repurposing opportunities for autoimmune diseases. These methods can uncover hidden relationships between drugs and autoimmune targets (Lipinski et al. 2019; Gupta et al. 2021).

5.2.8 FDA Drug Repurposing Initiatives

Organizations like the U.S. FDA have initiatives and programs to identify new uses for approved drugs. These efforts can involve collaborations with academic researchers and pharmaceutical companies. Investigating drugs that have previously undergone clinical trials but have failed for their intended indications can be valuable. Such drugs may be revisited and repurposed for autoimmune disorders if they exhibit safety and efficacy profiles (Pushpakom et al. 2019).

5.2.9 Patient-Derived Insights

Patient communities and advocacy groups sometimes play a role in identifying potential repurposing candidates based on anecdotal evidence or personal experiences with existing drugs. Analysis of patient data and electronic health records can uncover potential drug repurposing candidates. Researchers can identify patterns in medication use and disease outcomes, providing insights into drugs that may be effective for specific autoimmune disorders (Vivarelli et al. 2020).

5.2.10 Drug Combinations and Drug-Drug Interactions

Evaluating how drugs interact can lead to discovering new therapeutic combinations or repurposing opportunities. Autoimmune disorders frequently require a multifaceted treatment approach. Combining existing medications with complementary mechanisms of action can enhance therapeutic outcomes. For instance, combining an immunosuppressant with an anti-inflammatory drug can benefit autoimmune patients synergistically (Parisi et al. 2020).

5.2.11 Indication Expansion

Expanding the use of a drug to treat a broader range of conditions or patient populations based on emerging scientific evidence.

5.2.12 Disease-Centric Approach

This approach begins with a specific disease or patient population and identifies existing drugs that may potentially treat that disease or population. This can be done by looking at the disease's biology, the drug's mechanism of action, and any existing data on the drug's off-label use (Parisi et al. 2020).

5.2.13 Target-Centric Approach

This approach begins with a specific molecular target, such as a protein or gene involved in a disease's development or progression. Once the target has been identified, researchers can screen existing drug libraries for drugs that bind to or inhibit the target (Parisi et al. 2020).

5.2.14 Drug-Centric Approach

This approach begins with a specific existing drug and identifies new diseases or patient populations that the drug may treat effectively. This can be done by looking at the drug's pharmacology, toxicology, and clinical data (Parisi et al. 2020; Liu et al. 2013).

5.2.15 Natural Product Screening

Investigating natural compounds, such as plant extracts or marine organisms, for potential therapeutic properties can lead to drug repurposing candidates. Compounds derived from plants, aquatic organisms, or other sources can also be evaluated for their potential in autoimmune disorders. These compounds may possess anti-inflammatory or immunomodulatory properties that could be repurposed for therapeutic use.

5.3 Autoimmune Disorders with Inadequate Treatments

A broad spectrum of illnesses known as autoimmune disorders are defined by immune system abnormalities that result in abnormal host constituent reactivity on the components of B and T cells. These conditions can affect people of any age and almost any organ system, although they are more prevalent in women. These can include mild abnormalities that are readily missed or abrupt, life-threatening organ failure (Pisetsky 2023). For severe autoimmune illnesses, more than 50% and occasionally 70% or 80% of monozygotic twins are discordant. This discrepancy is especially noticeable because twins, at least while young, share much of the same environment. The hunt for environmental variables influencing the emergence of autoimmune illnesses has been the focus of several studies (Bach 2005).

The mechanisms that control the equilibrium between identifying infections and avoiding self-attacks are disrupted in systemic autoimmune disorders. While the significance of the innate immune system in the pathophysiology of systemic autoimmunity is still being extensively studied, adaptive immunity's ability to provide immunological memory remains of primary importance (Zhernakova et al. 2009). There are two theories on systemic autoimmune inflammatory diseases. First, disrupting the barrier control between innate and adaptive immunity may lead to a positive feedforward loop that perpetuates inflammation, aligning with interferons' effects. Second, abnormalities in tolerance checkpoints and prolonged inflammation may arise from compromised adaptive immunity with reactivated (auto)reactive memory by lymphocytes. Both innate and adaptive pathways are involved in systemic autoimmunity. Therefore, these two theories are not mutually incompatible (Wahren-Herlenius and Dörner 2013). Figure 5.1 illustrates the signaling pathways and mechanisms of various organ-specific and systemic autoimmune disorders, which are utilized in selecting the appropriate drug-repurposing approaches.

Various organs may experience inflammation, injury, or malfunction due to the aberrant or uncontrolled immune response. There are many different autoimmune diseases, and each one has its unique symptoms, organs that are affected, and underlying causes. These disorders continue to challenge the medical community with inadequate or suboptimal treatments. These disorders include, but are not limited to:

5.3.1 Myasthenia Gravis (MG)

MG is a chronic autoimmune neuromuscular disorder that causes weakness in the body's skeletal muscles. This condition occurs when the immune system mistakenly targets and attacks the acetylcholine receptors at the neuromuscular junction, where nerve cells communicate with muscles. Acetylcholine is a neurotransmitter that plays a crucial role in muscle contraction. One key symptom of myasthenia gravis is muscle weakness. It can affect various muscles, including eye movements, facial expressions, swallowing, and limb movements. Fatigue can lead to drooping eyelids (ptosis), difficulty in speaking, double vision (diplopia), and difficulty in swallowing (dysphagia). Although MG is a mostly treatable condition, it might cause severe

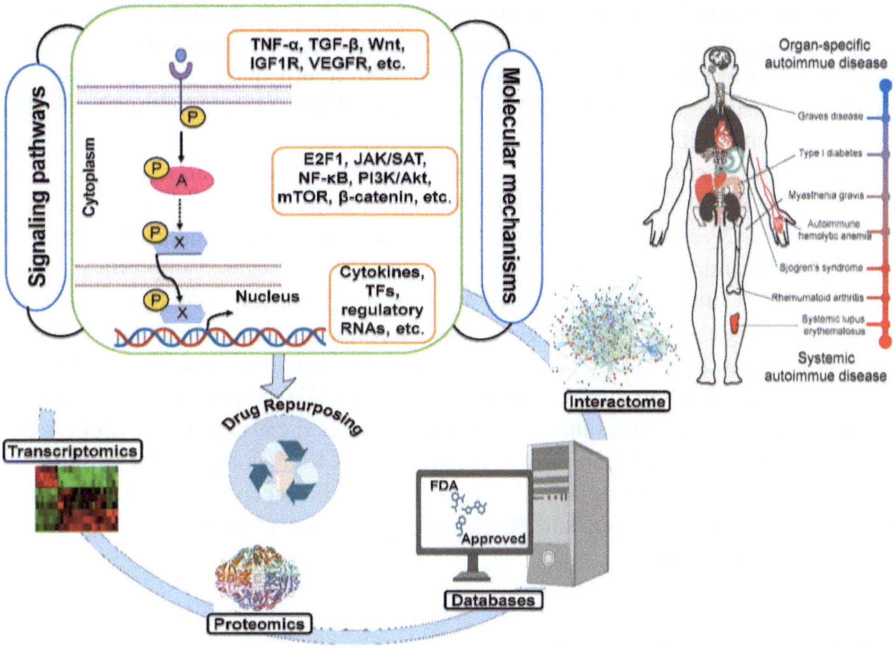

Fig. 5.1 Signaling pathways and mechanisms of various organ-specific and systemic autoimmune disorders utilized in drug repurposing approaches

complications and even death. MG is one of the most well-known cases of an autoimmune illness mediated by antibodies. Most of the time, this may be avoided with early detection and effective illness management. The illness peaks in younger adult women and older males, with onset ranging from early childhood to late adulthood (McGrogan et al. 2010). Current therapies often provide only partial relief, and significant side effects are common. Research indicates that the most common treatments for myasthenia gravis, such as anticholinesterase drugs and corticosteroids, are only partially effective and often cause significant side effects (Hehir 2nd and Li 2022; Schneider-Gold and Gilhus 2021). Few people have been identified and have a positive test result for a specific antibody in the blood serum against agrin, low-density lipoprotein receptor-related protein 4 (Lrp4), acetylcholine receptors (AChRs), or muscle-specific kinase (MuSK) in most MG patients. Antibodies are primarily employed as markers for the severity of the disease and are frequently seen in individuals with thymoma or late-onset MG (Szczudlik et al. 2014). Subtypes of MG are as follows:

- MG Due to Antibodies against AChR (AChR-MG)
- MuSK Antibody-Associated MG (MuSK-MG)
- Double-Seronegative Generalized MG
- Lrp4 Antibody-Associated MG (Lrp4-MG)

Since acetylcholinesterase inhibitors, intravenous immunoglobulins, immunosuppressants, and improved respiratory care became available in the early twentieth century, the mortality rate of MG has decreased significantly. Still, the disease's death rate is between 5% and 9% (Szczudlik et al. 2014). Most individuals with MG are female, and while the age at onset varies, most people present before the age of 50. Although patients with generalized or ocular MG may appear, it is thought that the overall symptom severity is less in this subset. Only 20% of patients experience eye symptoms 2 years into the illness. On the other hand, individuals who have both anti-Lrp4 and anti-agrin antibodies may experience more severe symptoms (Dresser et al. 2021). Complement inhibitors are one of the novel immunotherapies for MG. Following the binding of IgG to AChR epitopes, the classical and common complement cascades are initiated. The cascade's last stages lead to the synthesis of C5 convertase, which divides C5 into C5a and C5b. The membrane attack complex (MAC), which incorporates into the cell membrane and causes damage and lysis of the cell, is created when C5b joins forces with the C6-C9 components (Howard Jr. 2018). Eculizumab, Zilucoplan, Ravulizumab, and Fc receptor inhibitors, such as Efgartigimod, Nipocalimab, and Rozanolixizumab, are examples of complement inhibitors used in MG.

5.3.2 Sjögren's Syndrome

A chronic autoimmune condition called Sjögren's syndrome typically affects exocrine glands, especially the salivary and lacrimal glands. In Sjögren's syndrome, the immune system accidentally assaults and destroys these glands, resulting in dry mouth and eyes and several other symptoms, including the production of saliva and tears. It may also impact other body systems, resulting in various systemic symptoms. Treatment focuses on symptom management but does not address the underlying immune dysfunction. There is no cure for Sjögren's syndrome; however, current treatments focus on managing symptoms (Fox et al. 2021; Zeng et al. 2022; Song et al. 2024).

5.3.3 Addison's Disease

The primary cause of Addison's disease in industrialized nations is autoimmunity against the adrenal cortex; estimates of prevalence in Europe encompass nearly 220 per million. Adrenocortical cells are targeted by the immune system, which renders them incapable of synthesizing essential steroid hormones and requires lifelong hormone replacement treatment. In Addison's disease (AAD), also known as primary adrenocortical insufficiency, the adrenal cortex fails to generate adequate levels of mineralocorticoids and glucocorticoids. Women have a greater risk than men, which typically first appears between the second and fourth decades of life (Hellesen et al. 2018). Patients with Addison's disease may exhibit a variety of nonspecific symptoms, including tiredness, sickness, fainting, and weight loss. Still, they are frequently recognized by specific warning signs, such as skin and mucosal surface hyperpigmentation and a desire to consume salt (Betterle et al. 2002). Hormone replacement therapy in the form of fludrocortisone (mineralocorticoid substitution) and hydrocortisone (glucocorticoid substitution) is the current recommended treatment regimen (Bornstein et al. 2016). As with most other organ-specific autoimmune diseases, HLA class II alleles have the highest genetic correlations with AAD. The HLA DR3-B8 haplotype was shown to be linked to an elevated risk of AAD in a study of families with several afflicted people (Baker et al. 2010). The histological appearance of ADD has been characterized by lymphocytic infiltration of the adrenal cortex. The three cell layers in the cortex eventually break down and are replaced by fibrous tissue as the illness progresses, causing the cortex to lose its integrity (Imam et al. 1988). Adrenocortical cell loss may trigger AAD to stay subclinical for extended periods before developing the disease. Adrenocortical failure may not show symptoms until 90% of the cells die (Rosenthal et al. 1978). The duration of treatment and dosage of glucocorticoids taken are critical factors in determining a decent quality of life for patients with Addison's disease. A patient's risk of morbidity and death from an Addisonian crisis is heightened if they do not receive an adequate dosage of glucocorticoids at the appropriate time. To prevent such a catastrophe, most doctors have erred on the side of excess replacement in the past, but the link between high glucocorticoid doses and cardiovascular mortality is alarming (Choudhury and Meeran 2018).

5.3.4 Inflammatory Bowel Disease (IBD)

Chronic inflammation in the intestines is a hallmark of inflammatory bowel diseases (IBDs), among which are Crohn's disease (CD) and ulcerative colitis (UC). These conditions are often regarded as multifactorial illnesses (Baumgart and Sandborn 2007). These diseases differentiate themselves according to their affected gastrointestinal area, the depth and severity of the inflammation in the intestinal wall, and the peculiarity of their pathophysiology. The prevalence or occurrence of IBD is highest in the second to third decades of life, with another peak in the 60–70-year-old age group. At the onset and during the progression of IBD, the association between genetic factors (predisposing the patient to develop the disease), environmental factors (modulating the inflammatory pathways), and composition of the microbiota occurs. The impacted gastrointestinal regions, the extent of the intestinal wall inflammation, and the unique pathophysiology of these illnesses differ. The second and third decades of life have the highest prevalence of IBD, with a second peak observed in the 60–70-year-old age range. There are correlations between the hereditary factors, the environmental variables (which affect the inflammatory pathways), and the microbiota composition at the beginning and during the disease (Silva et al. 2016). Continuous follow-up and epidemiological inception cohorts allow medical professionals to analyze the different patterns of illness courses of IBD patients, ranging from aggressive to indolent stages. Before biological treatment was used to treat IBD, hospitalizations and surgeries were the primary factors of healthcare expenditures. However, new research shows that this cost trend has shifted, with biological agents now being the primary cause of healthcare costs (Burisch et al. 2015; Ullman and Itzkowitz 2011). Individuals with IBD who have significant, long-standing colitis are more prone to develop colorectal cancer (CRC) than people in the general population. Patients with IBD are more likely to develop colorectal cancer since immunosuppression and inflammation are the two key factors that cause carcinogenesis (Ananthakrishnan et al. 2014). Increased median levels of CRP and ERS have been linked to an increased risk of colorectal cancer (CRC) in individuals with IBD, according to previous research that suggested a relationship between the severity of histological inflammation and CRC risk in IBD patients (Wehkamp et al. 2016). There are no significant data to support basing therapy escalations on endoscopic results; instead, the treatment should always be customized depending on each patient's site, extent, and stage of the illness, with further changes determined by the clinical response. The recommendations do not recommend a quick, maximum immunosuppressive approach, including the delivery of antibodies, at the time of diagnosis. This could lead to overtreatment in many individuals and frequently result in severe consequences. Clinical follow-up needs to be regular because each patient's illness history is unique, and medication should be increased if there has been an inadequate response. This approach prevents the needless and expensive administration of antibody treatment to individuals with manageable conditions while facilitating the early diagnosis of illness consequences (Stabler 2013).

5.3.5 Pernicious Anemia (PA)

Cobalamin deficiency (CD) and autoimmune chronic atrophic gastritis (CAG) are two hallmarks of PA, an autoimmune illness. Worldwide, PA is the most frequent cause of CD (Htut et al. 2021). The incidence varies significantly with age, from 0.1% in the general population to 1.9% in people over 60 years. PA ranges between 20% and 50% of the etiology of CD in adults. While PA occurrence increases with age, regional differences exist. PA is frequently said to be widespread in Scandinavian nations. When Thomas Addison founded Pennsylvania in 1849, the state's history officially started. In 1860, Austin Flint documented the correlation between anemia and CAG, leading to the subsequent naming of PA. Remarkably, the cooked liver was able to cure anemia. This led to the hypothesis that the liver contained the intrinsic factor (IF) and the stomach discharge contained the extrinsic factor (cobalamin). Even though the anemia was originally believed to be pernicious, cobalamin supplementation has helped to keep it under control (Toh 2017). Gastric parietal cells produce two important biological products: HCl and intrinsic factors. Deficiency of these is triggered by intestinal metaplasia replacing the lost gastric parietal cells in chronic atrophic gastritis. Cobalamin malabsorption in the terminal ileum is caused by the loss of intrinsic factors and the emergence of neutralizing intrinsic factor antibodies, which results in cobalamin-deficient megaloblastic pernicious anemia. More recently, the finding that iron deficiency anemia, recognized as an achlorhydria consequence, primarily affects women and develops approximately 20 years before cobalamin-deficient pernicious anemia (Korstanje et al. 2002). Chronic atrophic gastritis can be consistently detected by screening tests that combine pepsinogen 1 (P1), pepsinogen 2, the pepsinogen 1/pepsinogen 2 ratio, and gastrin-17 (Rustgi et al. 2021). Early stage autoimmune gastritis is asymptomatic and can only be identified with serological antibody detection against stomach parietal cells. Only 10%–15% of the population with autoimmune gastritis develops pernicious anemia, and it only appears when vitamin B12 levels are imbalanced or low. Pernicious anemia has an underlying period of 20–30 years. Despite being "silent" until the end, immunologic markers unique to gastric autoimmunity can predict the gastric lesion years before it emerges clinically (Peterson and Fujinami 2007).

5.3.6 Multiple Sclerosis (MS)

Although the traditional definition of MS is a chronic inflammation of the central nervous system (CNS), a more accurate description would be one of a multicomponent illness involving a balance of pathophysiological pathways. In the primary progressive phenotype, there is a higher component of chronic neuronal, axonal, or myelin loss and failure of CNS repair processes, which is represented clinically by increased impairment as the illness progresses to secondary progressive MS without relapses (Compston and Coles 2008). MS describes CNS plaques associated with axonal damage, loss, demyelination, and inflammation. These plaques can be discovered in the gray matter as well as the white matter of the brain and spinal cord, primarily in the subpial area of the spinal cord and brainstem, corpus callosum, optic nerves and tracts, cerebellar peduncles, long tracts, and white matter surrounding the ventricles (Oksenberg and Baranzini 2010). It is considered that there are multiple factors contributing to the immune dysregulation in MS, including genetic predisposition, epigenetic and post-genomic events, and environmental factors like viruses (like Epstein–Barr virus (EBV)), chemicals, smoking, diet (obesity), and vitamin D levels from sun exposure (Dobson and Giovannoni 2019). The course of multiple sclerosis includes being at risk, as well as the asymptomatic, prodromal, and symptomatic stages of the illness. There are several less common presentations, including cortical presentations such as dominant parietal lobe syndromes, but the most often observed ones include optic neuritis, brainstem, and spinal cord syndromes. Approximately ten "asymptomatic" lesions are seen on magnetic resonance imaging (MRI) during every clinical incident (Huang et al. 2017). MS treatment is challenging and requires a variety of medications that work in different ways. While there is currently no established treatment for the primary progressive form, several medications may enhance the secondary progressive form and help to alter the disease's activity when the relapsing-remitting course predominates. The U.S. FDA and the European Medicines Agency (EMA) have licenses for 12 goods. Among them, since 2013, the following have been created and licensed: dimethyl fumarate (Tecfidera), alemtuzumab (Lemtrada), pegylated interferon β

(Plegridy), and glatiramer acetate (Copaxone) 40 mg. For the recurrent-remitting type of MS, interferon β-1a (Avonex) and glutamate acetate have been used as first-line therapies for over 20 years. Despite this, few studies have shown that vitamin D has positive therapeutic benefits (Dalakas 1991).

5.3.7 Dermatomyositis (DM)

DM involves subacute progressive proximal muscle weakness, a skin rash, or both (Findlay et al. 2015). DM in kids may sometimes emerge as a fever sickness. Amyopathic dermatomyositis (ADM) is a condition that affects 20% of DM patients who have typical histopathologic characteristics on muscle biopsy. These individuals may develop a rash but never have muscular weakening. According to specific theories, there are five categories into which DM skin symptoms can be categorized: pathognomonic, extremely distinctive, characteristic, more prevalent in juvenile DM, and rare in DM (DeWane et al. 2020). The etiology of DM is intricate, multifaceted, and not fully understood. Immunological, environmental, and genetic factors are believed to contribute to adult DM and JDM (Sontheimer 2004). Progressive, symmetric proximal muscular weakness and pathognomonic cutaneous signs are the hallmarks of classic dermatomyositis (CDM). In 30–50% of people in general, cutaneous illness develops within 3–6 months before myositis does, and 10% of patients first experience muscle soreness before developing skin abnormalities (Choy and Isenberg 2002). For idiopathic inflammatory myositis, corticosteroids are typically the first-line therapy; however, their effectiveness has not been thoroughly examined in randomized, placebo-controlled studies and is unlikely to be investigated in the same way in the future (Waldman et al. 2020; Pisetsky 2023). Hydroxychloroquine is recommended in conventional treatment algorithms as the first-line systemic treatment for cutaneous DM. Nevertheless, current data indicate that individuals receiving hydroxychloroquine are more likely to experience a cutaneous disease recurrence than to receive adequate disease management with hydroxychloroquine monotherapy. Drugs like Methotrexate, Azathioprine, Tacrolimus, Cyclosporine A, Cyclophosphamide, and Rituximab were already used to treat Dermatomyositis (Sasaki and Kohsaka 2018).

5.4 Limitations and Challenges of Current Treatment Strategies

Autoimmune disorders present a formidable challenge to healthcare providers and researchers in the field of pharmaceutical sciences and patients alike due to their complex and often debilitating nature. The current treatment strategies for these disorders are fraught with limitations and challenges that necessitate the exploration of alternative therapeutic approaches. When one's immune system is hyperactive, it attacks and harms the body's tissues, which results in autoimmune illnesses. In the normal course, the immune system produces proteins known as antibodies to defend against potentially hazardous elements, including viruses, cancer cells, and poisons. That's why some new weapons are needed to treat these autoimmune disorders. Autoimmune disorders encompass a wide range of conditions, including rheumatoid arthritis, multiple sclerosis, systemic lupus erythematosus, and Crohn's disease, among others. Despite the diversity of these disorders, they share common challenges in their treatment approaches:

5.4.1 Immunosuppression-Related Complications

Many current treatments rely on broad immunosuppression, which can lead to increased susceptibility to infections and long-term complications. A recent study by the National Institute of Health (NIH) found that patients with autoimmune disorders who are on immunosuppressive therapy are at an increased risk of developing severe infections, including pneumonia, tuberculosis, and sepsis. The study by Hartono et al. also found that immunosuppressive therapy can increase the risk of developing certain types of cancer (Hartono et al. 2013). Another study found that patients with rheumatoid arthritis who are on long-term immunosuppressive therapy are at an increased risk of developing osteoporosis and fractures (Staa et al. 2006; Fernåndez-Ruiz et al. 2019).

5.4.2 Ineffectiveness in Some Patients

Some patients do not respond well to standard treatments, leaving them with uncontrolled symptoms and disease progression. A study found that up to 40% of patients with rheumatoid arthritis do not respond well to standard treatments (Bredemeier et al. 2015a, b). Another study found that up to 30% of patients with multiple sclerosis do not respond well to first-line treatments (Hapfelmeier et al. 2023; Singh et al. 2016).

5.4.3 Adverse Effects

Existing therapies often come with a host of adverse effects, ranging from mild discomfort to severe organ damage. Studies have reported gastrointestinal problems, such as nausea, vomiting, and diarrhea, to be the most common adverse effects of immunosuppressive therapy. Other common adverse effects include fatigue, hair loss, and weight gain (Wiontzek 1946; Solomon et al. 2020; Ruiz and Kirk 2015).

5.5 Drug Repurposing as a Solution to Autoimmune Diseases

The concept of drug repurposing emerges as a promising avenue of exploration, potentially impacting the lives of those living with autoimmune disorders significantly. The unique appeal of drug repurposing for autoimmune diseases lies in the possibility of identifying new uses of existing pharmaceutical agents that have already undergone rigorous safety assessments. This accelerates the drug development timeline and reduces the inherent risks associated with novel drug discovery. Drug repurposing has the potential to accelerate the development of new treatments for autoimmune disorders (Pushpakom et al. 2019). Autoimmune disorders encompass various conditions, from rheumatoid arthritis and multiple sclerosis to lupus and inflammatory bowel disease. Each of these diseases presents its own set of challenges and intricacies. Therefore, the approaches to drug repurposing for autoimmune disorders must be as diverse and adaptable as the diseases themselves. Several autoimmune disorders have already benefited from drug repurposing efforts. A few examples are:

Thalidomide Originally developed as a sedative, thalidomide is now used to treat multiple myeloma and complications of leprosy. It has shown promise in managing the symptoms of systemic lupus erythematosus and other autoimmune conditions. Thalidomide was originally developed as a sedative in the 1950s, but it was withdrawn from the market after it was found to cause severe birth defects. However, thalidomide was later repurposed to treat multiple myeloma and complications of leprosy (Wimmelbücker and Kar 2023).

Methotrexate Initially developed as a cancer drug, methotrexate is now widely used to treat rheumatoid arthritis, psoriasis, and other autoimmune disorders. Methotrexate was originally developed as a cancer drug in the 1950s. However, it was later repurposed to treat rheumatoid arthritis and other autoimmune disorders. Methotrexate is now one of the most widely used drugs for treating autoimmune disorders (Alqarni and Zeidler 2020; Bedoui et al. 2019).

Rituximab Originally approved for certain types of lymphoma, rituximab is now used to treat various autoimmune conditions, including rheumatoid arthritis and lupus. Rituximab was initially approved to treat certain types of lymphoma in 1997. However, it was later repurposed to treat various autoimmune conditions, including rheumatoid arthritis and lupus. Rituximab is now one of the most effective treatments for certain autoimmune disorders (Bredemeier et al. 2015a; Froissart et al. 2015; Buch et al. 2011). Table 5.1 details a few other drugs repurposed for autoimmune disorders.

5.6 Nanotechnology in Drug Repurposing

Nanotechnology, a rapidly advancing field that manipulates materials at the nanoscale (typically at the level of molecules and atoms), has emerged as a powerful tool in drug repurposing. This innovative approach harnesses the unique properties of nanomaterials to enhance and streamline the discovery of new therapeutic applications for existing drugs, particularly in the context of autoimmune disorders. Nanoparticles for delivering repurposed drugs in treating autoimmune disorders have gained significant interest in recent years. Table 5.2 details a few nanocarrier systems of drugs repurposed to manage autoimmune disorders.

Nanocarrier systems, such as nanoparticles, liposomes, and micelles, have gained significant traction in drug repurposing. Their popularity can be attributed to several critical factors, including their size, composition, structure, versatility in carrying multiple drugs, and their high surface area-to-volume ratio, which makes them well-suited for targeted drug delivery (Rao et al. 2022). Nanocarrier systems as carriers for repurposed drugs have many advantages, including:

- Improving solubility and bioavailability: Nanocarrier systems can help improve the solubility and bioavailability of poorly soluble drugs, leading to increased efficacy and reduced toxicity.
- Reducing toxicity: Nanocarrier systems can help reduce drug toxicity by targeting specific tissues or cells, controlling their release, and preventing them from reaching healthy tissues. This is especially important for immunosuppressive drugs, which can have serious side effects if they are not used carefully.
- Improving targeting: Nanocarrier systems can help improve drug targeting to specific tissues or cells, increasing efficacy and reducing side effects.
- Improving permeability: Nanocarrier systems can help improve drug permeability across biological barriers, such as the blood-brain barrier and mucus layer.
- Reducing inflammation: Nanocarrier systems can help to reduce inflammation by targeting anti-inflammatory drugs to specific sites of inflammation.
- Improving targeting: Nanocarrier systems can help to improve the drug targeting to specific tissues or cells in the body, such as the joints in rheumatoid arthritis or the gut lining in inflammatory bowel disease. This can lead to increased efficacy and reduced side effects (Séguy et al. 2022; Mitchell et al. 2021; Tang and Li 2024; Nasra et al. 2022).

5.7 Future Perspective and Conclusion

The future of drug repurposing in autoimmune disorders holds considerable promise, driven by advancements in precision medicine, bioinformatics, and a deeper understanding of autoimmune pathophysiology. Integrating advanced computational tools and multi-omics approaches promises to further accelerate the identification of promising drug candidates. These technologies can analyze vast datasets to uncover hidden drug-disease relationships, predict drug efficacy, optimize

Table 5.1 Drugs repurposed for the management of autoimmune disorders

Drug	Original indication	New indication	Approval for original/ repurposed indication	Repurposing strategies	Findings	References
Edaravone	Cerebral infarction	Hashimoto thyroiditis	2001/–	Pharmacological analysis and exploration of the mechanism involved in the HO-1 and STAT3/PI3K/Akt pathway	Significantly decreased the severity score of thyroiditis	Li et al. (2018), Bailly (2019)
Pioglitazone	Type 2 diabetes	Addison's disease	1999/–	Observed changes in mRNA expression	Despite increasing CYP11B2 mRNA levels, pioglitazone suppresses aldosterone production	Pan et al. (2014)
Eculizumab	Paroxysmal nocturnal hemoglobinuria	Myasthenia gravis	2007/2017	Clinical analysis and pharmacologic analysis	FDA-approved for acetylcholine receptor	Dhillon (2018)
Ruxolitinib	Myelofibrosis	Inflammatory bowel disease	2011/–	Pharmacological analysis	Ruxolitinib alleviates colitis by inhibiting NF-κb	Overstreet et al. (2018)
Metformin	Type 2 diabetes	Psoriasis	2009/–	Pharmacological analysis	In clinical use for the repositioning Indication	Singh and Bhansali (2017), Jain et al. (2021)
Fingolimod	Transplantation rejection	Multiple sclerosis	–/2010	Structural and pharmacological analysis	First oral, disease-modifying therapy for multiple sclerosis to be approved	Terzi et al. (2023)
Ixazomib	Multiple myeloma	Dermatomyositis	2015/–	*Preclinical* in vivo *and* in vitro *models*	For the repurposed indication in clinical usage	Debrut et al. (2023)

Rituximab	Cancers	Rheumatoid disease	1997/2006	Retrospective clinical study (rituximab-treated non-Hodgkin lymphoma patients with concurrent rheumatoid arthritis in remission)	FDA-approved for the treatment of rheumatoid arthritis	Buch et al. (2011)
Leflunomide	Rheumatoid arthritis	Sjogren syndrome	1998/–	Placebo-controlled, double-blinded, phase 2A randomized clinical trial	Used in combination with hydroxychloroquine to treat Sjogren syndrome	van der Heijden et al. (2020)
Chloroquine	Antimalarial	Type I diabetes	1947/–	Retrospective clinical and Pharmacological analysis	Real-time PCR showed that IL-1β, IFNG, IL12, TNF-α, and IL-10 expression decreased in chloroquine-treated	de Almeida Júnior et al. (2020)
Mesalamine	Inflammatory bowel disease	Alopecia areata	1993/–	Retrospective clinical and Pharmacological analysis	For the newly designated indication, in clinical usage	Kiszewski et al. (2018)

Table 5.2 Application of nanocarrier systems in drug repurposing

Nanocarrier system	Drug	Original indication	Repurposed indication	Role of the nanocarrier system	References
Solid Lipid Nanoparticles (SLNs)	Metformin	Type 2 diabetes	Alzheimer's disease	Improve targeting to the brain and reduce side effects	Ebrahimi et al. (2022)
Micelles	Ibuprofen	Analgesic and anti-inflammatory	Inflammatory bowel disease	Improve targeting to the gut and reduce side effects	Felder et al. (2000), Klein and Eliakim (2010)
SLNs	Curcumin	Antioxidant	Arthritis	Improve bioavailability and reduce inflammation	Arora et al. (2015)
Liposomes	Methotrexate (MTX)	Anti-cancer	Psoriasis	Improve targeting and reduce toxicity	Ali et al. (2008)
Poly(lactic-co-glycolic acid) (PLGA) nanoparticles	Cyclosporine A	Psoriasis	Inflammatory bowel disease	Improve targeting and reduce toxicity	Melero et al. (2017)
Dendrimer-encapsulated infliximab	Infliximab	Crohn's disease	Multiple sclerosis	Improve targeting and reduce toxicity	Mignani et al. (2021)
Liposomes	Ibuprofen	Analgesic and anti-inflammatory	Rheumatoid arthritis	Improve targeting and reduce side effects	Chopra et al. (2008)
Nanoemulsion	MTX	Anti-cancer	Rheumatoid Arthritis (RA)	Enhanced targeting to inflamed joints, reduced systemic side effects	Suresh et al. (2022)
Liposomes	Hydroxychloroquine	Malaria	Systemic Lupus Erythematosus (SLE)	Improved targeting to immune cells, reduced toxicity	Dima et al. (2022), Stevens et al. (2020)
Albumin Nanoparticles	Dexamethasone	Anti-allergic	Rheumatoid Arthritis (RA)	Improved targeting to inflamed tissues, reduced systemic side effects	Wang et al. (2021)
Squalenyl Nanoparticles	Tofacitinib	Alopecia	Inflammatory Skin Diseases	Enhanced mucosal targeting, improved drug absorption	Christmann et al. (2020)

PLGA Nanoparticles	Cyclosporine A	Immunosuppressant	Psoriasis	Increased solubility and bioavailability	Kagawa et al. (2024)
Hyaluronic-coated Eudragit S100 Nanoparticles	Budesonide	Anti-inflammatory	Crohn's Disease	Targeted delivery to the intestinal tract, reduced systemic side effects	Seoudi et al. (2023)
Lentinan Nanoparticles	Budesonide	Anti-inflammatory	Ulcerative colitis	Macrophage-targeted delivery reduced systemic side effects	Lin et al. (2021)
Liposomes	Mycophenolic Acid	Immunosuppressant	Systemic Lupus Erythematosus (SLE)	Improved solubility and bioavailability, reduced systemic toxicity	Look et al. (2013)
Chitosan Nanoparticles	5-Aminosalicylic Acid	Analgesic and anti-inflammatory	Inflammatory Bowel Disease (IBD)	Targeted delivery to the colon, enhanced mucosal adhesion	Nalinbenjapun and Ovatlarnporn (2020)
Polycaprolactone Nanoparticles	Tacrolimus	Immunosuppressant	Atopic Dermatitis	Enhanced skin penetration, reduced systemic exposure	Camargo et al. (2021)

treatment regimens tailored to individual patients, and enhance our ability to pinpoint molecular targets and biomarkers specific to autoimmune diseases. This holistic view of disease mechanisms will facilitate the identification of repurposed drugs that can modulate these targets with high precision.

Collaboration between academic institutions, pharmaceutical companies, and regulatory agencies will be crucial in streamlining the drug repurposing process. Establishing robust frameworks for sharing clinical data, intellectual property management, and regulatory approval can significantly reduce the time and cost of bringing repurposed drugs to the market. Moreover, exploring novel drug delivery systems, such as nanotechnology and targeted delivery mechanisms, could enhance repurposed drugs' therapeutic efficacy and safety profiles. These advancements may open new avenues for treating autoimmune disorders with increased specificity and reduced side effects.

Drug repurposing presents a unique and valuable strategy in the quest to combat autoimmune disorders. By capitalizing on the known safety profiles and existing clinical data of established drugs, repurposing can expedite the development of effective treatments, providing patients with timely and cost-efficient therapeutic options. This chapter has highlighted key examples of successful drug repurposing in autoimmune diseases, demonstrating the potential for a significant clinical impact.

However, several challenges remain, including regulatory hurdles, economic considerations, and the need for robust clinical validation. Addressing these challenges will require concerted efforts from all stakeholders involved. Ultimately, the future of drug repurposing in autoimmune disorders looks bright, potentially transforming the treatment landscape and improving the quality of life of millions of patients worldwide. By continuing to harness innovative technologies and collaborative frameworks, we can unlock new therapeutic potentials and make significant strides in managing and potentially curing autoimmune diseases.

References

Ali MF et al (2008) Liposomal methotrexate hydrogel for treatment of localized psoriasis: preparation, characterization and laser targeting. Med Sci Monit 14(12):Pi66-74

Alqarni AM, Zeidler MP (2020) How does methotrexate work? Biochem Soc Trans 48(2):559–567

Ananthakrishnan AN et al (2014) Serum inflammatory markers and risk of colorectal cancer in patients with inflammatory bowel diseases. Clin Gastroenterol Hepatol 12(8):1342–8.e1

Arora R et al (2015) Curcumin loaded solid lipid nanoparticles ameliorate adjuvant-induced arthritis in rats. Eur J Pain 19(7):940–952

Ashburn TT, Thor KB (2004) Drug repositioning: identifying and developing new uses for existing drugs. Nat Rev Drug Discov 3(8):673–683

Bach JF (2005) Infections and autoimmune diseases. J Autoimmun 25(Suppl):74–80

Bailly C (2019) Potential use of edaravone to reduce specific side effects of chemo-, radio- and immuno-therapy of cancers. Int Immunopharmacol 77:105967

Baker PR et al (2010) Haplotype analysis discriminates genetic risk for DR3-associated endocrine autoimmunity and helps define extreme risk for Addison's disease. J Clin Endocrinol Metab 95(10):E263–E270

Baumgart DC, Sandborn WJ (2007) Inflammatory bowel disease: clinical aspects and established and evolving therapies. Lancet 369(9573):1641–1657
Bedoui Y et al (2019) Methotrexate an old drug with new tricks. Int J Mol Sci 20(20):5023
Begley CG et al (2021) Drug repurposing: misconceptions, challenges, and opportunities for academic researchers. Sci Transl Med 13(612):eabd5524
Betterle C et al (2002) Autoimmune adrenal insufficiency and autoimmune polyendocrine syndromes: autoantibodies, autoantigens, and their applicability in diagnosis and disease prediction. Endocr Rev 23(3):327–364
Bhagat RT, Butle SR (2021) Drug repurposing: a review. J Pharm Res Int 33(31B):161–169
Bornstein SR et al (2016) Diagnosis and treatment of primary adrenal insufficiency: an Endocrine Society clinical practice guideline. J Clin Endocrinol Metab 101(2):364–389
Breckenridge A, Jacob R (2019) Overcoming the legal and regulatory barriers to drug repurposing. Nat Rev Drug Discov 18(1):1–2
Bredemeier M, Campos GG, de Oliveira FK (2015a) Updated systematic review and meta-analysis of randomized controlled trials comparing low-versus high-dose rituximab for rheumatoid arthritis. Clin Rheumatol 34:1801–1805
Bredemeier M, Campos GG, de Oliveira FK (2015b) Updated systematic review and meta-analysis of randomized controlled trials comparing low- versus high-dose rituximab for rheumatoid arthritis. Clin Rheumatol 34(10):1801–1805
Buch MH et al (2011) Updated consensus statement on the use of rituximab in patients with rheumatoid arthritis. Ann Rheum Dis 70(6):909–920
Burisch J et al (2015) Costs and resource utilization for diagnosis and treatment during the initial year in a European inflammatory bowel disease inception cohort: an ECCO-EpiCom study. Inflamm Bowel Dis 21(1):121–131
Camargo GDA et al (2021) Characterization and in vitro and in vivo evaluation of Tacrolimus-loaded poly(ε-Caprolactone) nanocapsules for the management of atopic dermatitis. Pharmaceutics 13(12):2013
Cha Y et al (2018) Drug repurposing from the perspective of pharmaceutical companies. Br J Pharmacol 175(2):168–180
Chopra D et al (2008) Use of liposomal drugs in the treatment of rheumatoid arthritis. Curr Rheumatol Rev 4(1):63–70
Choudhury S, Meeran K (2018) Glucocorticoid replacement in Addison disease. Nat Rev Endocrinol 14(9):562
Choy EH, Isenberg DA (2002) Treatment of dermatomyositis and polymyositis. Rheumatology (Oxford) 41(1):7–13
Christmann R et al (2020) Tofacitinib loaded Squalenyl nanoparticles for targeted follicular delivery in inflammatory skin diseases. Pharmaceutics 12(12):1131
Compston A, Coles A (2008) Multiple sclerosis. Lancet 372(9648):1502–1517
Dalakas MC (1991) Polymyositis, dermatomyositis and inclusion-body myositis. N Engl J Med 325(21):1487–1498
de Almeida Júnior RF et al (2020) Chloroquine as a promising adjuvant therapy for type 1 Diabetes Mellitus. Sci Rep 10(1):12098
Debrut L et al (2023) OP0117 proteasome inhibition as a new treatment for dermatomyositis: results of a drug repurposing analysis based on the transcriptomic signature of patients' perifascicular fibers validated in pre-clinical models. Ann Rheumatic Dis 82(Suppl 1):77
DeWane ME, Waldman R, Lu J (2020) Dermatomyositis: clinical features and pathogenesis. J Am Acad Dermatol 82(2):267–281
Dey G (2019) An overview of drug repurposing. J Med Sci Clin Res 7:55–56
Dhillon S (2018) Eculizumab: a review in generalized Myasthenia gravis. Drugs 78(3):367–376
Dima A et al (2022) Hydroxychloroquine in systemic lupus erythematosus: overview of current knowledge. Ther Adv Musculoskelet Dis 14:1759720x211073001
DiMasi JA, Grabowski HG, Hansen RW (2015) The cost of drug development. N Engl J Med 372(20):1972
Dobson R, Giovannoni G (2019) Multiple sclerosis - a review. Eur J Neurol 26(1):27–40

Dresser L et al (2021) Myasthenia gravis: epidemiology, pathophysiology and clinical manifestations. J Clin Med 10(11):2235
Ebrahimi H et al (2022) Design and optimization of metformin-loaded solid lipid nanoparticles for neuroprotective effects in a rat model of diffuse traumatic brain injury: a biochemical, behavioral, and histological study. Eur J Pharm Biopharm 181:122–135
Felder JB et al (2000) Effects of nonsteroidal antiinflammatory drugs on inflammatory bowel disease: a case-control study. Am J Gastroenterol 95(8):1949–1954
Fernández-Ruiz M et al (2019) Monitoring of alphatorquevirus DNA levels for the prediction of immunosuppression-related complications after kidney transplantation. Am J Transplant 19(4):1139–1149
Findlay AR, Goyal NA, Mozaffar T (2015) An overview of polymyositis and dermatomyositis. Muscle Nerve 51(5):638–656
Fox RI et al (2021) Treatment of Sjögren's syndrome: current therapy and future directions. Rheumatology (Oxford) 60(5):2066–2074
Froissart A et al (2015) Rituximab in autoimmune thrombotic thrombocytopenic purpura: a success story. Eur J Intern Med 26(9):659–665
Grammer AC, Lipsky PE (2017) Drug repositioning strategies for the identification of novel therapies for rheumatic autoimmune inflammatory diseases. Rheum Dis Clin N Am 43(3):467–480
Gupta R et al (2021) Artificial intelligence to deep learning: machine intelligence approach for drug discovery. Mol Divers 25(3):1315–1360
Hapfelmeier A et al (2023) Retrospective cohort study to devise a treatment decision score predicting adverse 24-month radiological activity in early multiple sclerosis. Ther Adv Neurol Disord 16:17562864231161892
Hartono C, Muthukumar T, Suthanthiran M (2013) Immunosuppressive drug therapy. Cold Spring Harb Perspect Med 3(9):a015487
Hehir MK 2nd, Li Y (2022) Diagnosis and Management of Myasthenia Gravis. Continuum (Minneap Minn) 28(6):1615–1642
Hellesen A, Bratland E, Husebye ES (2018) Autoimmune Addison's disease - an update on pathogenesis. Ann Endocrinol (Paris) 79(3):157–163
Howard JF Jr (2018) Myasthenia gravis: the role of complement at the neuromuscular junction. Ann NY Acad Sci 1412(1):113–128
Htut TW, Thein KZ, Oo TH (2021) Pernicious anemia: pathophysiology and diagnostic difficulties. J Evid Based Med 14(2):161–169
Huang WJ, Chen WW, Zhang X (2017) Multiple sclerosis: pathology, diagnosis and treatments. Exp Ther Med 13(6):3163–3166
Ilg MM et al (2022) Phenotypic screening of 1,953 FDA-approved drugs reveals 26 hits with potential for repurposing for Peyronie's disease. PLoS One 17(12):e0277646
Imam K, Abdullah M, Felicetta JV (1988) Alopecia universalis as a feature of polyglandular autoimmunity type I. West J Med 149(3):338–341
Jain H et al (2021) Repurposing approved therapeutics for new indication: addressing unmet needs in psoriasis treatment. Curr Res Pharmacol Drug Discov 2:100041
Jourdan J-P et al (2020a) Drug repositioning: a brief overview. J Pharm Pharmacol 72(9):1145–1151
Jourdan JP et al (2020b) Drug repositioning: a brief overview. J Pharm Pharmacol 72(9):1145–1151
Kagawa A et al (2024) Therapeutic effects of 30 nm Cyclosporin A-loaded nanoparticles using PLGA-PEG-PLGA triblock copolymer for transdermal delivery in Mouse Models of Psoriasis. Appl Sci 14(9):3791
Kiszewski AE, Bevilaqua M, De Abreu LB (2018) Mesalazine in the treatment of extensive Alopecia Areata: a new therapeutic option? Int J Trichology 10(3):99–102
Klein A, Eliakim R (2010) Non steroidal anti-inflammatory drugs and inflammatory bowel disease. Pharmaceuticals (Basel) 3(4):1084–1092
Korstanje A et al (2002) The serological gastric biopsy: a non-endoscopical diagnostic approach in management of the dyspeptic patient: significance for primary care based on a survey of the literature. Scand J Gastroenterol Suppl 236:22–26

Krishnamurthy N et al (2022) Drug repurposing: a systematic review on root causes, barriers and facilitators. BMC Health Serv Res 22(1):1–17
Lee DSW, Rojas OL, Gommerman JL (2021) B cell depletion therapies in autoimmune disease: advances and mechanistic insights. Nat Rev Drug Discov 20(3):179–199
Li H et al (2018) Edaravone ameliorates experimental autoimmune thyroiditis in rats through HO-1-dependent STAT3/PI3K/Akt pathway. Am J Transl Res 10(7):2037–2046
Lin M et al (2021) Lentinan-based oral nanoparticle loaded Budesonide with macrophage-targeting ability for treatment of ulcerative colitis. Front Bioeng Biotechnol 9:702173
Lipinski CF et al (2019) Advances and perspectives in applying deep learning for drug design and discovery. Front Robot AI 6:108
Liu Z et al (2013) In silico drug repositioning: what we need to know. Drug Discov Today 18(3–4):110–115
Look M et al (2013) Nanogel-based delivery of mycophenolic acid ameliorates systemic lupus erythematosus in mice. J Clin Invest 123(4):1741–1749
Luo H et al (2016) Molecular docking for identification of potential targets for drug repurposing. Curr Top Med Chem 16(30):3636–3645
McGrogan A, Sneddon S, de Vries CS (2010) The incidence of myasthenia gravis: a systematic literature review. Neuroepidemiology 34(3):171–183
Melero A et al (2017) Targeted delivery of Cyclosporine A by polymeric nanocarriers improves the therapy of inflammatory bowel disease in a relevant mouse model. Eur J Pharm Biopharm 119:361–371
Mignani S et al (2021) In vivo therapeutic applications of phosphorus dendrimers: state of the art. Drug Discov Today 26(3):677–689
Mitchell MJ et al (2021) Engineering precision nanoparticles for drug delivery. Nat Rev Drug Discov 20(2):101–124
Mohammad Jafari R et al (2018) Drug repositioning: a review. J Iran Med Council 1(1):7–10
Nalinbenjapun S, Ovatlarnporn C (2020) Chitosan-5-aminosalicylic acid conjugates for colon-specific drug delivery: methods of preparation and in vitro evaluations. J Drug Deliv Sci Technol 57:101397
Nasra S, Bhatia D, Kumar A (2022) Recent advances in nanoparticle-based drug delivery systems for rheumatoid arthritis treatment. Nanoscale Adv 4(17):3479–3494
Ni Y et al (2023) A target-based discovery from a parasitic helminth as a novel therapeutic approach for autoimmune diseases. EBioMedicine 95:104751
Oksenberg JR, Baranzini SE (2010) Multiple sclerosis genetics--is the glass half full, or half empty? Nat Rev Neurol 6(8):429–437
Oprea TI et al (2011) Drug repurposing from an academic perspective. Drug Discov Today Therap Strateg 8(3–4):61–69
Overstreet AM et al (2018) The JAK inhibitor ruxolitinib reduces inflammation in an ILC3-independent model of innate immune colitis. Mucosal Immunol 11(5):1454–1465
Padhy B, Gupta Y (2011) Drug repositioning: re-investigating existing drugs for new therapeutic indications. J Postgrad Med 57(2):153
Pan ZQ et al (2014) The effect of pioglitazone on aldosterone and cortisol production in HAC15 human adrenocortical carcinoma cells. Mol Cell Endocrinol 394(1–2):119–128
Parisi D et al (2020) Drug repositioning or target repositioning: a structural perspective of drug-target-indication relationship for available repurposed drugs. Comput Struct Biotechnol J 18:1043–1055
Parvathaneni V et al (2019) Drug repurposing: a promising tool to accelerate the drug discovery process. Drug Discov Today 24(10):2076–2085
Peterson LK, Fujinami RS (2007) Inflammation, demyelination, neurodegeneration and neuroprotection in the pathogenesis of multiple sclerosis. J Neuroimmunol 184(1–2):37–44
Pisetsky DS (2023) Pathogenesis of autoimmune disease. Nat Rev Nephrol 19(8):509–524
Pushpakom S et al (2019) Drug repurposing: progress, challenges and recommendations. Nat Rev Drug Discov 18(1):41–58

Rajan JR et al (2023) An AI approach to identifying novel therapeutics for rheumatoid arthritis. J Pers Med 13(12):1633
Rao N et al (2022) Drug repurposing: a shortcut to new biological entities. Pharm Chem J 56(9):1203–1214
Reaume AGJDDTTS (2011) Drug repurposing through nonhypothesis driven phenotypic screening. Drug Discov Today Therap Strateg 8(3–4):85–88
Rosenthal FD, Davies MK, Burden AC (1978) Malignant disease presenting as Addison's disease. Br Med J 1(6127):1591–1592
Rudrapal M, Khairnar SJ, Jadhav AG (2020) Drug repurposing (DR): an emerging approach in drug discovery. In: Drug repurposing-hypothesis, molecular aspects and therapeutic applications, p 10
Ruiz R, Kirk AD (2015) Long-term toxicity of immunosuppressive therapy. Transpl Liver:1354–1363. https://doi.org/10.1016/B978-1-4557-0268-8.00097-X. Epub 2015 Apr 3
Rustgi SD, Bijlani P, Shah SC (2021) Autoimmune gastritis, with or without pernicious anemia: epidemiology, risk factors, and clinical management. Ther Adv Gastroenterol 14:17562848211038771
Sasaki H, Kohsaka H (2018) Current diagnosis and treatment of polymyositis and dermatomyositis. Mod Rheumatol 28(6):913–921
Schneider-Gold C, Gilhus NE (2021) Advances and challenges in the treatment of myasthenia gravis. Ther Adv Neurol Disord 14:17562864211065406
Séguy L, Groo A-C, Malzert-Fréon A (2022) How nano-engineered delivery systems can help marketed and repurposed drugs in Alzheimer's disease treatment? Drug Discov Today 27(6):1575–1589
Seoudi SS et al (2023) Targeted delivery of budesonide in acetic acid induced colitis: impact on miR-21 and E-cadherin expression. Drug Deliv Transl Res 13(11):2930–2947
Shi H et al (2022) Mechanism of action of Tripterygium wilfordii for treatment of idiopathic membranous nephropathy based on network pharmacology. Ren Fail 44(1):116–125
Silva FA et al (2016) The immunological basis of inflammatory bowel disease. Gastroenterol Res Pract 2016:2097274
Singh S, Bhansali A (2017) Randomized placebo control study of metformin in psoriasis patients with metabolic syndrome (systemic treatment Cohort). Indian J Endocrinol Metab 21(4):581–587
Singh JA et al (2016) 2015 American College of Rheumatology Guideline for the treatment of rheumatoid arthritis. Arthritis Rheumatol 68(1):1–26
Sleigh SH, Barton CL (2010) Repurposing strategies for therapeutics. Pharm Med 24(3):151–159
Solomon DH et al (2020) Adverse effects of low-dose methotrexate: a randomized trial. Ann Intern Med 172(6):369–380
Song W, Wang H, Wang X (2024) Research hotspots and emerging trends in the treatment of Sjogren's syndrome: a bibliometric analysis from 1900 to 2022. Heliyon 10(1):e23216
Sontheimer RD (2004) The management of dermatomyositis: current treatment options. Expert Opin Pharmacother 5(5):1083–1099
Staa TPV et al (2006) Clinical assessment of the long-term risk of fracture in patients with rheumatoid arthritis. Arthritis Rheumat 54(10):3104–3112
Stabler SP (2013) Vitamin B12 deficiency. N Engl J Med 368(2):149–160
Stevens DM, Crist RM, Stern ST (2020) Nanomedicine reformulation of Chloroquine and Hydroxychloroquine. Molecules 26(1):175
Sun W, Sanderson PE, Zheng W (2016) Drug combination therapy increases successful drug repositioning. Drug Discov Today 21(7):1189–1195
Suresh P et al (2022) Development of a novel methotrexate-loaded nanoemulsion for rheumatoid arthritis treatment with site-specific targeting subcutaneous delivery. Nanomaterials (Basel) 12(8):1299
Szczudlik P et al (2014) Antititin antibody in early- and late-onset myasthenia gravis. Acta Neurol Scand 130(4):229–233

Talevi A, Bellera CL (2020) Challenges and opportunities with drug repurposing: finding strategies to find alternative uses of therapeutics. Expert Opin Drug Discov 15(4):397–401

Tang Y, Li LJIJON (2024) The application of nanovaccines in autoimmune diseases, pp 367–388

Terzi M et al (2023) Clinical and demographic characteristics and two-year efficacy and safety data of 508 multiple sclerosis patients with Fingolimod treatment. Noro Psikiyatr Ars 60(1):23–27

Toh BH (2017) Pathophysiology and laboratory diagnosis of pernicious anemia. Immunol Res 65(1):326–330

Ullman TA, Itzkowitz SH (2011) Intestinal inflammation and cancer. Gastroenterology 140(6):1807–1816

Unal U et al (2022) Drug repurposing for rheumatoid arthritis: identification of new drug candidates via bioinformatics and text mining analysis. Autoimmunity 55(3):147–156

van der Heijden EHM et al (2020) Leflunomide–hydroxychloroquine combination therapy in patients with primary Sjögren's syndrome (RepurpSS-I): a placebo-controlled, double-blinded, randomised clinical trial. Lancet Rheumatol 2(5):e260–e269

Vivarelli S et al (2020) Patient-derived tumor organoids for drug repositioning in cancer care: a promising approach in the era of tailored treatment. Cancers (Basel) 12(12):3636

Wahren-Herlenius M, Dörner T (2013) Immunopathogenic mechanisms of systemic autoimmune disease. Lancet 382(9894):819–831

Waldman R, DeWane ME, Lu J (2020) Dermatomyositis: diagnosis and treatment. J Am Acad Dermatol 82(2):283–296

Wang Q et al (2021) Nanomedicines for the treatment of rheumatoid arthritis: state of art and potential therapeutic strategies. Acta Pharm Sin B 11(5):1158–1174

Wehkamp J et al (2016) Inflammatory bowel disease. Dtsch Arztebl Int 113(5):72–82

Wimmelbücker L, Kar A (2023) A history of thalidomide in India. Med Hist 67(3):228–246

Wiontzek H (1946) Adverse effects of immunosuppressive therapy. Dtsch Med Wochenschr 1969. 94(49):2568

Xue H et al (2018) Review of drug repositioning approaches and resources. Int J Biol Sci 14(10):1232–1244

Zeng W et al (2022) The future of targeted treatment of primary Sjögren's syndrome: a focus on extra-glandular pathology. Int J Mol Sci 23(22):14135

Zhernakova A, van Diemen CC, Wijmenga C (2009) Detecting shared pathogenesis from the shared genetics of immune-related diseases. Nat Rev Genet 10(1):43–55

COVID-19 Management Using Drug Repurposing

6

Saima, Ruchika Sharma, S. Latha, and Anoop Kumar

Abstract

Drug repurposing is a well-known technique in which already approved drugs are explored for alternate indications. It saves time and money which are involved in the drug development process. This technique is widely used to identify promising available drugs in the treatment of various diseases including COVID-19. This chapter offers a comprehensive exploration of the multifaceted approach in the management of COVID-19 patients through drug repurposing. This chapter starts with an overview of COVID-19 landscape, followed by the concept of drug repurposing, elucidating its significance as a swift and cost-effective method for identifying potential drug candidates. The discussion regarding repurposing candidates identified so far was also included along with their possible antiviral mechanisms. Furthermore, this chapter also highlights the transformative impact of innovative approaches such as network pharmacology, machine learning, artificial intelligence, and computational methods in the identification of repurposed drugs. Ultimately, this chapter serves as a valuable resource, offering both a comprehensive understanding of COVID-19 management through drug repurposing and an exploration of cutting-edge methodologies that hold promise in our ongoing battle against emerging infectious diseases.

Saima
Department of Pharmacology, Magadh Institute of Pharmacy, Patna, India

R. Sharma
Centre for Precision Medicine and Pharmacy, Delhi Pharmaceutical Science & Research University (DPSRU), New Delhi, India

S. Latha · A. Kumar (✉)
Department of Pharmacology, Delhi Institute of Pharmaceutical Sciences and Research (DIPSAR), Delhi Pharmaceutical Science & Research University (DPSRU), New Delhi, India
e-mail: anoopdpsru@dpsru.edu.in

R. K. Kaundal et al. (eds.), *Drug Repurposing: Exploring Newer Therapeutic Potentials*,
https://doi.org/10.1007/978-981-95-4522-3_6

Keywords

Drug repurposing · Molecular docking · Network pharmacology · Molecular dynamics · COVID-19

6.1 Introduction

The novel coronavirus that emerged in 2019, initially named Severe Acute Respiratory Syndrome Coronavirus-2 (SARS-CoV-2) by the International Committee on Taxonomy of Viruses, falls within the beta-CoV genus (Pal et al. 2020). This group also encompasses other significant coronaviruses such as the Severe Acute Respiratory Syndrome CoV (SARS-CoV) and Middle Eastern Respiratory Syndrome CoV (MERS-CoV) (Petrosillo et al. 2020). On February 11, 2020, the World Health Organization (WHO) introduced a new designation for the illness caused by SARS-CoV-2, officially termed Coronavirus Disease (COVID-19) (World Health Organization 2024a). A month later, on March 11, 2020, COVID-19 was declared a pandemic. SARS-CoV-2 made its first appearance in Wuhan, China, in December 2019, with the initial case connected to a seafood market. Subsequently, the virus spread rapidly, and as of January 17, 2021, it had infected over 93 million individuals and claimed more than 2 million lives (World Health Organization 2021). Although the mortality rate of SARS-CoV-2 appears lower than that of SARS-CoV and MERS-CoV, it spreads more swiftly and widely, resulting in a global healthcare and financial crisis.

In this chapter, we provide an overview of the current knowledge regarding the geographic distribution, virology, clinical manifestations, diagnosis, and management of COVID-19. Coronaviruses have been responsible for significant outbreaks of fatal pneumonia in the twenty-first century. Severe acute respiratory syndrome coronavirus (SARS-CoV) emerged in 2003, spreading to five continents with a 10% fatality rate (Cheng et al. 2007). Both SARS-CoV and MERS-CoV are zoonotic, originating from bats and civets, and dromedaries, respectively. As of now, no specific therapeutic drug or vaccine has been approved for treating human coronaviruses, which are considered a significant threat to human health.

Coronaviruses are enveloped viruses with a positive RNA genome, belonging to the Coronaviridae family of the order Nidovirales. They are divided into four genera (a, b, g, and d), with SARS-CoV-2 falling into the b genus. CoVs possess at least four structural proteins: spike (S) protein, envelope (E) protein, membrane (M) protein, and nucleocapsid (N) protein. Spike plays a crucial role in host attachment and virus–cell membrane fusion during infection, partially determining the host range (Fehr and Perlman 2015; Kumar 2021).

Potential therapies for combating coronaviruses can be categorized based on their target: those acting on the human immune system or human cells or acting directly on the virus itself (Wu et al. 2020). Enhancing the innate immune response, particularly through interferons, can help control coronavirus replication. Additionally, blocking the signal pathways necessary for virus replication in human cells may have antiviral effects. Coronaviruses often bind to receptor proteins on the cell surface, such as the SARS virus binding to the angiotensin-converting enzyme 2 (ACE2) receptor and MERS binding to the DPP4 receptor (Pushpakom et al. 2019; Scialo et al. 2020; Sharma et al. 2022). Therapies acting directly on the virus include preventing viral RNA synthesis, inhibiting virus replication through critical enzymes, and interfering with the virus binding to human cell receptors or inhibiting the virus's self-assembly process by acting on certain structural proteins.

Drug repurposing, also known as drug repositioning or reprofiling, is a strategy aimed at discovering new uses for approved or investigational drugs beyond their original medical indications. This approach offers several advantages over developing entirely new drugs for specific medical conditions (Low et al. 2020). Numerous researchers have employed drug repurposing to identify new therapeutic targets, pathways, and interventions.

This approach offers several advantages compared to the development of entirely new drugs for specific uses. First, and most significantly, it involves a lower risk of failure. Repurposed drugs have already undergone safety assessments in preclinical models and humans, making them less likely to fail, especially from a safety perspective, in subsequent efficacy trials. Second, the drug development timeline can be significantly shortened because much of the preclinical testing, safety evaluation, and, sometimes, formulation work has already been completed. Third, it generally requires less investment, although the extent of savings varies based on the development stage of the repurposed candidate. Regulatory and phase III costs may be similar for a repurposed drug compared to a new drug in the same indication, but substantial savings can be achieved in preclinical and phase I and II costs (Low et al. 2020; Talevi and Bellera 2020; Weth et al. 2023). Collectively, these advantages can lead to a less risky and faster return on investment in the development of repurposed drugs, resulting in lower average costs once failures are considered. In fact, bringing a repurposed drug to market is estimated to cost around $300 million on average, in contrast to the estimated $2–3 billion for a new chemical entity (Talevi and Bellera 2020). Furthermore, repurposed drugs may uncover new targets and pathways that can be explored for further therapeutic opportunities.

Amid the global challenge of COVID-19, drug repurposing has become a critical strategy. By identifying existing drugs with established safety profiles, this approach enables a swift and cost-effective response to the pandemic. It significantly reduces the risk associated with the development of entirely new treatments and expedites the availability of potential therapies to combat the virus. With preclinical and safety assessments already completed for approved drugs, repurposed candidates are more likely to succeed in clinical trials, focusing efforts on efficacy evaluations. This approach not only accelerates the drug development process but also leverages the existing infrastructure of the pharmaceutical industry, ensuring that effective treatments can be efficiently produced and widely distributed. Drug repurposing serves as a cornerstone in the fight against COVID-19, providing a practical and life-saving solution to address the evolving challenges of the pandemic (Garg and Kumar 2021; Wang et al. 2020).

6.2 Repurposed Drugs for COVID-19

Repurposing drugs for the management of COVID-19 involves the exploration of existing medications that were initially created for different medical purposes to assess their potential in treating or alleviating the symptoms of COVID-19. This strategy seeks to accelerate the development of potential treatments by utilizing drugs that are already known for their safety and established mechanisms of action. Numerous drugs have been investigated for repurposing in the context of COVID-19, although their efficacy varies, and research in this area continues. Table 6.1 depicts a range of repurposed drugs utilized in the treatment of COVID-19. Some of the notable drugs being considered for repurposing in the treatment of COVID-19 include the following.

6.2.1 Antimalarial Drugs

Antimalarial drugs have been repurposed for the treatment of COVID-19. The details are mentioned in the following sections.

6.2.1.1 Chloroquine and Derivatives

Chloroquine is a medication primarily used to prevent and treat malaria. It belongs to a class of drugs known as antimalarials. Additionally, chloroquine has been used in the treatment of certain autoimmune conditions, such as rheumatoid arthritis and lupus erythematosus, due to its immunomodulatory properties. During the early stages of the COVID-19 pandemic, chloroquine and its derivative, hydroxychloroquine, gained attention as potential treatments for COVID-19. Laboratory studies suggested that chloroquine and hydroxychloroquine might have antiviral effects against SARS-CoV-2, the virus that causes COVID-19. These effects were thought to involve interference with viral entry into cells and inhibition of viral replication. Hydroxychloroquine, despite being primarily known as an antimalarial drug, gained attention early in the COVID-19 pandemic as a potential treatment option. Its proposed mechanism of action in COVID-19 was based on its ability to modulate the immune response and inhibit viral replication (Lei et al. 2020).

A study conducted in France by Raoult et al. attracted attention for proposing that a combination of hydroxychloroquine and azithromycin exhibited encouraging outcomes in diminishing the viral load. The research indicated that hydroxychloroquine effectively reduced the nasopharyngeal carriage of SARS-CoV-2 in COVID-19 patients within a relatively short period, ranging from 3 to 6 days for most individuals. A notable distinction emerged between patients treated with hydroxychloroquine and the control group, evident as early as day 3 post-inclusion. Nonetheless, criticisms were raised against the study for its methodology and the limited size of the sample. Following larger-scale trials such as the Recovery Trial in the UK, the Solidarity Trial conducted by the World Health Organization (WHO), and the RECOVERY Trial in the United States, no substantial advantages of hydroxychloroquine were identified in relation to decreasing mortality or enhancing clinical

Table 6.1 Drugs repurposed for COVID-19

Name and class of drugs	Earlier indication	Mechanism in COVID-19	Ongoing research	References
Remdesivir (anti-viral)	Originally developed for Ebola virus	Inhibits viral RNA polymerase, preventing viral replication Reduces viral load and severity of illness	Ongoing trials to evaluate efficacy in different stages of COVID-19 and in combination with other treatments	Wang et al. (2020), Beigel et al. (2020a), Spinner et al. (2020)
Ritonavir (anti-viral)	Initially used for HIV/AIDS	Protease inhibitor targeting viral replication May inhibit SARS-CoV-2 replication by interfering with viral protease activity	Limited clinical evidence; ongoing studies to assess effectiveness in COVID-19	Cao et al. (2020a), Yamamoto et al. (2020)
Favipiravir (anti-viral)	Approved for influenza treatment	RNA polymerase inhibitor, inhibiting viral replication May reduce viral load and duration of symptoms	Ongoing research to determine efficacy and optimal dosing in COVID-19	Cai et al. (2020), Chen et al. (2020a)
Molnupiravir (anti-viral)	Investigational antiviral	Oral nucleoside analog prodrug Incorporates into viral RNA, causing errors during replication and hindering viral replication May reduce viral load and transmission	Early clinical trials underway to evaluate efficacy and safety in COVID-19	Painter and Holman (2022), Kabinger et al. (2021a)
Ritonavir-boosted Nirmatrelvir (Paxlovid) (anti-viral)	N/A—Newly developed for COVID-19	Protease inhibitor combination targeting viral replication Reduces viral replication and severity of illness	Clinical trials ongoing to evaluate efficacy and safety in treating COVID-19, particularly in early stages of infection	A Randomized, Double-blind, Placebo-controlled (2022a, b)

(continued)

Table 6.1 (continued)

Name and class of drugs	Earlier indication	Mechanism in COVID-19	Ongoing research	References
Ibuprofen (Anti-inflammatory)	Pain relief, fever reduction, inflammation reduction	Non-selective inhibitor of cyclooxygenase (COX), reducing production of prostaglandins involved in inflammation and pain	Limited evidence; conflicting reports on potential effects in COVID-19	Little et al. (2020), Abassi et al. (2020)
Mefenamic Acid (Anti-inflammatory)	Pain relief, fever reduction, inflammation reduction	Non-selective inhibitor of cyclooxygenase (COX), reducing production of prostaglandins involved in inflammation and pain	Limited evidence; role in COVID-19 management not well-established	Gluth et al. (2020), Dostan et al. (2020)
Indomethacin (Anti-inflammatory)	Pain relief, fever reduction, inflammation reduction	Non-selective inhibitor of cyclooxygenase (COX), reducing production of prostaglandins involved in inflammation and pain Potential antiviral effects by inhibiting viral replication	Limited clinical data; ongoing research to evaluate efficacy and safety in COVID-19	Scarsi et al. (2020), Porreca and Di Febbo (2020)
Dexamethasone (Corticosteroids)	Various inflammatory conditions, autoimmune diseases	Suppression of pro-inflammatory cytokines Modulation of immune response Reduction of lung inflammation and edema	Ongoing studies to evaluate efficacy, optimal dosage, and timing of administration in different stages of COVID-19	RECOVERY Collaborative Group (2020a), Villar et al. (2020)

(continued)

Table 6.1 (continued)

Name and class of drugs	Earlier indication	Mechanism in COVID-19	Ongoing research	References
Hydrocortisone (Corticosteroids)	Adrenal insufficiency, allergic reactions, inflammatory conditions	Suppression of pro-inflammatory cytokines Modulation of immune response Reduction of lung inflammation and edema	Clinical trials investigating efficacy and safety in severe COVID-19 cases, including the RECOVERY trial	Angus et al. (2020), Dequin et al. (2020a)
Prednisolone (Corticosteroids)	Asthma, autoimmune diseases, inflammatory conditions	Suppression of pro-inflammatory cytokines Modulation of immune response Reduction of lung inflammation and edema	Limited clinical data available, ongoing research to assess efficacy and optimal dosing in COVID-19	Jeronimo et al. (2021), WHO Rapid Evidence Appraisal for COVID-19 Therapies (REACT) Working Group et al. (2020)
Methylprednisolone (Corticosteroids)	Asthma, autoimmune diseases, inflammatory conditions	Suppression of pro-inflammatory cytokines Modulation of immune response Reduction of lung inflammation and edema	Ongoing studies to evaluate effectiveness and safety in severe COVID-19 cases	Tomazini et al. (2020), Corral et al. (2020)
Chloroquine and derivatives (Antimalarials)	Malaria treatment and prophylaxis	Proposed inhibition of viral entry and replication Immunomodulatory effects	Limited efficacy and safety data in COVID-19; controversial findings from early studies	Lewis et al. (2021), Junqueira and Rowe (2021)
Artemisinin and derivatives (artesunate, artemether) (Antimalarials)	Malaria treatment	Laboratory studies suggest potential direct antiviral effects against coronaviruses, including SARS-CoV-2 Mechanism may involve inhibition of viral replication and modulation of host immune response	Limited clinical data in COVID-19; ongoing research to evaluate efficacy and safety	Agrawal et al. (2022), Fuzimoto (2021)

(continued)

Table 6.1 (continued)

Name and class of drugs	Earlier indication	Mechanism in COVID-19	Ongoing research	References
Azithromycin (Antibiotics)	Bacterial Infections	Proposed to have antiviral and anti-inflammatory properties	Clinical trials assessing its role in combination therapy and prevention	Mangkuliguna et al. (2021)
Doxycycline (Antibiotics)	Bacterial Infections	Broad-spectrum antibiotic with potential antiviral and anti-inflammatory effects	Studies investigating its efficacy and safety in COVID-19	Dhar et al. (2023)
Tocilizumab (Immunomodulator)	Rheumatoid arthritis, cytokine release syndrome	Interleukin-6 (IL-6) receptor antagonist, reducing inflammation and cytokine storm Potential to modulate immune response in severe COVID-19 cases	Ongoing trials to evaluate efficacy and safety, particularly in severe COVID-19 cases	Stone et al. (2020), REMAP-CAP Investigators et al. (2021)
Baricitinib (Immunomodulator)	Rheumatoid arthritis	Janus kinase (JAK) inhibitor, reducing inflammation and cytokine production Potential antiviral effects by inhibiting viral entry	Clinical trials investigating efficacy and safety, especially in combination with other treatments	Kalil et al. (2021), Marconi et al. (2021)
Anakinra (Immunomodulator)	Rheumatoid arthritis, cytokine release syndrome	Interleukin-1 (IL-1) receptor antagonist, reducing inflammation and cytokine storm Potential to modulate immune response in severe COVID-19 cases	Limited clinical data; ongoing research to assess efficacy and optimal dosing	Huet et al. (2020), Cavalli et al. (2020)

outcomes among COVID-19 patients admitted to hospitals. Ultimately, the WHO terminated the hydroxychloroquine arm of the Solidarity Trial (Tang et al. 2021; World Health Organization 2024b). Numerous meta-analyses were undertaken to compile information from various studies. These analyses frequently failed to reveal consistent evidence endorsing the effectiveness of hydroxychloroquine. Some meta-analyses indicated potential safety issues, particularly concerning cardiac-related side effects (Chandler et al. 2020; Elavarasi et al. 2020; Lei et al. 2020; Lucchetta et al. 2023; Shamshirian et al. 2020; Ullah et al. 2020).

Hydroxychloroquine may inhibit viral replication by affecting the endocytic pathway of the virus. It is thought to interfere with the ability of the virus to enter host cells by increasing the pH of endosomes and lysosomes, where the virus particles are taken up. This alteration in pH can impair the virus's ability to release its genetic material into the host cell (Chen et al. 2020b; Lei et al. 2020). Besides this, hydroxychloroquine has immunomodulatory properties and can affect the immune response. It may reduce the production of pro-inflammatory cytokines, such as interleukin-6 (IL-6), and suppress certain immune reactions. In COVID-19, where an overactive immune response can lead to severe symptoms, this immunomodulatory effect was considered beneficial (Chandler et al. 2020).

However, subsequent clinical trials and real-world evidence did not consistently support the use of chloroquine or hydroxychloroquine in the treatment of COVID-19. Some studies found no significant benefit in terms of reducing mortality or improving clinical outcomes in COVID-19 patients treated with these medications. Additionally, concerns were raised about their safety, particularly regarding the risk of cardiac side effects (Singh et al. 2021).

6.2.1.2 Artemisinin and Its Derivatives

Artemisinin and its derivatives, such as artemether and artesunate, are compounds commonly used in the treatment of malaria. They have also been investigated for their potential antiviral properties against various viruses, including coronaviruses like SARS-CoV and MERS-CoV.

Laboratory studies suggest that artemisinin and its derivatives may possess direct antiviral effects against coronaviruses, including SARS-CoV-2, potentially by targeting specific stages of the viral life cycle and inhibiting key viral proteins crucial for replication or entry into host cells. Additionally, these compounds have demonstrated the ability to modulate the host immune response, regulating cytokine production and inflammation. Given the role of excessive inflammation in COVID-19 severity, these properties may help mitigate immune dysregulation. Furthermore, artemisinin derivatives exhibit various other effects, such as anti-inflammatory, antioxidant, and immunomodulatory actions, which could further contribute to their potential efficacy against COVID-19 by reducing inflammation and oxidative stress associated with the disease (Badraoui et al. 2022; Fuzimoto 2021; Götte 2021; Uckun et al. 2021; Zhou et al. 2021). Figure 6.1 shows artemisinin's properties and mechanism of action in COVID-19.

6.2.2 Antiviral Drugs

The existing antiviral drugs have also been repurposed for the treatment of COVID-19. The detail is explained below.

6.2.2.1 Remdesivir

Originally developed for Ebola, remdesivir is an antiviral drug that has received Emergency Use Authorization (EUA) from the FDA for the treatment of COVID-19.

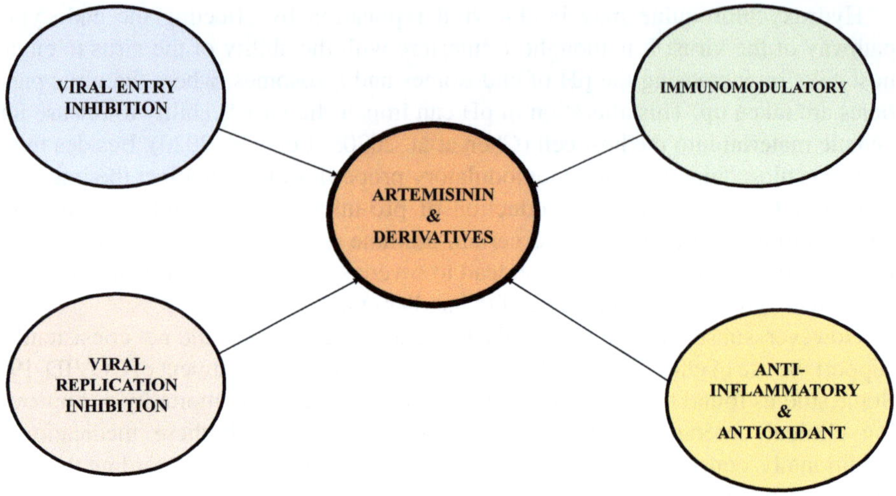

Fig. 6.1 Artemisinin's properties and mechanism of action in COVID-19

It has shown some effectiveness in reducing the recovery time for hospitalized patients (Eastman et al. 2020). Remdesivir is used for the treatment of COVID-19 because it has shown promise as an antiviral medication that can help reduce the severity and duration of symptoms in some patients. Its mechanism of action against COVID-19 involves inhibiting viral replication, and it is considered one of the potential treatment options for COVID-19. However, its effectiveness varies depending on the stage of the disease and other factors (Götte 2021).

Beigel JH et al. conducted the Adaptive COVID-19 Treatment Trial (ACTT-1), funded by the National Institute of Allergy and Infectious Diseases (NIAID), which stands as a crucial investigation into remdesivir. The trial revealed that, in comparison to a placebo, remdesivir reduced the time to recovery for hospitalized adults with COVID-19. According to the findings, there was a moderate decrease in the time to recovery, shifting from a median of 15 days in the placebo group to 10 days in the remdesivir group (Beigel et al. 2020b).

Various meta-analyses endeavored to amalgamate information from numerous studies. Outcomes diverged, with certain analyses suggesting a possible advantage in terms of clinical enhancement, while others concluded there was no noteworthy effect on mortality. The challenges in interpreting these findings were underscored by differences in study designs, patient demographics, and outcome measures across the analyzed studies. Todd C. Lee et al. conducted a comprehensive review and meta-analysis on remdesivir, assessing its effectiveness in hospitalized inpatients based on their oxygen requirements. Employing both frequentist random-effects meta-analysis on the risk ratio scale and Bayesian random-effects meta-analysis on the risk difference scale to provide a contextualized understanding of probabilistic benefits, the study considered a $\geq 1\%$ absolute risk reduction as clinically significant. The results of this meta-analysis strongly suggest that remdesivir significantly reduces mortality among COVID-19 patients who require supplemental oxygen but are not on ventilators. The findings recommend a reassessment of treatment guidelines in light of these outcomes (Lee et al. 2022). Similarly, Alain Amstutz et al. conducted a systematic review and individual patient data meta-analysis of randomized controlled trials to assess the impact of remdesivir on patients hospitalized with COVID-19. The meta-analysis of individual patient data indicated that remdesivir was effective in reducing mortality among hospitalized COVID-19 patients who required no or conventional oxygen support. However, the analysis lacked sufficient power to evaluate the impact of remdesivir on patients who were already on ventilators. Further research is needed to determine the effectiveness of remdesivir in patients requiring more respiratory support or those with acquired immunity. Additionally, the cost-effectiveness of remdesivir warrants further investigation (Amstutz et al. 2023).

Remdesivir is a nucleotide analog prodrug, which means it can be incorporated into the viral RNA during replication. Its active form interferes with the virus's genetic material, preventing it from replicating properly. Specifically, remdesivir inhibits the action of the viral RNA-dependent RNA polymerase (RdRp), an essential enzyme for the replication of the virus. By disrupting the viral replication process, remdesivir can slow down the spread of the virus within the body (Götte 2021).

The effectiveness of remdesivir in the treatment of COVID-19 has been a subject of clinical trials and research. It has been authorized for emergency use in some countries, including the United States, based on its potential to reduce the time to recovery in hospitalized patients. However, its efficacy in reducing mortality or preventing severe illness varies (Eastman et al. 2020). Remdesivir appears to be most beneficial when administered in the early stages of the disease or for patients with moderate symptoms. Its effectiveness in severely ill patients is less clear. Multiple clinical trials have established its effectiveness in reducing hospitalization time, mortality rates, and symptom alleviation. Regulatory approvals, such as FDA's Emergency Use Authorization, underscore its significance in the fight against the pandemic.

6.2.2.2 Ritonavir-Boosted Nirmatrelvir (Paxlovid)

Ritonavir-boosted nirmatrelvir, marketed under the brand name Paxlovid, is an antiviral medication developed by Pfizer for the treatment of COVID-19. It consists of two components: nirmatrelvir, a novel protease inhibitor, and ritonavir, a pharmacokinetic enhancer that increases the plasma half-life of nirmatrelvir. Nirmatrelvir inhibits the activity of the SARS-CoV-2 main protease, also known as the 3CL protease. This protease plays a critical role in the replication of the virus by cleaving viral polyproteins into functional proteins necessary for viral replication (Hashemian et al. 2023).

Ritonavir inhibits cytochrome P450 3A4 (CYP3A4), an enzyme involved in the metabolism of nirmatrelvir. By inhibiting CYP3A4, ritonavir increases the plasma concentration and half-life of nirmatrelvir, enhancing its antiviral efficacy. Paxlovid received Emergency Use Authorization (EUA) from regulatory authorities, including the U.S. Food and Drug Administration (FDA), for the treatment of COVID-19 in high-risk individuals (U.S. Food and Drug Administration 2024).

The medication is administered orally as a combination of nirmatrelvir (300 mg) and ritonavir (100 mg) tablets, taken together twice daily for 5 days. It is currently available for use in eligible patients under EUA provisions and is being distributed by healthcare providers and pharmacies. Clinical trials evaluating Paxlovid have shown promising results in reducing the risk of hospitalization and death in nonhospitalized adults with mild-to-moderate COVID-19 who are at high risk of progressing to severe disease (National institute of health (NIH) 2024).

In the EPIC-HR (Evaluation of Protease Inhibition for COVID-19 in High-Risk Patients) trial, Paxlovid demonstrated an 89% reduction in the risk of hospitalization or death compared to a placebo in high-risk individuals with COVID-19. The study also showed a significant reduction in viral load and duration of symptoms in patients treated with Paxlovid compared to placebo (Hammond et al. 2022).

6.2.2.3 Ritonavir

Ritonavir is a protease inhibitor that was initially developed for the treatment of HIV/AIDS. It inhibits the activity of the HIV protease enzyme, preventing the cleavage of viral polyproteins and the production of infectious viral particles (Talha and Dhamoon 2024). It has been investigated for its potential antiviral activity against SARS-CoV-2, the virus responsible for COVID-19. It is often used in combination with other antiviral medications, such as lopinavir, in an attempt to enhance their efficacy. However, clinical trials evaluating ritonavir and lopinavir as treatments for COVID-19 have shown mixed results. Some studies have demonstrated limited efficacy in reducing the viral load or improving clinical outcomes, while others have shown no significant benefit compared to standard care or placebo (Cao et al. 2020b; Li et al. 2020; RECOVERY Collaborative Group 2020b; U.S. Food and Drug Administration 2024).

Ritonavir has also been studied as a pharmacokinetic enhancer for other antiviral medications, such as nirmatrelvir in the combination medication Paxlovid, where it increases the plasma concentration and half-life of the active antiviral agent. Ritonavir inhibits the activity of the HIV protease enzyme by binding to its active site, preventing the cleavage of viral polyproteins and the formation of mature infectious viral particles. This inhibition disrupts the replication cycle of HIV. While ritonavir was initially developed for HIV/AIDS treatment, it has been investigated for its potential antiviral activity against SARS-CoV-2, the virus responsible for COVID-19. Ritonavir has been shown to inhibit the replication of SARS-CoV-2 in laboratory studies, leading to interest in its potential therapeutic use for COVID-19 (Hung et al. 2022).

Clinical trials evaluating ritonavir, often in combination with the protease inhibitor lopinavir (sold as Kaletra), for the treatment of COVID-19 have yielded mixed results (RECOVERY Collaborative Group 2020b). Some studies have reported modest reductions in viral load or improvements in clinical outcomes, particularly when initiated early in the course of illness. However, other trials have shown no significant benefit of ritonavir/lopinavir compared to standard care or placebo. In addition to its direct antiviral activity, ritonavir is also used as a pharmacokinetic enhancer for other medications. Ritonavir inhibits cytochrome P450 3A4 (CYP3A4), an enzyme involved in drug metabolism, thereby increasing the plasma concentration and half-life of co-administered medications. This pharmacokinetic boosting effect is utilized in combination therapies, such as Paxlovid (ritonavir-boosted nirmatrelvir), for COVID-19 treatment (Loos et al. 2023).

6.2.2.4 Favipiravir

Favipiravir is an antiviral medication that inhibits the replication of RNA viruses by targeting the viral RNA-dependent RNA polymerase enzyme. It is approved for the treatment of influenza in several countries and has shown broad-spectrum activity against other RNA viruses. It has been investigated as a potential treatment option due to its mechanism of action and antiviral activity against SARS-CoV-2 in laboratory studies. Favipiravir is typically administered orally and is most effective when initiated early in the course of illness, ideally within the first few days of symptom onset. It is generally well-tolerated, with common adverse effects including gastrointestinal symptoms and reversible elevation of liver enzymes. It is a broad-spectrum antiviral medication that inhibits the replication of RNA viruses by selectively targeting the viral RNA-dependent RNA polymerase enzyme. By blocking viral RNA synthesis, favipiravir disrupts the replication cycle of RNA viruses, including influenza and other respiratory viruses (Batool et al. 2023; Bosaeed et al. 2022a; Furuta et al. 2017).

Clinical trials evaluating favipiravir for the treatment of COVID-19 have yielded mixed results. Some studies have reported a reduction in the time to viral clearance and improvement in clinical outcomes, while others have shown no significant benefit compared to standard care or placebo (Bosaeed et al. 2022b; Chuah et al. 2022; McMahon et al. 2022; Shinkai et al. 2021; Udwadia et al. 2021). Favipiravir has been investigated for its potential antiviral activity against SARS-CoV-2. Laboratory studies have shown that favipiravir can inhibit the replication of SARS-CoV-2 in cell culture, suggesting its potential utility as a treatment for COVID-19 (Chiba et al. 2021).

6.2.2.5 Molnupiravir

Molnupiravir (also known as MK-4482 or EIDD-2801) is an antiviral medication that has gained attention for its potential use in the treatment of COVID-19. It is a prodrug that is metabolized into its active form, NHC (β-D-N4-hydroxycytidine), within the body (Tian et al. 2022). NHC is a nucleoside analogue that inhibits viral RNA synthesis by introducing errors into the viral RNA during replication. This leads to the accumulation of mutations in the viral genome, ultimately resulting in the inhibition of viral replication and spread.

Laboratory studies have shown that molnupiravir has potent antiviral activity against SARS-CoV-2, the virus responsible for COVID-19. It has demonstrated efficacy in inhibiting viral replication and reducing viral load in cell culture and animal models (Kabinger et al. 2021b; Painter et al. 2021; Singla and Goyal 2022; Tian et al. 2022).

Clinical trials evaluating molnupiravir for the treatment of COVID-19 have shown promising results. In a phase 2/3 trial involving nonhospitalized patients with mild-to-moderate COVID-19, molnupiravir significantly reduced the risk of hospitalization or death compared to placebo.

Early treatment with molnupiravir (within 5 days of symptom onset) was associated with a greater reduction in the risk of hospitalization or death. The medication was generally well-tolerated, with no safety concerns identified in the trial (Chang et al. 2023; Sinha et al. 2022). It has received Emergency Use Authorization (EUA) from regulatory authorities, including the U.S. Food and Drug Administration (FDA), for the treatment of COVID-19 in certain high-risk patients. The EUA allows for the emergency use of molnupiravir in nonhospitalized patients aged 18 years and older with mild-to-moderate COVID-19 who are at a high risk of progression to severe disease, including those with underlying health conditions or immunocompromised status (National Institute of Health (NIH) 2024b).

6.2.3 Antiparasitic Drugs

Antiparasitic drugs have also been tested for the treatment of COVID-19. The details are explained below.

6.2.3.1 Ivermectin

Ivermectin, a medication primarily used to combat parasitic infections, had garnered attention and been the subject of research due to its possible antiviral effects against SARS-CoV-2. Nevertheless, its efficacy remained a topic of ongoing discussion and required further investigation (Popp et al. 2022). Certain early in vitro studies suggested that ivermectin exhibited antiviral activity against SARS-CoV-2. These initial findings sparked interest in investigating ivermectin as a potential treatment for COVID-19. In order to assess the antiviral effects of ivermectin against SARS-CoV-2, Vero/hSLAM cells were infected with the virus at a 0.1 MOI (multiplicity of infection) for 2 h, followed by the addition of 5 μM ivermectin. Analysis of supernatant and cell pellets collected from days 0 to 3 through RT-PCR revealed the replication status of SARS-CoV-2 RNA. Additionally, to further evaluate the efficacy of ivermectin, cells infected with SARS-CoV-2 were treated with various dilutions of ivermectin 2 h post-infection, and the supernatant and cell pellets were collected for real-time RT-PCR at 48 h. Collectively, these results indicate that ivermectin possesses antiviral activity against the SARS-CoV-2 clinical isolate in vitro, with a single dose capable of controlling viral replication within 24–48 h in this experimental system. The hypothesis is that this effect may be attributed to inhibiting IMPα/β1-mediated nuclear import of viral proteins, as observed in other RNA viruses. However, further confirmation of this mechanism specific to SARS-CoV-2 and identification of the impacted SARS-CoV-2 and/or host component(s) are necessary (Caly et al. 2020).

Inhibition of SARS-CoV-2 replication has also been demonstrated in vitro. A 72-patient trial in Dhaka, Bangladesh, compared three groups: oral ivermectin alone

(12 mg daily for 5 days), ivermectin with doxycycline, and a placebo. Virological clearance was quicker in the 5-day ivermectin group (9.7 days) than the placebo (12.7 days; $p = 0.02$). The ivermectin + doxycycline group showed no significant difference (11.5 days; $p = 0.27$). No severe adverse events were reported. A 5-day ivermectin course was deemed safe and effective for mild COVID-19, but larger trials are needed for confirmation (Ahmed et al. 2021).

The exact mechanism of action of ivermectin in COVID-19 is not yet fully understood and is a subject of ongoing research. However, there are several proposed mechanisms by which ivermectin might potentially have antiviral effects against SARS-CoV-2. Some studies suggest that ivermectin may inhibit the replication of SARS-CoV-2 in infected cells. It may interfere with the virus's ability to replicate its genetic material, thus reducing viral load. They might modulate the host immune response, reducing the exaggerated inflammatory response observed in severe COVID-19 cases (Castillejos-López et al. 2022). By dampening the immune overreaction, it could potentially help prevent damage to lung tissue. Ivermectin has anti-inflammatory properties, which could help mitigate the cytokine storm seen in severe COVID-19 cases. By reducing the release of pro-inflammatory cytokines, it may contribute to less severe lung inflammation. But, it is important to note that while there is ongoing research and some promising data, ivermectin's use for COVID-19 remains a topic of debate, and it has not been universally accepted as a standard treatment.

6.2.3.2 Niclosamide

Niclosamide is an anthelmintic drug that has garnered some interest in the context of COVID-19 due to its potential antiviral properties. Some studies have suggested that niclosamide could inhibit SARS-CoV-2, the virus that causes COVID-19, by interfering with viral replication (Singh et al. 2022).

Niclosamide exerts its potential antiviral effects against COVID-19 through a multifaceted mechanism. First, it disrupts endosomal acidification, crucial for viral entry, inhibiting SARS-CoV-2 from fusing its membrane with the host cell and releasing its genetic material (Pindiprolu and Pindiprolu 2020). Second, niclosamide interferes with the processing of viral proteins, particularly viral proteases, essential for replication, thus impeding the production of new viral particles. Additionally, it modulates host cell signaling pathways, such as inhibiting the Wnt/β-catenin pathway, critical for viral replication and immune evasion, ultimately hampering SARS-CoV-2's ability to replicate and spread within the host (Xu et al. 2020).

6.2.3.3 Nitazoxanide

Nitazoxanide is a broad-spectrum antiparasitic and antiviral medication that has been investigated for its potential in treating various viral infections, including COVID-19. Its antiviral activity stems from its ability to inhibit the replication of several RNA and DNA viruses, including coronaviruses like SARS-CoV-2 (Mahmoud et al. 2020).

Nitazoxanide's mechanism of action against viruses involves interfering with various stages of the viral replication cycle. It inhibits viral replication by blocking the maturation of viral particles and interfering with viral protein synthesis. Additionally, nitazoxanide has been shown to modulate host cell immune responses, potentially enhancing the body's ability to combat viral infections (Xu et al. 2020). Several studies and clinical trials have evaluated the efficacy of nitazoxanide in treating COVID-19. While some preliminary data have shown promising results, larger clinical trials are needed to establish its safety and effectiveness fully (Abuelazm et al. 2022; Martins-Filho et al. 2022).

6.2.4 Corticosteroids

Corticosteroids are widely used in the management of COVID-19 patients to reduce the severe inflammation induced by SARS-Cov-2 viral infection (El-Saber Batiha et al. 2022). The details are mentioned below.

6.2.4.1 Dexamethasone

Dexamethasone is a corticosteroid medication that has been used in the treatment of severe cases of COVID-19. Its use is based on clinical trials and studies that have shown its effectiveness in specific situations (Ahmed and Hassan 2020).

The RECOVERY Collaborative Group conducted a clinical trial on dexamethasone in hospitalized patients with COVID-19. In this controlled and open-label trial, which compared various treatment options for patients hospitalized with COVID-19, participants were randomly assigned to receive oral or intravenous dexamethasone (6 mg once daily) for a maximum of 10 days or to receive standard care alone. The primary outcome assessed was the 28-day mortality rate. The findings revealed that, among patients hospitalized with COVID-19, the administration of dexamethasone led to reduced 28-day mortality, particularly for those who were receiving either invasive mechanical ventilation or oxygen alone at the time of randomization. However, there was no significant impact on 28-day mortality for patients not requiring respiratory support. This study was funded by the Medical Research Council, National Institute for Health Research, and others, with the trial registered under ClinicalTrials.gov number NCT04381936 and ISRCTN number 50189673 (Fuzimoto 2021). There is a consistent body of evidence indicating that the advantages of corticosteroids, particularly dexamethasone, are more prominent in severe cases of COVID-19. Corticosteroids have demonstrated effectiveness in regulating the inflammatory response, mitigating cytokine storms, and enhancing outcomes for individuals experiencing acute respiratory distress syndrome (ARDS) (RECOVERY Collaborative Group et al. 2021).

A widely available corticosteroid, dexamethasone, has been shown to be effective in reducing mortality in severe cases of COVID-19, particularly in patients requiring mechanical ventilation or supplemental oxygen. Dexamethasone also acts as an immunosuppressive agent and suppresses the immune system's response (Sharun et al. 2020). This can help reduce the inflammation and damage to the lungs and other organs caused by the immune system's reaction to the virus. By reducing inflammation in the lungs, dexamethasone can help alleviate breathing difficulties and lung damage, which are common in severe COVID-19 cases. In severe cases of COVID-19, the body's immune response can become overly active and lead to an inflammatory condition known as a cytokine storm. A cytokine storm is an excessive and uncontrolled immune response involving the release of a large number of pro-inflammatory cytokines. In the context of COVID-19, a cytokine storm is a severe immune reaction that can occur in some patients. It is often associated with the worsening of symptoms and can lead to acute respiratory distress syndrome (ARDS) and other life-threatening complications. Dexamethasone helps reduce this excessive inflammation, which can contribute to the severity of the disease (Villar et al. 2020).

Dexamethasone has also been found to improve oxygenation in patients with severe respiratory distress related to COVID-19. It helps make it easier for patients to breathe by reducing lung inflammation (RECOVERY Collaborative Group 2020a). It is typically reserved for severe COVID-19 cases where the patient requires supplemental oxygen or mechanical ventilation. It is not recommended for mild cases, as suppressing the immune response may not be beneficial in the early stages of infection.

6.2.4.2 Hydrocortisone

Hydrocortisone is a corticosteroid medication commonly used for its anti-inflammatory and immunosuppressive properties (Coutinho and Chapman 2011). In the context of COVID-19 treatment, hydrocortisone has been utilized to address the excessive inflammation and immune response associated with severe cases of the disease.

The mechanism of action of hydrocortisone in COVID-19 treatment is similar to other corticosteroids such as prednisolone and methylprednisolone. It works by suppressing the inflammatory response and modulating the immune system, thereby reducing inflammation and tissue damage caused by the virus. Specifically, hydrocortisone inhibits the production of pro-inflammatory cytokines and chemokines, which are elevated in severe cases of COVID-19 and contribute to the development of a cytokine storm. By dampening this exaggerated immune response, hydrocortisone helps prevent the immune system from causing excessive damage to the body's tissues, particularly in the lungs (World Health Organization 2020).

Several studies and clinical trials have investigated the efficacy and safety of hydrocortisone in managing the cytokine storm and reducing inflammation associated with severe COVID-19. A study published in JAMA Internal Medicine evaluated the use of corticosteroids, including hydrocortisone, in critically ill COVID-19 patients. The study found that corticosteroid treatment was associated with lower 28-day mortality among critically ill patients with COVID-19, further supporting the use of hydrocortisone in managing severe cases of the disease (Dequin et al. 2020b; Taher et al. 2023).

6.2.4.3 Prednisolone

Prednisolone is a corticosteroid used in the treatment of various inflammatory conditions, including severe cases of COVID-19. In COVID-19 treatment, prednisolone is utilized to mitigate the inflammatory response and reduce lung inflammation, particularly in patients with severe disease requiring hospitalization and supplemental oxygen (Ghanei et al. 2021).

The mechanism of action of prednisolone in COVID-19 treatment revolves around its anti-inflammatory and immunosuppressive properties. Prednisolone suppresses the inflammatory response by inhibiting the production of pro-inflammatory cytokines and chemokines. In COVID-19, where an excessive immune response can lead to a cytokine storm, prednisolone helps dampen this response, reducing inflammation and tissue damage. Furthermore, prednisolone modulates the activity of various immune cells, including T cells and macrophages, which play critical roles in the immune response to viral infections. By suppressing the activity of these immune cells, prednisolone helps prevent the immune system from causing excessive damage to the body's tissues in response to the virus. In addition to its anti-inflammatory effects, prednisolone reduces lung inflammation, allowing for better gas exchange in the lungs and potentially improving oxygenation in severe cases of COVID-19. However, the use of prednisolone in COVID-19 treatment should be carefully considered and monitored, as its benefits must be weighed against potential side effects, such as immunosuppression, hyperglycemia, and increased risk of secondary infections (Mehta et al. 2022; National Institute of Health (NIH) 2024c).

6.2.4.4 Methylprednisolone

Methylprednisolone, another corticosteroid, is also utilized in the treatment of severe cases of COVID-19 to reduce inflammation and improve patient outcomes. Similar to prednisolone, methylprednisolone acts through its anti-inflammatory and immunosuppressive properties to mitigate the inflammatory response associated with severe COVID-19.

The mechanism of action of methylprednisolone in COVID-19 treatment involves suppression of the inflammatory response by inhibiting the production of pro-inflammatory cytokines and chemokines. By dampening the cytokine storm observed in severe cases of COVID-19, methylprednisolone helps reduce inflammation and tissue damage, thereby improving patient outcomes. Additionally, methylprednisolone modulates the activity of various immune cells involved in the immune response to viral infections, such as T cells and macrophages. By suppressing the activity of these immune cells, methylprednisolone prevents excessive damage to the body's tissues caused by the immune response to the virus. Furthermore, methylprednisolone reduces lung inflammation, allowing for better gas exchange in the lungs and improved oxygenation in severe cases of COVID-19. However, as with prednisolone, the use of methylprednisolone in COVID-19 treatment should be carefully considered and monitored due to potential side effects, such as immunosuppression, hyperglycemia, and increased risk of secondary infections (Jeronimo et al. 2021; Taher et al. 2023). Figure 6.2 shows the mechanism of action of corticosteroids (dexamethasone, hydrocortisone, prednisolone, and methylprednisolone) in COVID-19.

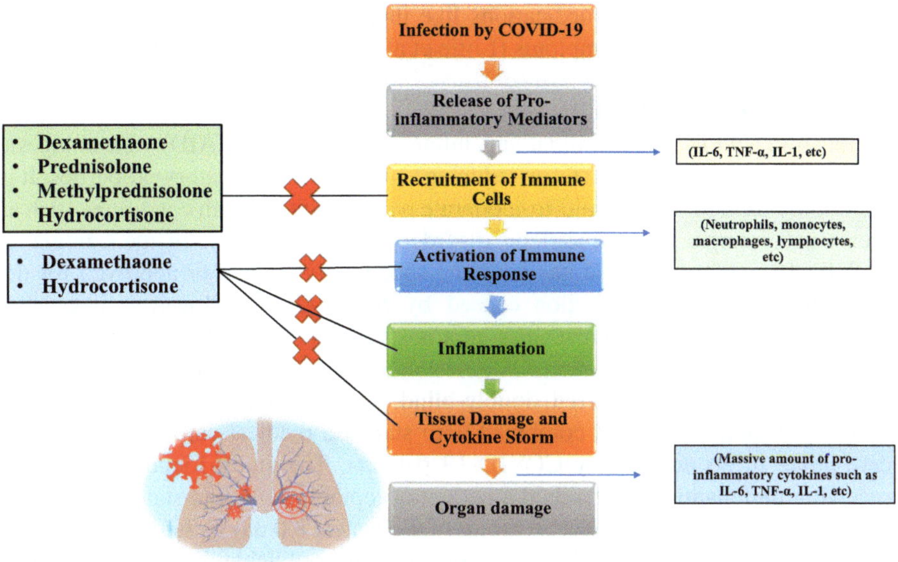

Fig. 6.2 Mechanism of action of corticosteroids (dexamethasone, hydrocortisone, prednisolone, and methylprednisolone) in COVID-19

6.2.5 Anti-Inflammatory Agents

Anti-inflammatory agents are widely used in the management of mild COVID-19 patients. The details are mentioned below.

6.2.5.1 Ibuprofen

Ibuprofen is a widely used nonsteroidal anti-inflammatory drug (NSAIDs) known for its analgesic (pain-relieving), antipyretic (fever-reducing), and anti-inflammatory properties. It works by inhibiting the production of prostaglandins, which are chemicals in the body that promote inflammation, pain, and fever. The use of ibuprofen in the context of COVID-19 primarily revolves around its anti-inflammatory and antipyretic (fever-reducing) properties rather than direct antiviral activity against the SARS-CoV-2 virus. COVID-19, caused by the SARS-CoV-2 virus, can lead to symptoms such as fever, muscle aches, headache, and general discomfort, which are often managed with medications like ibuprofen (Rinott et al. 2020a).

The mechanism of action of ibuprofen involves inhibiting the enzyme cyclooxygenase (COX), which is responsible for the production of prostaglandins, lipid compounds involved in inflammation, fever, and pain. By inhibiting COX, ibuprofen reduces the production of prostaglandins, leading to decreased inflammation, pain relief, and reduction of fever. In COVID-19 patients, ibuprofen may be used to alleviate symptoms such as fever and muscle aches. Fever, in particular, is a common symptom of viral infections, including COVID-19. While fever is part of the body's immune response to fight off infections, it can also cause discomfort and may be managed with antipyretic medications like ibuprofen to improve patient comfort and reduce fever-related symptoms (Little et al. 2020; Rinott et al. 2020b).

6.2.5.2 Mefenamic Acid

Mefenamic acid is a nonsteroidal anti-inflammatory drug (NSAID) with analgesic (pain-relieving), antipyretic (fever-reducing), and anti-inflammatory properties. It is commonly used to relieve mild-to-moderate pain associated with conditions such as menstrual cramps, musculoskeletal disorders, and headaches. In the context of COVID-19, mefenamic acid has not been specifically studied or approved for the treatment of the viral infection caused by SARS-CoV-2. However, like other NSAIDs, it may be used to manage symptoms such as fever and body aches in individuals with mild-to-moderate COVID-19 symptoms. Ivan Delgado-Enciso et al. (2022) conducted a randomized, double-blind, placebo-controlled trial to assess the efficacy of mefenamic acid combined with standard medical care versus standard care alone in 36 ambulatory COVID-19 patients. The study found that patients receiving mefenamic acid reached a patient-acceptable symptom state (PASS) significantly earlier than the placebo group (4.4 days vs. 8.0 days). Additionally, mefenamic acid-treated patients had a higher probability of achieving PASS by day 8. Symptoms such as headache, retro-orbital eye pain, and sore throat also resolved faster in the mefenamic acid group. No severe adverse effects were reported with the experimental treatment. These results suggest that mefenamic acid could be beneficial in reducing symptomatology and expediting recovery in COVID-19 patients, possibly due to its anti-inflammatory and potential antiviral properties. Further studies are needed to validate these findings and explore potential combinations with other antiviral drugs like remdesivir, molnupiravir, or favipiravir (Guzman-Esquivel et al. 2022).

6.2.5.3 Indomethacin

In the ongoing quest for effective treatments against COVID-19, researchers have explored the potential of repurposing existing drugs. Among them, indomethacin, a nonsteroidal anti-inflammatory drug (NSAID), has garnered attention due to its known mechanism of action in alleviating pain, swelling, and inflammation by inhibiting prostaglandin production. Although indomethacin has not been specifically studied or approved for COVID-19 caused by the SARS-CoV-2 virus, it may offer symptom management benefits for individuals with mild-to-moderate cases. One pivotal study by Carla Amici et al. (2006) delved into the antiviral properties of indomethacin against severe acute respiratory syndrome coronavirus (SARS-CoV) and canine coronavirus (CCoV). Their findings revealed that indomethacin exhibited potent antiviral effects against both coronaviruses, acting by blocking viral RNA synthesis independently of its COX inhibitory activity. This significant discovery, demonstrated both in in vitro cell cultures and in vivo in CCoV-infected dogs, suggests a potential role for indomethacin in SARS therapy, thanks to its dual anti-inflammatory and antiviral properties (Amici et al. 2006). Building upon this foundation, Rajan Ravichandran et al. (2022) conducted an open-label randomized clinical trial to assess the efficacy and safety of indomethacin in hospitalized

COVID-19 patients. The study enrolled 210 RT-PCR-positive individuals, randomly assigning them to control or case arms. Notably, patients in the case arm received indomethacin instead of paracetamol alongside standard care. The primary endpoint focused on the development of hypoxia/desaturation (SpO2 ≤ 93), while secondary endpoints included time to become afebrile, cough, and myalgia resolution. The results of this trial were promising. None of the patients in the indomethacin arm developed desaturation, in stark contrast to 20 individuals in the paracetamol arm. Moreover, those treated with indomethacin experienced faster symptomatic relief, with most symptoms resolving in half the time compared to the paracetamol arm. Additionally, fewer patients in the indomethacin group had fever on the seventh day. Importantly, no adverse events were reported in either group, affirming the safety profile of indomethacin for COVID-19 management (Ravichandran et al. 2022). In conclusion, although indomethacin has not yet received dedicated approval for treating COVID-19, its proven effectiveness and safety in alleviating symptoms and averting desaturation in mild-to-moderate cases show potential. More investigations and clinical studies are necessary to fully understand its capabilities and refine treatment strategies for individuals with COVID-19. Figure 6.3 shows the

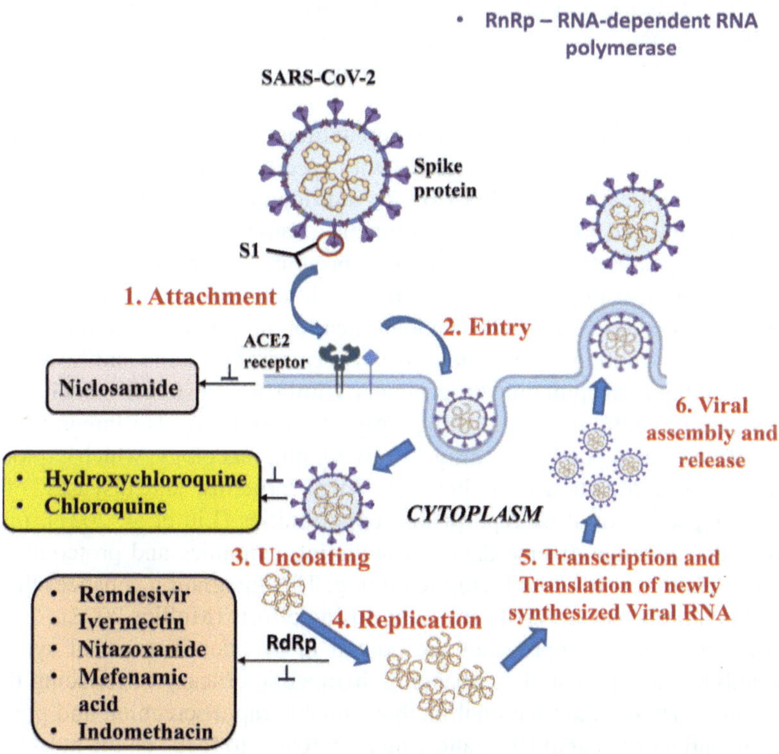

Fig. 6.3 Mechanism of action of niclosamide, hydroxychloroquine, chloroquine, remdesivir, ivermectin, nitazoxanide, mefenamic acid, and indomethacin against COVID-19

mechanism of action of niclosamide, hydroxychloroquine, chloroquine, remdesivir, ivermectin, nitazoxanide, mefenamic acid, and indomethacin against COVID-19.

6.3 Drug Repurposing Strategies and Techniques

New strategies and techniques for drug repurposing have emerged as critical components of the global response to the pandemic. These approaches leverage a combination of innovative methodologies, computational tools, and data analysis to identify potential drug candidates for repurposing in the context of COVID-19.

6.3.1 *In Silico* Drug Repurposing

Computational methods, including molecular docking, molecular dynamics simulations, and structural bioinformatics, are employed to screen existing drugs for their alternate indications. This approach allows for the rapid evaluation of a vast library of compounds. In the pursuit of finding effective treatments for COVID-19, researchers have turned to innovative approaches, one of which is *in silico* drug repurposing. This method involves sophisticated computational techniques, drawing on bioinformatics and artificial intelligence, to sift through existing drugs and identify potential candidates for repurposing against the novel coronavirus (Mullins 2022; Sencanski et al. 2020).

One key strategy in this approach is computational docking studies. This involves predicting how a drug molecule interacts with specific viral proteins, such as those crucial for the replication of SARS-CoV-2. Additionally, molecular dynamics simulations are employed to model the movement of atoms and molecules over time, providing insights into the stability of drug–protein complexes (Pinzi and Rastelli 2019). Machine learning, another powerful tool, analyzes vast datasets to predict drug–target interactions, aiding in the identification of potential repurposing candidates. By understanding the interactions between viral proteins and host cell proteins, researchers can pinpoint drug targets within the host cell machinery (Issa et al. 2021). Predicting potential drug–drug interactions is also integral to avoid adverse effects when combining drugs. Network pharmacology, which explores the broader network of interactions between drugs, proteins, and diseases, offers a holistic perspective on drug repurposing opportunities (Liu et al. 2021). *In silico* methods also leverage diverse datasets, including genomics and proteomics data, electronic health records, and literature mining. This integration enhances the accuracy and comprehensiveness of drug repurposing efforts (Mullins 2022).

Moreover, *in silico* approaches are crucial in the identification of synergistic drug combinations, potentially enhancing therapeutic efficacy or reducing the risk of resistance. These computational methods enable rapid screening and prioritization of potential drug candidates, allowing researchers to focus on the most promising options for experimental validation. Ongoing research efforts, data integration, and global collaboration contribute to the dynamic nature of *in silico* drug repurposing. However, it is essential to acknowledge the challenges associated with *in silico* drug repurposing, including the need for rigorous experimental validation, the possibility of false-positives, and the dynamic nature of the virus, necessitating constant adaptation of computational models. In summary, *in silico* drug repurposing has emerged as a valuable strategy, offering a complementary and accelerated approach to identifying potential therapeutics for COVID-19. The integration of computational methods with traditional drug discovery processes has proven instrumental in the urgent quest for effective treatments (Park 2019).

6.3.2 High-Throughput Screening

This experimental technique involves the testing of a large number of drugs or compounds to identify potential candidates for repurposing. High-throughput screening assays specifically designed to evaluate drug effectiveness against SARS-CoV-2 have been used to rapidly identify potential treatments. In the realm of drug discovery, the collaboration between high-throughput screening (HTS) and drug repurposing has emerged as a potent strategy, particularly in the context of addressing urgent health challenges like COVID-19 (Jonguitud-Borrego et al. 2022). HTS provides a rapid and systematic way to sift through a diverse array of chemical compounds, offering an efficient avenue to test their biological activities. This opens the door for the exploration of existing drugs, originally developed for different purposes, for potential benefits against new diseases. Beyond its utility with existing drugs, HTS has the capacity to uncover entirely new compounds that exhibit promising effects against specific biological targets. This synergizes with drug repurposing initiatives, broadening the scope to include both familiar drugs and entirely new entities (Dey and Kumar 2024). HTS can be tailored to focus on specific targets or pathways associated with a disease. In the realm of drug repurposing, this tailored approach allows for a more focused exploration of existing drugs that might interact with these identified targets, forming a logical basis for repurposing.

6.3.3 Network Pharmacological Techniques

Network pharmacology integrates data on drug targets, disease-related proteins, and biological pathways to identify potential drug candidates (Saima et al. 2024). It helps uncover the complex interactions between drugs, proteins, and diseases. In the urgent pursuit of effective treatments for COVID-19, network pharmacology emerges as a transformative approach in drug repurposing. Unlike traditional methods, it comprehensively maps the intricate networks of genes, proteins, and pathways involved in the disease, allowing for the identification of key targets. By exploring interactions between existing drugs and multiple targets within the COVID-19 network, network pharmacology unveils potential repurposing candidates and considers the pleiotropic effects of these drugs. It aids in predicting synergistic drug combinations and leverages big data for a data-driven prioritization of candidates. This holistic approach not only accelerates drug development but also remains adaptable to emerging data, offering a promising avenue for innovative therapeutic solutions in the battle against COVID-19 (Bhati et al. 2023; de Anda-Jáuregui et al. 2018).

6.4 Current Challenges

Drug repurposing, a strategy with the potential to rapidly identify new treatments, is not without its array of challenges. In the context of emerging diseases such as COVID-19, where the underlying disease mechanisms are not fully understood, finding suitable candidates for repurposing can be daunting. A profound challenge lies in the availability of comprehensive data, spanning clinical, biological, and chemical information. This data is essential to identify potential candidates, and challenges often emerge in terms of data sharing, integration, and quality. The biological complexity of many diseases further complicates repurposing efforts. Diseases often involve multiple intricate pathways and mechanisms, making it challenging to find existing drugs that can effectively target and modulate all relevant aspects of the disease. Another significant challenge arises from the off-label use of drugs, which is often a necessity in repurposing. This practice can raise regulatory and liability concerns, as well as complicate issues related to dosing and safety, given that the drug was originally intended for a different purpose (Sharma et al. 2021).

Moreover, patent and intellectual property issues can loom large. The existence of patents on drugs can raise complex legal challenges and may limit the availability of certain drugs for repurposing. Repurposed drugs may also exhibit unexpected side effects when used in new contexts, which can limit their suitability for repurposing. Obtaining the necessary funding and investment for drug development is often a challenge, especially when compared to the development of entirely new drugs. Commercial viability is another issue to contend with, as market dynamics, competition, and pricing can significantly affect the success of repurposed drugs. Furthermore, ensuring the long-term safety and efficacy of repurposed drugs, especially when the original drug was not initially intended for chronic use, is a critical concern. Scaling up production and maintaining a stable supply chain are logistical challenges, which can be especially pronounced during a public health crisis, as seen in the COVID-19 pandemic. Lastly, ethical and equity considerations arise when making decisions regarding drug repurposing, availability, and access. Decisions must take into account the equitable distribution of treatments and the ethical implications of drug allocation. Successfully addressing these multifaceted challenges necessitates a comprehensive, multidisciplinary approach. Collaborative efforts among researchers, clinicians, regulators, and the pharmaceutical industry are essential to harness the full potential of drug repurposing in the management of diseases like COVID-19. These challenges underscore the importance of perseverance and innovation in the field of drug repurposing (Hua et al. 2022; Kapoor et al. 2020).

6.5 Future Prospects

Currently, repurposing existing drugs represents a valuable strategy in the battle against COVID-19. Scientists have delved into a diverse range of established pharmaceuticals, with the goal of finding candidates that can be repurposed to effectively combat the virus (Sharma et al. 2021). Noteworthy instances encompass antiviral medications like remdesivir and immunomodulatory agents like dexamethasone. These treatments, initially designed for different purposes, have exhibited varying levels of effectiveness in managing COVID-19. This success has led to their emergency use authorization and integration into treatment regimens. This strategy holds the advantage of capitalizing on well-documented safety profiles, potentially expediting regulatory approvals, and enhancing the availability of treatment options. Nevertheless, challenges, such as limited efficacy and the emergence of drug-resistant viral strains, remain salient concerns (Hua et al. 2022; Kapoor et al. 2020; Low et al. 2020).

Contemplating the future of managing COVID-19 through drug repurposing reveals several promising and dynamic aspects. Scientists will persist in their efforts to discover repurposed drugs that are more effective and specific in combating COVID-19 while minimizing adverse effects. This endeavor entails revisiting existing pharmaceutical compounds to pinpoint those capable of precisely targeting the virus's mechanisms. The exploration of combining drugs is set to gain prominence, with the goal of simultaneously using multiple drugs to address different phases of the virus's life cycle. Combining drugs can enhance treatment efficacy, potentially reduce the risk of drug resistance, and offer complementary modes of action for a more comprehensive approach to controlling the disease.

The ever-evolving landscape of COVID-19 is marked by the continuous emergence of new viral variants. As these variants may affect the effectiveness of existing treatments, adapting and repurposing drugs to maintain their efficacy against evolving strains will be a primary concern (Hossain et al. 2020). This adaptive approach is critical for ensuring that our therapeutic toolbox remains potent in light of changing viral characteristics.

Moreover, cutting-edge technologies, such as artificial intelligence and machine learning, will play an increasingly prominent role in identifying potential drug candidates. These technologies can rapidly analyze extensive datasets, forecast interactions between drugs and the virus, and streamline the process of drug repurposing. This approach offers a promising avenue for discovering fresh applications of existing drugs. With the accumulation of clinical data and a deeper understanding of COVID-19, the use of repurposed drugs may transition to a more individualized and tailored approach (Singh et al. 2020). Medications may find specific applications tailored to distinct patient groups or particular disease stages. Customizing treatments to individual patient profiles may optimize their effectiveness while minimizing potential side effects. While COVID-19 vaccination campaigns are pivotal in managing the pandemic, the strategy of drug repurposing maintains its significance. It serves as a crucial treatment option for individuals unable to receive vaccines due to contraindications, allergies, or vaccine availability limitations. Furthermore, repurposed drugs can continue to fulfill a role in treating breakthrough cases and managing post-vaccination situations.

6.6 Conclusions

The repurposing of drugs has emerged as a critical strategy in the fight against COVID-19, offering a rapid and cost-effective approach to identify potential treatments. Through the exploration of existing drugs in various therapeutic classes such as antivirals, antimalarials, anti-inflammatories, corticosteroids, and antiparasitic agents, researchers have been able to leverage their known mechanisms of action to target key pathways involved in the pathogenesis of the virus. This approach has not only expedited the drug discovery process but has also offered a promising avenue for the development of effective treatments amidst the urgent global health crisis. Furthermore, the utilization of advanced techniques such as *in silico* approaches, molecular docking, network pharmacology, machine learning, and high-throughput screening has revolutionized the field of drug repurposing, enabling researchers to efficiently sift through vast amounts of data and identify potential candidates for further investigation. These computational methods have played a pivotal role in predicting drug–target interactions, elucidating molecular mechanisms, and prioritizing compounds for experimental validation, thereby accelerating the pace of drug discovery and development. Moving forward, continued research efforts in drug repurposing hold immense potential for addressing the evolving challenges posed by COVID-19, including emerging variants and treatment-resistant strains. By leveraging existing knowledge and innovative methodologies, researchers can continue to identify new therapeutic options, optimize treatment regimens, and ultimately improve outcomes for patients affected by this devastating disease. Collaborative initiatives between academia, industry, and regulatory agencies will be crucial in advancing the field of drug repurposing and ultimately mitigating the impact of COVID-19 on a global scale.

References

A Randomized, Double-blind, Placebo-controlled, Phase 2/3 Study to Evaluate the Efficacy and Safety of Nirmatrelvir/ritonavir (PF-07321332/ritonavir) in Non-hospitalized High-risk Adults With Symptomatic COVID-19 Infection (2022a). ClinicalTrials.gov Identifier: NCT05012340

A Randomized, Double-blind, Placebo-controlled, Phase 2/3 Study to Evaluate the Efficacy and Safety of Nirmatrelvir/ritonavir (PF-07321332/ritonavir) in Non-hospitalized High-risk Children, Adolescents, and Adults With Symptomatic COVID-19 Infection (2022b). ClinicalTrials.gov Identifier: NCT05047601

Abassi Z, Knaney Y, Karram T, Heyman SN, Abassi A (2020) COVID-19: the intersection of conspiracy theory, uncertainty, and fear. Am J Trop Med Hyg 103(3):971–976

Abuelazm M, Ghanem A, Awad AK, Farahat RA, Labieb F, Katamesh BE, Abdelazeem B (2022) The effect of Nitazoxanide on the clinical outcomes in patients with COVID-19: a systematic review and meta-analysis of randomized controlled trials. Clin Drug Investig 42(12):1031–1047. https://doi.org/10.1007/s40261-022-01213-y

Agrawal PK, Agrawal C, Blunden G (2022) *Artemisia* extracts and Artemisinin-based Antimalarials for COVID-19 management: could these be effective antivirals for COVID-19 treatment? Molecules (Basel, Switzerland) 27(12):3828. https://doi.org/10.3390/molecules27123828. (Retraction published Molecules. 2023;28(21):)

Ahmed MH, Hassan A (2020) Dexamethasone for the treatment of Coronavirus disease (COVID-19): a review. SN Compr Clin Med 2(12):2637–2646. https://doi.org/10.1007/s42399-020-00610-8

Ahmed S, Karim MM, Ross AG et al (2021) A five-day course of ivermectin for the treatment of COVID-19 may reduce the duration of illness. Int J Infect Dis 103:214–216. https://doi.org/10.1016/j.ijid.2020.11.191

Amici C, Di Caro A, Ciucci A et al (2006) Indomethacin has a potent antiviral activity against SARS coronavirus. Antivir Ther 11(8):1021–1030

Amstutz A, Speich B, Mentré F et al (2023) Effects of remdesivir in patients hospitalised with COVID-19: a systematic review and individual patient data meta-analysis of randomised controlled trials. [Correction in: Lancet Respir Med. 2023;11(8):e77]. Lancet Respir Med 11(5):453–464. https://doi.org/10.1016/S2213-2600(22)00528-8

Angus DC, Derde L, Al-Beidh F, Annane D, Arabi Y, Beane A et al (2020) Effect of hydrocortisone on mortality and organ support in patients with severe COVID-19: the REMAP-CAP COVID-19 Corticosteroid Domain randomized clinical trial. JAMA 324(13):1317–1329

Badraoui R, Saoudi M, Hamadou WS, Elkahoui S, Siddiqui AJ, Alam JM, Jamal A, Adnan M, Suliemen AME, Alreshidi MM, Yadav DK, Naïli H, Ben-Nasr H (2022) Antiviral effects of Artemisinin and its derivatives against SARS-CoV-2 Main protease: computational evidences and interactions with ACE2 Allelic variants. Pharmaceuticals (Basel, Switzerland) 15(2):129. https://doi.org/10.3390/ph15020129

Batool S, Vuthaluru K, Hassan A, Bseiso O, Tehseen Z, Pizzorno G, Rodriguez Reyes Y, Saleem F (2023) Efficacy and safety of Favipiravir in treating COVID-19 patients: a meta-analysis of randomized control trials. Cureus 15(1):e33676. https://doi.org/10.7759/cureus.33676

Beigel JH, Tomashek KM, Dodd LE, Mehta AK, Zingman BS, Kalil AC et al (2020a) Remdesivir for the treatment of Covid-19—final report. N Engl J Med 383(19):1813–1826

Beigel JH, Tomashek KM, Dodd LE et al (2020b) Remdesivir for the treatment of Covid-19 - final report. N Engl J Med 383(19):1813–1826. https://doi.org/10.1056/NEJMoa2007764

Bhati V, Kumar A, Lather V, Sharma R, Pandita D (2023) Association of temozolomide with progressive multifocal leukoencephalopathy: a disproportionality analysis integrated with network pharmacology. Expert Opin Drug Saf:1–10. https://doi.org/10.1080/14740338.2023.2278682

Bosaeed M, Alharbi A, Mahmoud E, Alrehily S, Bahlaq M, Gaifer Z, Alturkistani H, Alhagan K, Alshahrani S, Tolbah A, Musattat A, Alanazi M, Jaha R, Sultana K, Alqahtani H, Al Aamer K, Jaser S, Alsaedy A, Ahmad A, Abalkhail M et al (2022a) Efficacy of favipiravir in adults with mild COVID-19: a randomized, double-blind, multicentre, placebo-controlled clinical trial. Clin Microbiol Infect 28(4):602–608. https://doi.org/10.1016/j.cmi.2021.12.026

Bosaeed M, Alharbi A, Mahmoud E et al (2022b) Efficacy of favipiravir in adults with mild COVID-19: a randomized, double-blind, multicentre, placebo-controlled clinical trial. Clin Microbiol Infect 28:602–608

Cai Q, Yang M, Liu D, Chen J, Shu D, Xia J et al (2020) Experimental treatment with favipiravir for COVID-19: an open-label control study. Engineering 6(10):1192–1198

Caly L, Druce JD, Catton MG, Jans DA, Wagstaff KM (2020) The FDA-approved drug ivermectin inhibits the replication of SARS-CoV-2 in vitro. Antivir Res 178:104787. https://doi.org/10.1016/j.antiviral.2020.104787

Cao B, Wang Y, Wen D, Liu W, Wang J, Fan G et al (2020a) A trial of lopinavir–ritonavir in adults hospitalized with severe Covid-19. N Engl J Med 382(19):1787–1799

Cao B, Wang Y, Wen D, Liu W, Wang J, Fan G et al (2020b) A trial of lopinavir-ritonavir in adults hospitalized with severe Covid-19. N Engl J Med 382:1787–1799

Castillejos-López M, Torres-Espíndola LM, Huerta-Cruz JC et al (2022) Ivermectin: a controversial focal point during the COVID-19 pandemic. Life (Basel) 12(9):1384. https://doi.org/10.3390/life12091384

Cavalli G, De Luca G, Campochiaro C, Della-Torre E, Ripa M, Canetti D et al (2020) Interleukin-1 blockade with high-dose anakinra in patients with COVID-19, acute respiratory distress syndrome, and hyperinflammation: a retrospective cohort study. Lancet Rheumatol 2(6):e325–e331

Chandler LC, Yusuf IH, McClements ME et al (2020) Immunomodulatory effects of hydroxychloroquine and Chloroquine in viral infections and their potential application in retinal gene therapy. Int J Mol Sci 21(14):4972. https://doi.org/10.3390/ijms21144972

Chang YC, Chen YC, Huang CC, Fu CM, Lee YT, Wu PJ, Lee WC, Lee CT, Liao SC, Tsai KF (2023) Clinical effectiveness of molnupiravir in patients with COVID-19 undergoing haemodialysis. Int J Antimicrob Agents 62(1):106834. https://doi.org/10.1016/j.ijantimicag.2023.106834

Chen C, Zhang Y, Huang J, Yin P, Cheng Z, Wu J et al (2020a) Favipiravir versus Arbidol for COVID-19: a randomized clinical trial. MedRxiv

Chen L, Chen H, Dong S, Huang W, Chen L, Wei Y, Shi L, Li J, Zhu F, Zhu Z, Wang Y, Lv X, Yu X, Li H, Wei W, Zhang K, Zhu L, Qu C, Hong J, Hu C et al (2020b) The effects of Chloroquine and Hydroxychloroquine on ACE2-related Coronavirus pathology and the cardiovascular system: an evidence-based review. Function 1(2):zqaa012. https://doi.org/10.1093/function/zqaa012

Cheng VC, Lau SK, Woo PC, Yuen KY (2007) Severe acute respiratory syndrome coronavirus as an agent of emerging and reemerging infection. Clin Microbiol Rev 20(4):660–694. https://doi.org/10.1128/CMR.00023-07

Chiba S, Kiso M, Nakajima N, Iida S, Maemura T, Kuroda M, Sato Y, Ito M, Okuda M, Yamada S, Iwatsuki-Horimoto K, Watanabe T, Imai M, Armbrust T, Baric RS, Halfmann PJ, Suzuki T, Kawaoka Y (2021) Co-administration of Favipiravir and the Remdesivir Metabolite GS-441524 effectively reduces SARS-CoV-2 replication in the lungs of the Syrian Hamster Model. MBio 13(1):e0304421. https://doi.org/10.1128/mbio.03044-21

Chuah CH, Chow TS, Hor CP et al (2022) Efficacy of early treatment with favipiravir on disease progression among high-risk patients with coronavirus disease 2019 (COVID-19): a randomized, open-label clinical trial. Clin Infect Dis 75:0–9

Corral L, Bahamonde A, Delasrevillas FA, Hernández R, García I, de la Fuente I et al (2020) GLUCOCOVID: a controlled trial of methylprednisolone in adults hospitalized with COVID-19 pneumonia. MedRxiv

Coutinho AE, Chapman KE (2011) The anti-inflammatory and immunosuppressive effects of glucocorticoids, recent developments and mechanistic insights. Mol Cell Endocrinol 335(1):2–13. https://doi.org/10.1016/j.mce.2010.04.005

de Anda-Jáuregui G, Guo K, McGregor BA, Hur J (2018) Exploration of the anti-inflammatory drug space through network pharmacology: applications for drug repurposing. Front Physiol 9:151. https://doi.org/10.3389/fphys.2018.00151

Dequin PF, Heming N, Meziani F, Plantefève G, Voiriot G, Badié J et al (2020a) Effect of hydrocortisone on 21-day mortality or respiratory support among critically ill patients with COVID-19: a randomized clinical trial. JAMA 324(13):1298–1306

Dequin PF, Heming N, Meziani F et al (2020b) Effect of hydrocortisone on 21-day mortality or respiratory support among critically ill patients with COVID-19: a randomized clinical trial. JAMA 324(13):1298–1306. https://doi.org/10.1001/jama.2020.16761

Dey D, Kumar A (2024) Unveiling the potential of recently FDA-approved drugs as quorum sensing inhibitors against P. Aeruginosa using high-performance computational techniques. J Biomol Struct Dyn:1–18. https://doi.org/10.1080/07391102.2024.2304682

Dhar R, Kirkpatrick J, Gilbert L, Khanna A, Modi MM, Chawla RK, Dalal S, Maturu VN, Stern M, Keppler OT, Djukanovic R, Gadola SD (2023) Doxycycline for the prevention of progression of COVID-19 to severe disease requiring intensive care unit (ICU) admission: a randomized, controlled, open-label, parallel group trial (DOXPREVENT.ICU). PLoS One 18(1):e0280745. https://doi.org/10.1371/journal.pone.0280745

Dostan J, Faramarzi M, Gholamnezhad Z (2020) Mefenamic acid and Serotonin reuptake inhibitors have similar analgesic effects in women with primary dysmenorrhea. Acta Med Iran 58(3):115–119

Eastman RT, Roth JS, Brimacombe KR et al (2020) Remdesivir: a review of its discovery and development leading to emergency use authorization for treatment of COVID-19. ACS Cent Sci 6(5):672–683. https://doi.org/10.1021/acscentsci.0c00489

Elavarasi A, Prasad M, Seth T et al (2020) Chloroquine and hydroxychloroquine for the treatment of COVID-19: a systematic review and meta-analysis. J Gen Intern Med 35(11):3308–3314. https://doi.org/10.1007/s11606-020-06146-w

El-Saber Batiha G, Al-Gareeb AI, Saad HM, Al-Kuraishy HM (2022) COVID-19 and corticosteroids: a narrative review. Inflammopharmacology 30(4):1189–1205. https://doi.org/10.1007/s10787-022-00987-z

Fehr AR, Perlman S (2015) Coronaviruses: an overview of their replication and pathogenesis. Methods Mol Biol 1282:1–23. https://doi.org/10.1007/978-1-4939-2438-7_1

Furuta Y, Komeno T, Nakamura T (2017) Favipiravir (T-705), a broad spectrum inhibitor of viral RNA polymerase. Proc Jpn Acad Ser B Phys Biol Sci 93(7):449–463. https://doi.org/10.2183/pjab.93.027

Fuzimoto AD (2021) An overview of the anti-SARS-CoV-2 properties of Artemisia annua, its antiviral action, protein-associated mechanisms, and repurposing for COVID-19 treatment. J Integr Med 19(5):375–388. https://doi.org/10.1016/j.joim.2021.07.003

Garg A, Kumar A (2021) Repurposing of drugs for COVID-19 infections. In: COVID-19: current challenges and future perspectives, vol 14, 12th edn, pp 146–159. https://doi.org/10.2174/9789811498640121010015

Ghanei M, Solaymani-Dodaran M, Qazvini A et al (2021) The efficacy of corticosteroids therapy in patients with moderate to severe SARS-CoV-2 infection: a multicenter, randomized, open-label trial. Respir Res 22:245. https://doi.org/10.1186/s12931-021-01833-6

Gluth S, Smith LE, Abbott JH (2020) Mefenamic acid and diclofenac have a similar safety and efficacy profile in patients with osteoarthritis. Ann Fam Med 18(2):162–168. https://doi.org/10.1370/afm.2514

Götte M (2021) Remdesivir for the treatment of Covid-19: the value of biochemical studies. Curr Opin Virol 49:81–85. https://doi.org/10.1016/j.coviro.2021.04.014

Guzman-Esquivel J, Galvan-Salazar HR, Guzman-Solorzano HP et al (2022) Efficacy of the use of mefenamic acid combined with standard medical care vs. standard medical care alone for the treatment of COVID-19: a randomized double-blind placebo-controlled trial. Int J Mol Med 49(3):29. https://doi.org/10.3892/ijmm.2022.5084

Hammond J, Leister-Tebbe H, Gardner A et al (2022) Oral Nirmatrelvir for high-risk, nonhospitalized adults with Covid-19. N Engl J Med 386(15):1397–1408. https://doi.org/10.1056/NEJMoa2118542

Hashemian SMR, Sheida A, Taghizadieh M, Memar MY, Hamblin MR, Bannazadeh Baghi H, Sadri Nahand J, Asemi Z, Mirzaei H (2023) Paxlovid (Nirmatrelvir/Ritonavir): a new approach to Covid-19 therapy? Biomedicine & pharmacotherapy = Biomedecine & pharmacotherapie 162:114367. https://doi.org/10.1016/j.biopha.2023.114367

Hossain MS, Hami I, Sawrav MSS, Rabbi MF, Saha O, Bahadur NM, Rahaman MM (2020) Drug repurposing for prevention and treatment of COVID-19: a clinical landscape. Discover 8(4):e121. https://doi.org/10.15190/d.2020.18

Hua Y, Dai X, Xu Y, Xing G, Liu H, Lu T et al (2022) Drug repositioning: Progress and challenges in drug discovery for various diseases. Eur J Med Chem 234:114239. https://doi.org/10.1016/j.ejmech.2022.114239

Huet T, Beaussier H, Voisin O, Jouveshomme S, Dauriat G, Lazareth I et al (2020) Anakinra for severe forms of COVID-19: a cohort study. Lancet Rheumatol 2(7):e393–e400

Hung YP, Lee JC, Chiu CW, Lee CC, Tsai PJ, Hsu IL, Ko WC (2022) Oral Nirmatrelvir/Ritonavir therapy for COVID-19: the Dawn in the dark? Antibiotics (Basel, Switzerland) 11(2):220. https://doi.org/10.3390/antibiotics11020220

Issa NT, Stathias V, Schürer S, Dakshanamurthy S (2021) Machine and deep learning approaches for cancer drug repurposing. Semin Cancer Biol 68:132–142. https://doi.org/10.1016/j.semcancer.2019.12.011

Jeronimo CMP, Farias MEL, Val FFA et al (2021) Methylprednisolone as adjunctive therapy for patients hospitalized with Coronavirus disease 2019 (COVID-19; Metcovid): a randomized, double-blind, phase IIb, placebo-controlled trial. Clin Infect Dis 72(9):e373–e381. https://doi.org/10.1093/cid/ciaa1177

Jonguitud-Borrego N, Malcı K, Anand M et al (2022) High-throughput and automated screening for COVID-19. Front Med Technol 4:969203. https://doi.org/10.3389/fmedt.2022.969203

Junqueira DR, Rowe BH (2021) Efficacy and safety outcomes of proposed randomized controlled trials investigating hydroxychloroquine and chloroquine during the early stages of the COVID-19 pandemic. Br J Clin Pharmacol 87(4):1758–1767. https://doi.org/10.1111/bcp.14598

Kabinger F, Stiller C, Schmitzová J, Dienemann C, Kokic G, Hillen HS et al (2021a) Mechanism of molnupiravir-induced SARS-CoV-2 mutagenesis. Nat Struct Mol Biol 28(9):740–746

Kabinger F, Stiller C, Schmitzová J, Dienemann C, Kokic G, Hillen HS et al (2021b) Mechanism of Molnupiravir-induced SARS-CoV-2 mutagenesis. Nat Struct Mol Biol 28(9):740–746. https://doi.org/10.1038/s41594-021-00651-0

Kalil AC, Patterson TF, Mehta AK, Tomashek KM, Wolfe CR, Ghazaryan V et al (2021) Baricitinib plus Remdesivir for hospitalized adults with Covid-19. N Engl J Med 384(9):795–807

Kapoor Y, Sharma R, Kumar A (2020) Repurposing of existing drugs for the bacterial infections: an in silico and in vitro study. Infect Disorders-Drug Targ (Formerly Current Drug Targets-Infectious Disorders) 20(2):182–197. https://doi.org/10.2174/1871526519666181126094244

Kumar A (2021) COVID-19 current challenges and future perspectives. Bentham Science Publishers

Lee TC, Murthy S, Del Corpo O et al (2022) Remdesivir for the treatment of COVID-19: a systematic review and meta-analysis. Clin Microbiol Infect 28(9):1203–1210. https://doi.org/10.1016/j.cmi.2022.04.018

Lei ZN, Wu ZX, Dong S et al (2020) Chloroquine and hydroxychloroquine in the treatment of malaria and repurposing in treating COVID-19. Pharmacol Ther 216:107672. https://doi.org/10.1016/j.pharmthera.2020.107672

Lewis K, Chaudhuri D, Alshamsi F, Carayannopoulos L, Dearness K, Chagla Z, Alhazzani W, GUIDE Group (2021) The efficacy and safety of hydroxychloroquine for COVID-19 prophylaxis: a systematic review and meta-analysis of randomized trials. PLoS One 16(1):e0244778. https://doi.org/10.1371/journal.pone.0244778

Li Y, Xie Z, Lin W, Cai W, Wen C, Guan Y et al (2020) Efficacy and safety of lopinavir/ritonavir or arbidol in adult patients with mild/moderate COVID-19: anexploratory randomized controlled trial. Med 1:105–113

Little P, Moore M, Kelly J, Williamson I, Leydon G, McDermott L et al (2020) Ibuprofen, paracetamol, and steam for patients with respiratory tract infections in primary care: pragmatic randomised factorial trial. BMJ 370:f6041

Liu DY, Liu JC, Liang S et al (2021) Drug repurposing for COVID-19 treatment by integrating network pharmacology and transcriptomics. Pharmaceutics 13(4):545. https://doi.org/10.3390/pharmaceutics13040545

Loos NHC, Beijnen JH, Schinkel AH (2023) The inhibitory and inducing effects of ritonavir on hepatic and intestinal CYP3A and other drug-handling proteins. Biomed Pharmacother 162:114636. https://doi.org/10.1016/j.biopha.2023.114636

Low ZY, Farouk IA, Lal SK (2020) Drug repositioning: new approaches and future prospects for life-debilitating diseases and the COVID-19 pandemic outbreak. Viruses 12(9):1058. https://doi.org/10.3390/v12091058

Lucchetta R, Matuoka JY, Oliveira Junior HA et al (2023) Hydroxychloroquine for non-hospitalized COVID-19 patients: a systematic review and meta-analysis of randomized clinical trials. Arq Bras Cardiol 120(4):e20220380. https://doi.org/10.36660/abc.20220380

Mahmoud DB, Shitu Z, Mostafa A (2020) Drug repurposing of nitazoxanide: can it be an effective therapy for COVID-19? J Genet Eng Biotechnol 18(1):35. https://doi.org/10.1186/s43141-020-00055-5

Mangkuliguna G, Glenardi N, Pramono LA (2021) Efficacy and safety of Azithromycin for the treatment of COVID-19: a systematic review and meta-analysis. Tubercul Resp Dis 84(4):299–316. https://doi.org/10.4046/trd.2021.0075

Marconi VC, Ramanan AV, de Bono S, Kartman CE, Krishnan V, Liao R et al (2021) Efficacy and safety of baricitinib for the treatment of hospitalised adults with COVID-19 (COV-BARRIER): a randomised, double-blind, parallel-group, placebo-controlled phase 3 trial. Lancet Respir Med 9(12):1407–1418

Martins-Filho PR, do Nascimento-Júnior EM, Barreto-Alves JA, Fakhouri R, Ferreira LC (2022) Efficacy and safety of nitazoxanide in treating SARS-CoV-2 infection: a systematic review and meta-analysis of blinded, placebo-controlled, randomized clinical trials. Eur J Clin Pharmacol 78(11):1813–1821. https://doi.org/10.1007/s00228-022-03380-5

McMahon JH, Lau JS, Coldham A et al (2022) Favipiravir in early symptomatic COVID-19, a randomised placebo-controlled trial. EClinicalMedicine 54:101703

Mehta J, Rolta R, Mehta BB, Kaushik N, Choi EH, Kaushik NK (2022) Role of Dexamethasone and Methylprednisolone Corticosteroids in Coronavirus disease 2019 hospitalized patients: a review. Front Microbiol 13:813358. https://doi.org/10.3389/fmicb.2022.813358

Mullins JGL (2022) Drug repurposing *in silico* screening platforms. Biochem Soc Trans 50(2):747–758. https://doi.org/10.1042/BST20200967

National institute of health (NIH), COVID-19 Treatment Guidelines, Ritonavir-Boosted Nirmatrelvir (Paxlovid) https://www.covid19treatmentguidelines.nih.gov. Assessed on 08/03/2024

National Institute of Health (NIH), COVID-19 Treatment Guidelines, https://www.covid19treatmentguidelines.nih.gov. Assessed on 08/03/2024b

National Institute of Health (NIH), COVID-19 Treatment Guidelines, National Institute of Health (NIH), COVID-19 Treatment Guidelines, https://www.covid19treatmentguidelines.nih.gov. Assessed on 08/03/2024c

Painter WP, Holman W (2022) Molnupiravir, a novel oral antiviral agent for the treatment of COVID-19. Curr Opin Virol 50:67–74

Painter WP, Holman W, Bush JA, Almazedi F, Malik H, Eraut N et al (2021) Human safety, tolerability, and pharmacokinetics of Molnupiravir, a novel broad-spectrum oral antiviral agent with activity against SARS-CoV-2. Antimicrob Agents Chemother 65(5):e02428–e02420. https://doi.org/10.1128/AAC.02428-20

Pal M, Berhanu G, Desalegn C, Kandi V (2020) Severe acute respiratory syndrome Coronavirus-2 (SARS-CoV-2): An update. Cureus 12(3):e7423. https://doi.org/10.7759/cureus.7423

Park K (2019) A review of computational drug repurposing. Transl Clin Pharmacol 27(2):59–63. https://doi.org/10.12793/tcp.2019.27.2.59

Petrosillo N, Viceconte G, Ergonul O, Ippolito G, Petersen E (2020) COVID-19, SARS and MERS: are they closely related? Clin Microbiol Infect 26(6):729–734. https://doi.org/10.1016/j.cmi.2020.03.026

Pindiprolu SKSS, Pindiprolu SH (2020) Plausible mechanisms of Niclosamide as an antiviral agent against COVID-19. Med Hypotheses 140:109765. Advance online publication. https://doi.org/10.1016/j.mehy.2020.109765

Pinzi L, Rastelli G (2019) Molecular docking: shifting paradigms in drug discovery. Int J Mol Sci 20(18):4331. https://doi.org/10.3390/ijms20184331

Popp M, Reis S, Schießer S et al (2022) Ivermectin for preventing and treating COVID-19. Cochrane Database Syst Rev 6(6):CD015017. https://doi.org/10.1002/14651858.CD015017.pub3

Porreca E, Di Febbo C (2020) Indomethacin as a possible option for managing cytokine storm in COVID-19 patients. Lancet Rheumatol 2(7):e379

Pushpakom S, Iorio F, Eyers PA et al (2019) Drug repurposing: progress, challenges and recommendations. Nat Rev Drug Discov 18(1):41–58. https://doi.org/10.1038/nrd.2018.168

Ravichandran R, Mohan SK, Sukumaran SK et al (2022) An open label randomized clinical trial of Indomethacin for mild and moderate hospitalised Covid-19 patients. Sci Rep 12:6413. https://doi.org/10.1038/s41598-022-10370-1

RECOVERY Collaborative Group (2020a) Dexamethasone in hospitalized patients with Covid-19. N Engl J Med 384(8):693–704

RECOVERY Collaborative Group (2020b) Group Lopinavir-ritonavir in patients admitted to hospital with COVID-19 (RECOVERY): a randomised, controlled, open-label, platform trial. Lancet 396:1345–1352

RECOVERY Collaborative Group, Horby P, Lim WS et al (2021) Dexamethasone in hospitalized patients with Covid-19. N Engl J Med 384(8):693–704. https://doi.org/10.1056/NEJMoa2021436

REMAP-CAP Investigators, Gordon AC, Mouncey PR, Al-Beidh F, Rowan KM, Nichol AD et al (2021) Interleukin-6 receptor antagonists in critically ill patients with Covid-19. N Engl J Med 384(16):1491–1502

Rinott E, Kozer E, Shapira Y, Bar-Haim A, Youngster I (2020a) Ibuprofen use and clinical outcomes in COVID-19 patients. Clin Microbiol Infect 26(9):1259.e5–1259.e7. https://doi.org/10.1016/j.cmi.2020.06.003

Rinott E et al (2020b) Ibuprofen use and clinical outcomes in COVID-19 patients. Clin Microbiol Infect 2020:261259

Saima L, Sharma R, Kumar A (2024) Role of network pharmacology in prediction of mechanism of neuroprotective compounds. In: Neuroprotection: method and protocols. Springer USNew York, pp 159–179

Scarsi M, Piantoni S, Colombo E, Airó P (2020) COVID-19: immunopathology and its implications for therapy. Swiss Med Wkly 150(1112):w20221

Scialo F, Daniele A, Amato F, Pastore L, Matera MG, Cazzola M, Castaldo G, Bianco A (2020) ACE2: the major cell entry receptor for SARS-CoV-2. Lung 198(6):867–877. https://doi.org/10.1007/s00408-020-00408-4

Sencanski M, Perovic V, Pajovic SB, Adzic M, Paessler S, Glisic S (2020) Drug repurposing for candidate SARS-CoV-2 main protease inhibitors by a novel *In Silico* method. Molecules 25(17):3830. https://doi.org/10.3390/molecules25173830

Shamshirian A, Hessami A, Heydari K et al (2020) The role of hydroxychloroquine in COVID-19: a systematic review and meta-analysis. Ann Acad Med Singap 49(10):789–800

Sharma N, Singh A, Sharma R, Kumar A (2021) Repurposing of auranofin against bacterial infections: an in silico and in vitro study. Curr Comput Aided Drug Des 17(5):687–701. https://doi.org/10.2174/1386207323666200717155640

Sharma R, Kumar A, Majeed J, Thakur AK, Aggarwal G (2022) Drugs acting on the renin–angiotensin–aldosterone system (RAAS) and deaths of COVID-19 patients: a systematic review and meta-analysis of observational studies. Egypt Heart J 74(1):64

Sharun K, Tiwari R, Dhama J, Dhama K (2020) Dexamethasone to combat cytokine storm in COVID-19: clinical trials and preliminary evidence. Int J Surg 82:179–181. https://doi.org/10.1016/j.ijsu.2020.08.038

Shinkai M, Tsushima K, Tanaka S et al (2021) Efficacy and safety of favipiravir in moderate COVID-19 pneumonia patients without oxygen therapy: a randomized, phase III clinical trial. Infect Dis Ther 10:2489–2509

Singh TU, Parida S, Lingaraju MC, Kesavan M, Kumar D, Singh RK (2020) Drug repurposing approach to fight COVID-19. Pharmacol Rep 72(6):1479–1508. https://doi.org/10.1007/s43440-020-00155-6

Singh H, Chauhan P, Kakkar AK (2021) Hydroxychloroquine for the treatment and prophylaxis of COVID-19: the journey so far and the road ahead. Eur J Pharmacol 890:173717. https://doi.org/10.1016/j.ejphar.2020.173717

Singh S, Weiss A, Goodman J, Fisk M, Kulkarni S, Lu I, Gray J, Smith R, Sommer M, Cheriyan J (2022) Niclosamide-A promising treatment for COVID-19. Br J Pharmacol 179(13):3250–3267. https://doi.org/10.1111/bph.15843

Singla S, Goyal S (2022) Antiviral activity of molnupiravir against COVID-19: a schematic review of evidences. Bull Natl Res Cent 46(1):62. https://doi.org/10.1186/s42269-022-00753-9

Sinha S, Suram VK, Chary SS, Naik S, Singh VB, Jain MK, Suthar CP, Borthakur S, Sawardekar V, Sk N, Reddy N, Talluri L, Thakur P, Reddy M, Panapakam M, Vattipalli R (2022) Efficacy and safety of Molnupiravir in mild COVID-19 patients in India. Cureus 14(11):e31508. https://doi.org/10.7759/cureus.31508

Spinner CD, Gottlieb RL, Criner GJ, Arribas López JR, Cattelan AM, Soriano Viladomiu A et al (2020) Effect of Remdesivir vs standard care on clinical status at 11 days in patients with moderate COVID-19. JAMA 324(11):1048–1057

Stone JH, Frigault MJ, Serling-Boyd NJ, Fernandes AD, Harvey L, Foulkes AS et al (2020) Efficacy of tocilizumab in patients hospitalized with Covid-19. N Engl J Med 383(24):2333–2344

Taher A, Lashkari M, Keramat F et al (2023) Comparison of the efficacy of equivalent doses of dexamethasone, methylprednisolone, and hydrocortisone for treatment of COVID-19-related acute respiratory distress syndrome: a prospective three-arm randomized clinical trial. Wien Med Wochenschr 173:140–151. https://doi.org/10.1007/s10354-022-00993-4

Talevi A, Bellera CL (2020) Challenges and opportunities with drug repurposing: finding strategies to find alternative uses of therapeutics. Expert Opin Drug Discov 15(4):397–401. https://doi.org/10.1080/17460441.2020.1704729

Talha B, Dhamoon AS (2024) Ritonavir. [Updated 2023 Aug 8]. In: StatPearls [Internet]. StatPearls Publishing, Treasure Island. Available from: https://www.ncbi.nlm.nih.gov/books/NBK544312/

Tang W, Khalili L, Giles J, Gartshteyn Y, Kapoor T, Guo C, Chen T, Theodore D, Askanase A (2021) The rise and fall of Hydroxychloroquine with the COVID-19 pandemic: narrative review of selected data. Rheumatol Ther 8(2):681–691. https://doi.org/10.1007/s40744-021-00315-x

Tian L, Pang Z, Li M, Lou F, An X, Zhu S, Song L, Tong Y, Fan H, Fan J (2022) Molnupiravir and its antiviral activity against COVID-19. Front Immunol 13:855496. https://doi.org/10.3389/fimmu.2022.855496

Tomazini BM, Maia IS, Cavalcanti AB, Berwanger O, Rosa RG, Veiga VC et al (2020) Effect of dexamethasone on days alive and ventilator-free in patients with moderate or severe acute respiratory distress syndrome and COVID-19: the CoDEX randomized clinical trial. JAMA 324(13):1307–1316

U.S. Food and Drug Administration (2024) FDA revises letter of authorization for the emergency use authorization for Paxlovid. https://www.fda.gov/drugs/drug-safety-and-availability/fda-revises-letter-authorization-emergency-use-authorization-paxlovid. Accessed on 8/03/2024

Uckun FM, Saund S, Windlass H, Trieu V (2021) Repurposing anti-malaria phytomedicine artemisinin as a COVID-19 drug. Front Pharmacol 12:649532. Published 2021 Mar 19. https://doi.org/10.3389/fphar.2021.649532

Udwadia ZF, Singh P, Barkate H et al (2021) Efficacy and safety of favipiravir, an oral RNA-dependent RNA polymerase inhibitor, in mild-to-moderate COVID-19: a randomized, comparative, open-label, multicenter, phase 3 clinical trial. Int J Infect Dis 103:62–71

Ullah W, M Abdullah H, Roomi S et al (2020) Safety and efficacy of hydroxychloroquine in COVID-19: a systematic review and meta-analysis. J Clin Med Res 12(8):483–491. https://doi.org/10.14740/jocmr4233

Villar J, Ferrando C, Martínez D, Ambrós A, Muñoz T, Soler JA et al (2020) Dexamethasone treatment for the acute respiratory distress syndrome: a multicentre, randomised controlled trial. Lancet Respir Med 8(3):267–276

Wang M, Cao R, Zhang L et al (2020) Remdesivir and chloroquine effectively inhibit the recently emerged novel coronavirus (2019-nCoV) in vitro. Cell Res 30:269–271. https://doi.org/10.1038/s41422-020-0282-0

Weth FR, Hoggarth GB, Weth AF et al (2023) Unlocking hidden potential: advancements, approaches, and obstacles in repurposing drugs for cancer therapy. Br J Cancer. https://doi.org/10.1038/s41416-023-02502-9

WHO Rapid Evidence Appraisal for COVID-19 Therapies (REACT) Working Group, Sterne JAC, Murthy S, Diaz JV, Slutsky AS, Villar J et al (2020) Association between administration of systemic corticosteroids and mortality among critically ill patients with COVID-19: a meta-analysis. JAMA 324(13):1330–1341

World Health Organization (2020) Corticosteroids for COVID-19: living guidance, 2 September 2020. World Health Organization. https://apps.who.int/iris/handle/10665/334125

World Health Organization (2021) COVID-19 Weekly epidemiological update - January 17, 2021. https://www.who.int/docs/default-source/coronaviruse/situation-reports/weekly_epidemiological_update_23.pdf?sfvrsn=6ac0778_5&download=true. Accessed 19 Jan 2021

World Health Organization (2024a) Novel Coronavirus (2019-nCoV) situation reports. https://www.who.int/emergencies/diseases/novel-coronavirus-2019. Assessed on 5/02/2024

World Health Organization (Year) Coronavirus disease (COVID-19): Solidarity Trial and hydroxychloroquine. https://www.who.int/news-room/questions-and-answers/item/coronavirus-disease-covid-19-hydroxychloroquine. Accessed on 8/03/2024b

Wu C, Liu Y, Yang Y, Zhang P, Zhong W, Wang Y, Wang Q, Xu Y, Li M, Li X, Zheng M, Chen L, Li H (2020) Analysis of therapeutic targets for SARS-CoV-2 and discovery of potential drugs by computational methods. Acta Pharm Sin B 10(5):766–788. https://doi.org/10.1016/j.apsb.2020.02.008

Xu J, Shi PY, Li H, Zhou J (2020) Broad Spectrum antiviral agent Niclosamide and its therapeutic potential. ACS Infect Dis 6(5):909–915. https://doi.org/10.1021/acsinfecdis.0c00052

Yamamoto N, Matsuyama S, Hoshino T, Yamamoto N, Nishiura K, Kawaoka Y, Matsuyama S (2020) Ritonavir blocks replication of severe acute respiratory syndrome coronavirus 2 (SARS-CoV-2). bioRxiv

Zhou Y, Gilmore K, Ramirez S et al (2021) In vitro efficacy of artemisinin-based treatments against SARS-CoV-2. Sci Rep 11:14571. https://doi.org/10.1038/s41598-021-93361-y

Drug Repurposing: Exploring Newer Therapeutic Potential

7

Bhaskaranand Pancholi, Amarjeet Rao, and Debapriya Garabadu

Abstract

New drug formulation and development is costly and can take 10–15 years. The drug repurposing approach focuses on re-evaluating known drugs previously approved for one medical condition to explore their potential effectiveness in treating different diseases or conditions for which they were not initially intended. This approach has many significant aspects, such as cost and time efficiency, safety, expanded treatment options, rare diseases, and accelerated drug development. Conversely, there are many challenges to re-evaluating the drugs, such as regulatory challenges, financial issues, clinical trials, intellectual property, and demands for market analysis. It minimizes drug failure, maximizes the drug's therapeutic potential, and enhances the success rate. The study also shows various experimental and computational hypothesis-based strategies that are being used in the drug repurposing approaches. The study enlists various repurposed drugs of central nervous system, respiratory system, gastrointestinal system, and other diseases for newer indications. It is observed that the drug repurposing approach fastens the drug development process for several diseases. Hence, drug repurposing approach could be considered an alternative approach to traditional drug discovery systems.

Keywords

Drug repurposing · Central nervous system · Respiratory system · Gastrointestinal system · Therapeutic potential

---

B. Pancholi · A. Rao · D. Garabadu (✉)
Department of Pharmacology, School of Health Sciences, Central University of Punjab, Bathinda, India
e-mail: debapriya.garabadu@cup.edu.in

R. K. Kaundal et al. (eds.), *Drug Repurposing: Exploring Newer Therapeutic Potentials*,
https://doi.org/10.1007/978-981-95-4522-3_7

7.1 Introduction

New drug formulation and development is costly and can take 10–15 years. Due to the lengthy discovery process, drug repurposing (or "repositioning") is now an option for reducing the time needed to create a medicine. Drug repurposing, or drug repositioning, entails investigating current medications for potential new therapeutic use. Those drugs have already been developed and approved by different regulatory bodies, such as the FDA, MHRA, and EMA, for the other diseases for new signals. Many pharmaceutical companies use drug repurposing to regenerate some of their FDA-approved and previously unsuccessful pipeline molecules as novel medicines for various illness conditions due to the enormous promise of a reduced development cycle (Parvathaneni et al. 2019). The idea of repurposing "old" medications to treat joint and uncommon diseases is gaining traction since it allows for the use of compounds that have been de-risked and may result in cheaper overall development costs and shorter development times. Drug repurposing, sometimes referred to as drug repositioning or drug reprofiling, is a significant topic with several ramifications for pharmaceutical research and healthcare (Langedijk et al. 2015a). This approach focuses on re-evaluating known drugs previously approved for one medical condition to explore their potential effectiveness in treating different diseases or conditions for which they were not initially intended (Pushpakom et al. 2019a).

7.1.1 Significance

The following salient features highlight the importance of medication repurposing:

(a) *Cost and Time Efficiency:* Creating a novel medication from the ground up is time-consuming and costly, requiring many years and significant financial resources. Compared to lengthy traditional research and development processes, drug repurposing offers patients effective medications more quickly and less expensively. Additionally, this strategy helps combat rising drug development costs, cutting patients' out-of-pocket expenses, and lowering the actual cost of therapy (Parvathaneni et al. 2019; Naylor et al. 2015).

(b) *Safety:* The repurposed medication has already been shown to be safe in Phase II and III studies; it is unlikely that safety concerns will surface in research. During their initial development, repurposed medications frequently undergo thorough safety testing (Sisignano et al. 2016).

(c) *Expanded Treatment Options:* The range of therapies for various medical diseases is expanded via drug repurposing. It can provide novel therapeutic possibilities, especially for conditions with few other effective treatments. Some of the best examples of developed medicine, such as sildenafil, minoxidil, aspirin, and valproic acid, have arisen from unorganized ("field") discovery methods. These processes frequently depend on the known pharmacology of medicine (e.g., an off-target side effect) to address a clinical issue from a different domain (Pushpakom et al. 2019b).

(d) *Rare Diseases:* Approximately 7000 rare diseases have been found, impacting 10% of the population. Still, less than 6% of all rare disorders have a recognized course of treatment. Drug repurposing is the new approach to treating various disease conditions compared to developing novel orphan drugs. It can be helpful for time-saving, cost-efficient, and higher success rates (Roessler et al. 2021a).

(e) *Accelerated Drug Development:* The traditional drug development process is lengthy, slow, and time-consuming. It is imperative to find new approaches to shorten the drug discovery process. Numerous techniques are available which are approved by the FDA for using the existing drug for another purpose. We do not need a human population because clinical studies have already been performed (Parvathaneni et al. 2019).

7.1.2 Drug Repurposing Difficulties

There are many advantages to repurposing drugs, such as cost-effectiveness, lesser time development, and physiochemical properties, while simultaneously, there are many drawbacks and challenges of repurposing drugs.

The following are some points that highlight the challenges of repurposing drugs:

(a) Regulatory challenges: No hard and fast regulatory guidelines for repurposing drug individuals (Gatta et al. 2011).
(b) Financial issues: Pharmaceutical businesses face the difficulty of inadequate funding for research and lack of financial incentives (Gatta et al. 2011).
(c) There may be further challenges to overcome when conducting clinical trials with repurposed candidates. Previously performed clinical trials do not justify the significant use of the repurposed drug, so there might be a phase I clinical trial (Padhy and Gupta 2011; Shineman et al. 2014).
(d) Problems with intellectual property prevent the repositioned chemical from being commercialized (Padhy and Gupta 2011).
(e) Demands market analysis: Getting a new indication for an already-approved medication does not always provide the same market exclusivity as creating an entirely new drug, even if a patent protects the original medication. This may create a more competitive market and fewer financial incentives (Novac 2013).

The benefits of drug repurposing are becoming increasingly apparent despite these obstacles, mainly as it can result in shorter development times and lower risks than new drug development. Efforts to address these challenges, such as creating regulatory pathways and financial incentives for repurposing, could potentially enhance the economic motivations for exploring new uses for existing drugs (Fig. 7.1).

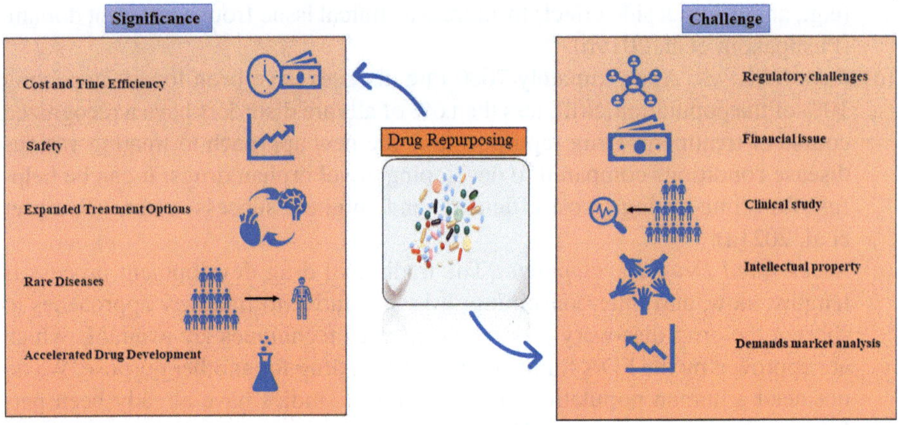

Fig. 7.1 Significance and challenges of drug repurposing

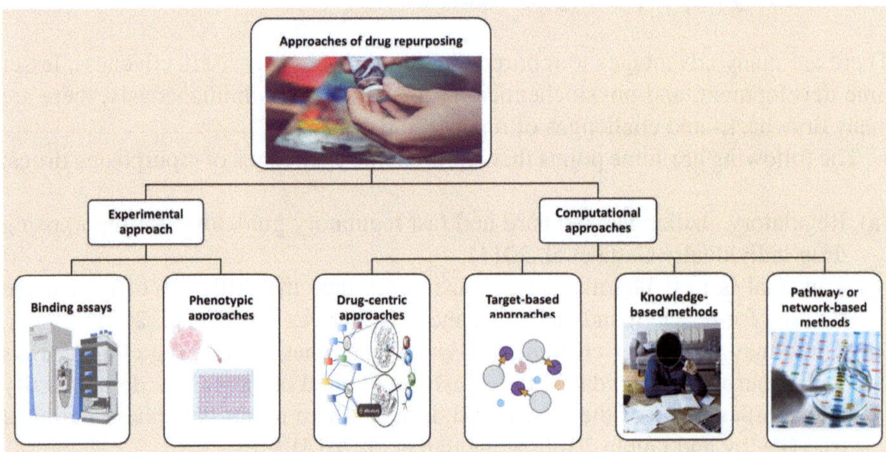

Fig. 7.2 Methods of drug repurposing

7.2 Approaches to Drug Repurposing

Drug repurposing approaches depend upon the hypothesis. Hypothesis-based strategies for repurposed drugs depend upon experimental and computational methods that can help evaluate the new therapeutic use of existing drugs and understand the mechanisms and pathways of pathogens (Parvathaneni et al. 2019) (Fig. 7.2).

I. *Experimental approach:* There are a few approaches includeing the following:

   (a) *Binding assays:* Some techniques help to identify the various compounds' targets, such as mass spectroscopy (MS), and proteomics. In the case of thermal stability, the cellular thermostability assay is helpful to predict the target protein stability through binding compounds with the highest cellular affinity.

(b) *Phenotypic approaches:* This approach is based on the cell-line study, employing cellular models to find substances that affect particular disease-related traits like cell division, death, or proliferation (Lage et al. 2018). Using in vivo disease models to find substances that exhibit whole-body therapeutic effectiveness and animal models is also effective in identifying medications or drug-like compounds initially used in phenotypic assays related to disease to choose and validate candidates for repositioning. Instead, other approaches are beneficial to detecting the new use of old molecules, such as robotic screening, in silico methods (Wilkinson and Pritchard 2015), and genome editing technologies such as the CRISPR/Cas-9 system, which helps to systematically delete or alter genes to find those that, when changed, result in the desired phenotype, thus offering possible targets for pharmaceuticals (Nishimura and Hara 2018).

II. *Computational approaches:* There are some other approaches for repurposing the drugs which are as follows:

(a) *Drug-centric approaches:* Discovery of molecules for newer therapeutic use can be determined by studying the drug–protein binding of its off-target hits. It helps to search for a new target for the previously approved drug. Polypharmacological approaches are part of drug-centric approaches; they can help identify the latest targets of a single pill. It also found out the unrevealed off-target of a previously approved drug. Nevertheless, this strategy must incorporate all the information obtained from clinical research, in vitro and in vivo pharmaceutical testing, and computer modeling (Reddy and Zhang 2013).

(b) *Target-based approaches:* This approach helps to determine the new indication by associating a medication with a particular ailment through its protein targets. The primary target and off-target proteins can be used to define a further indication for a specific drug (Sawada et al. 2015).

(c) *Knowledge-based methods:* Knowledge-based approaches give valid justification for the drug except for unidentified pharmacological targets, unexpected drug–drug similarities, and novel biomarkers, among other previously unrecognized mechanisms. Knowledge-based approaches increase the prediction certainty by incorporating significant information into drug repurposing (Jin and Wong 2014).

(d) *Pathway- or network-based methods:* Pathway- or network-based methods utilize the advantage of disease omics data, including a practical medication path and interaction between the drug and target. These techniques use biological systems' interconnectedness to comprehend the complex interactions between genes, proteins, and other molecular elements (Parvathaneni et al. 2019).

7.3 How Can We Reuse the Old Drug for Different Targets?

(a) *Identify the New Target:* Start by identifying a specific disease or medical condition for which there is a need for a new treatment. The "New Target" in drug repurposing refers to a novel application of an existing drug for a medical condition different from its original intended use. Identifying new targets involves exploring the potential of drugs to treat diseases or conditions beyond their initial therapeutic purposes, often driven by advances in our understanding of biology and disease mechanisms. This approach can lead to more efficient drug development and the discovery of new treatment options. Research the molecular and biological pathways associated with the target to better understand the disease (Roessler et al. 2021b).

(b) *Disease or Medical Condition Selection*: The first step in repurposing old drugs is to choose a specific disease, medical condition, or therapeutic need for which a new treatment is required. This decision should be based on factors like the inadequacy of current therapies or the presence of unmet medical needs. The initial step in repurposing existing drugs is selecting a particular disease, medical condition, or therapeutic requirement where a new treatment is urgently needed. Various factors, such as the ineffectiveness of current therapies or the existence of unaddressed medical needs, should guide this choice. In essence, it involves identifying areas where there is room for improvement or where existing medications do not adequately serve patients. This decision serves as the foundation for the subsequent drug repurposing process, focusing research and development efforts on addressing these unmet medical needs (Roessler et al. 2021b).

(c) *Understand Disease Biology:* Once a target disease or condition is selected, it is essential to delve into the underlying biology and pathophysiology of the state. This involves studying the molecular and cellular mechanisms responsible for the disease and identifying key pathways, proteins, or targets involved in its progression. Understanding the disease biology is a critical step in drug development and repurposing. It consists of gaining in-depth knowledge about the underlying mechanisms and pathways that drive a specific disease or medical condition. This understanding encompasses factors such as the root causes of the disease, the cellular and molecular processes involved, the key molecules and genes implicated, and the interactions within the biological systems (Law et al. 2013).

(d) *Consider Genetic and Biomarker Information:* To gain insights into potential drug targets, consider any available genetic and biomarker information related to the disease. Genetic data, such as mutations or gene expression patterns, can provide valuable clues regarding potential targets (Munj et al. 2022).

(e) *Review Existing Literature:* Thoroughly review the scientific literature related to the selected disease. This involves studying research findings on disease mechanisms, recent discoveries, and potential therapeutic targets. A comprehensive literature review helps understand the current knowledge and treatment options (Langedijk et al. 2015b).

(f) *Consult Experts:* Engage with experts in the field, including clinicians, researchers, and medical professionals specializing in the disease area. Their expertise and insights can provide a deeper understanding of the disease's complexities and help identify potential targets (Cha et al. 2018).
(g) *Prioritize Targets:* After gathering relevant information, prioritize potential drug targets based on their relevance to the disease biology and their potential to be influenced by existing drugs. Look for well-validated marks with a clear connection to the disease (Arshad et al. 2020).
(h) *Validate the Target:* Before proceeding further, it is crucial to confirm that the selected target is a promising candidate for drug repurposing. This may involve conducting initial experiments or using available biological data to ensure the target's role in the disease (Park 2019).
(i) *Assess the Drug Library:* Compile a list of existing drugs that have previously been approved for other indications. Consider both patented and generic drugs, as repurposing can apply to both (Luo et al. 2011).
(j) *Data Mining and Bioinformatics:* Utilize bioinformatics and data mining tools to analyze large datasets of drug compounds and their effects on various biological targets. Look for drugs that have demonstrated activity against proteins or pathways related to the new target (Liu et al. 2020).
(k) In Vitro *and* In Vivo *Testing:* Conduct initial in vitro experiments to test the effects of selected drugs on the new target or disease model. If positive results are obtained, advance to in vivo preclinical testing in animal models to assess safety and efficacy (Ju et al. 2022).
(l) *Clinical Trials:* If preclinical results are promising, plan and execute clinical trials to evaluate the drug's safety and effectiveness for the new indication. This may involve Phase I, II, and III clinical trials, as required by regulatory agencies (Wang and Guan 2021).
(m) *Post-Market Surveillance:* Continuously monitor and gather data on the drug's performance and safety after it is on the market for the new indication. Address any emerging safety or efficacy concerns (Mower et al. 2022).
(n) *Combination Therapies:* Consider combining the repurposed drug with other medications or treatments to enhance its therapeutic effects, particularly in complex diseases (Liu et al. 2021).
(o) *Collaborations and Partnerships:* Collaborate with academic institutions, pharmaceutical companies, and research organizations to leverage expertise and resources in repurposing (Kiriiri et al. 2020).

7.4 Newer Therapeutic Potentials

Drug repositioning, referred to as drug reprofiling, is the active study area of repurposing existing medications for novel therapeutic purposes. It entails finding novel applications for medications currently licensed for use in other contexts.

7.4.1 Newer Therapeutic Approach for CNS

Some drugs have been re-evaluated for the other disease conditions; these are as follows (Table 7.1):

Gabapentin and Pregabalin: Originally developed as antiepileptic medications, these drugs are now commonly used to manage neuropathic pain and have applications in treating conditions like fibromyalgia.

Aripiprazole: Initially designed as an atypical antipsychotic, aripiprazole is now used in the treatment of bipolar disorder, major depressive disorder, and schizophrenia.

Amantadine: Originally used as an antiviral agent, it has been repurposed to manage symptoms of Parkinson's disease and to treat fatigue in multiple sclerosis.

Memantine: Originally developed for Alzheimer's disease, memantine is also used to treat moderate-to-severe Alzheimer's and other cognitive disorders.

Baclofen: This muscle relaxant was repurposed to treat spasticity in conditions like multiple sclerosis and spinal cord injuries.

Methylphenidate: Known for its use in attention-deficit hyperactivity disorder (ADHD), it can also be applied in narcolepsy and fatigue associated with multiple sclerosis.

Lithium: Used in bipolar disorder, lithium has mood-stabilizing properties and is an example of drug repurposing within the psychiatric field.

Dalfampridine: Originally developed to improve walking in multiple sclerosis patients, dalfampridine (also known as fampridine) is a potassium channel blocker that may enhance nerve conduction, potentially improving the walking speed and function.

Riluzole: Initially developed for treating amyotrophic lateral sclerosis (ALS), riluzole has been investigated for its potential neuroprotective effects. It may have implications for conditions affecting the motor system.

Botulinum toxin: Originally used for cosmetic purposes and treating certain neurological disorders, injections have been repurposed for managing spasticity and dystonia in various conditions, including cerebral palsy.

Levodopa/Carbidopa: While levodopa is a standard treatment for Parkinson's disease, it can also be considered a repurposed drug. It is converted into dopamine in the brain, helping alleviate motor symptoms associated with Parkinson's.

Tizanidine: Originally a muscle relaxant, tizanidine has been repurposed for managing spasticity associated with conditions such as multiple sclerosis, spinal cord injury, and stroke.

7.4.2 Newer Therapeutic Approach for Respiratory Diseases

Some drugs have been re-evaluated for other disease conditions, which are as follows (Table 7.2):

Table 7.1 Repurposed drugs used in CNS

Sr. no.	Name of the drug	Mechanism of action of the drug	Old use	Repurposed use	References
1	Gabapentin and Pregabalin	Gabapentin modulates the action of the GABA synthetic enzyme, glutamic acid decarboxylase (GAD), and the glutamate synthesizing enzyme, branched-chain amino acid transaminase	Developed initially as antiepileptic medications	It is now commonly used to manage neuropathic pain and has applications in treating conditions like fibromyalgia	Di Cesare et al. (2023), Wiffen et al. (2017)
2	Aripiprazole	The efficacy of aripiprazole can be mainly attributed to this combination of partial agonism/antagonism at D2Rs and serotonin 5-HT1A receptors, together with antagonism at serotonin 5-HT2A receptors	Initially designed as an atypical antipsychotic	Aripiprazole is now used in the treatment of bipolar disorder, major depressive disorder, and schizophrenia	Shapiro et al. (2003), Eugene et al. (2021)
3	Amantadine	During replication, it may also inhibit the assembly of influenza A virus isolates from each subtype (H1N1, H2N2, and H3N2). Influenza B has a structurally different M2 protein, which makes amantadine ineffective. In the treatment of Parkinson's disease, studies have shown amantadine acts on dopamine neurons. Amantadine is a weak, noncompetitive antagonist of the NMDA receptor, which increases dopamine release and prevents dopamine reuptake	Originally used as an antiviral agent	It has been repurposed to manage symptoms of Parkinson's disease and to treat fatigue in multiple sclerosis	Staničová et al. (2001)

(continued)

Table 7.1 (continued)

Sr. no.	Name of the drug	Mechanism of action of the drug	Old use	Repurposed use	References
4	Memantine	Memantine is a clinically helpful drug in many neurological disorders, including Alzheimer's disease. The principal mechanism of action of memantine is believed to be the blockade of current flow through the channels of N-methyl-d-aspartate (NMDA) receptors—a glutamate receptor subfamily broadly involved in brain function	Originally developed for Alzheimer's disease	It has been repurposed to manage symptoms of Parkinson's disease and to treat fatigue in multiple sclerosis	Reisberg et al. (2003), Johnson and Kotermanski (2006)
5	Baclofen	Baclofen is an agonist for gamma-aminobutyric acid $(GABA)_B$ receptors on pre- and postsynaptic neurons in the central nervous system (CNS) and peripheral nervous system. Its action inhibits the transmission of mono- and polysynaptic reflexes at the spinal cord, relaxing spasticity	This is a muscle relaxant	Repurposed for the treatment of spasticity in conditions like multiple sclerosis and spinal cord injuries	Romito et al. (2021), Dario and Tomei (2004)

(continued)

Table 7.1 (continued)

Sr. no.	Name of the drug	Mechanism of action of the drug	Old use	Repurposed use	References
6	Methylphenidate	With the increase in dopamine levels, methylphenidate can provide neuroprotection in certain conditions like Parkinson's disease, which involves loss of dopaminergic neurons and methamphetamine abuse	Known for its use in attention-deficit hyperactivity disorder (ADHD)	It can also be applied in narcolepsy and fatigue associated with multiple sclerosis	Volkow et al. (2002), Ayache and Chalah (2017)
7	Lithium	Lithium inhibits therapeutic concentrations, and this inhibition leads to multiple pharmacological effects; GSK3 has been known to regulate gene expression, embryonic development, neuronal survival, and circadian rhythms, among others	Used in bipolar disorder, lithium has mood-stabilizing properties	Drug repurposed within the psychiatric field	Alda (2015), Lüscher Dias et al. (2020)
8	Dalfampridine	Dalfampridine increases the amplitude and duration of nerve conduction, resulting in enriched nerve conduction in demyelinated animal nerves	Originally developed to improve walking in multiple sclerosis patients	Dalfampridine (also known as fampridine) is a potassium channel blocker that may enhance nerve conduction, potentially improving the walking speed and function	Dunn and Blight (2011), Lamore III et al. (2010)

(continued)

Table 7.1 (continued)

Sr. no.	Name of the drug	Mechanism of action of the drug	Old use	Repurposed use	References
9	Riluzole	Riluzole, a glutamatergic antagonist, is a disease-modifying treatment shown to extend survival time in ALS	Initially developed for the treatment of amyotrophic lateral sclerosis (ALS)	Riluzole has been investigated for its potential neuroprotective effects. It may have implications for conditions affecting the motor system	Hinchcliffe and Smith (2017), Bellingham (2011)
10	Botulinum toxin	Intramuscular administration of botulinum toxin acts at the neuromuscular junction to cause muscle paralysis by inhibiting the release of acetylcholine from presynaptic motor neurons	Initially used for cosmetic purposes and treating certain neurological disorders	Botulinum toxin injections have been repurposed for managing spasticity and dystonia in various conditions, including cerebral palsy	Chatterjee (2004), Nigam and Nigam (2010), Multani et al. (2019)
11	Levodopa/ Carbidopa	Inhibitor of L-amino acid decarboxylase, the plasma enzyme metabolizes levodopa peripherally	Levodopa is a standard treatment for Parkinson's disease	It can also be considered a repurposed drug. It is converted into dopamine in the brain, helping alleviate motor symptoms associated with Parkinson's	Ovallath and Sulthana (2017), Elkouzi et al. (2019)
12	Tizanidine	TZN is an α_2-adrenergic receptor agonist that reduces spasticity by enhancing motor neuron presynaptic inhibition	Originally a muscle relaxant	Tizanidine has been repurposed for managing spasticity associated with conditions such as multiple sclerosis, spinal cord injury, and stroke	Mohsen et al. (2023), Kamen et al. (2008)

Table 7.2 Repurposed drugs used in respiratory diseases

Sr.no.	Name of the drug	Mechanism of action of the drug	Old use	Repurposed use	References
1	Dornase alfa (pulmozyme)	Dornase alfa is a recombinant human deoxyribonuclease I (rhDNase) that cleaves down extracellular long-chain DNA in the sputum, reducing mucous viscosity and preventing airway infections and damage to the lung parenchyma. It improves the lung function and quality of life	Originally developed for cystic fibrosis	It has been investigated for its potential benefits in other respiratory conditions characterized by mucus buildup	Yang and Montgomery (2021), Wagener and Kupfer (2012), Goetz et al. (2022)
2	Montelukast (Singulair)	Montelukast causes inhibition of airway cysteinyl leukotriene receptors, as demonstrated by the ability to inhibit bronchoconstriction due to inhaled LTD4 in asthmatics. Doses as low as 5 mg cause substantial blockage of LTD4-induced bronchoconstriction	Initially approved for treating asthma and allergic rhinitis, montelukast is a leukotriene receptor antagonist	Some studies have explored its use in chronic obstructive pulmonary disease (COPD) and eosinophilic bronchitis	Louie et al. (2013), Wermuth et al. (2023)

Table 7.2 (continued)

Sr.no.	Name of the drug	Mechanism of action of the drug	Old use	Repurposed use	References
3	Statins (e.g., simvastatin)	Using acetyl-CoA as a substrate, mevalonic acid is formed, and subsequent reactions lead to the formation of cholesterol	Commonly used to lower cholesterol	They were investigated for their anti-inflammatory properties in the context of respiratory diseases. Some studies suggest potential benefits in conditions such as asthma and COPD	McFarland et al. (2014), Marin et al. (2011)
4	Methotrexate	It inhibits the enzyme AICAR transformylase, leading to hindrance in adenosine and guanine metabolism	This drug, originally developed as a chemotherapy agent, has immunosuppressive properties	Methotrexate has been studied for its potential role in treating severe asthma	Cronstein and Aune (2020), Tian and Cronstein (2007), Davies et al. (1996)
5	Thalidomide	The complete mechanism is unknown	Initially developed as a sedative and later withdrawn due to teratogenic effects	Thalidomide has been explored for its anti-inflammatory and immunomodulatory properties in conditions such as sarcoidosis and certain interstitial lung diseases	Zhou et al. (2013), Lima et al. (2006) Dong et al. (2017)

Dornase alfa (Pulmozyme): Originally developed for cystic fibrosis, dornase alfa is an enzyme that helps thin mucus in the airways, making it easier to clear. It has been investigated for its potential benefits in other respiratory conditions characterized by mucus buildup.

Montelukast (Singulair): Originally approved for treating asthma and allergic rhinitis, montelukast is a leukotriene receptor antagonist. Some studies have explored its use in chronic obstructive pulmonary disease (COPD) and eosinophilic bronchitis.

Statins (e.g., *simvastatin):* Statins, commonly used to lower cholesterol, have been investigated for their anti-inflammatory properties in respiratory diseases. Some studies suggest potential benefits in conditions such as asthma and COPD.

Methotrexate: This drug, originally developed as a chemotherapy agent, has immunosuppressive properties and has been studied for its potential role in treating severe asthma.

Thalidomide: Initially developed as a sedative and later withdrawn due to teratogenic effects, thalidomide has been explored for its anti-inflammatory and immunomodulatory properties in the context of conditions such as sarcoidosis and certain interstitial lung diseases.

7.4.3 Newer Therapeutic Approaches for Other Therapeutic Conditions

The following drugs are used in rare disease conditions (Table 7.3):

Ivermectin: Ivermectin was developed as an antiparasitic medication for animals. It has gained attention for its potential antiviral properties and has been studied as a treatment for specific viral infections in humans, including COVID-19. However, its efficacy for this purpose is still a subject of debate and ongoing research.

Ketamine: Ketamine is an anesthetic and analgesic used in human and veterinary medicine. Ketamine has shown promise in the treatment of depression, particularly treatment-resistant depression. It is administered in lower doses than those used for anesthesia.

Colchicine: Colchicine is traditionally used to treat gout and familial Mediterranean fever. It has been investigated for its potential anti-inflammatory effects and is being studied for use in conditions such as cardiovascular disease.

Methylene Blue: Methylene blue has been used as a dye and in various medical procedures. It has been explored for its potential as an antidepressant and as an antimalarial agent.

Doxycycline: Doxycycline is an antibiotic used to treat bacterial infections. It has been studied for its potential anti-inflammatory and anti-cancer properties.

Table 7.3 Repurposed drugs used in rare diseases

Sr. no.	Name of the drug	Mechanism of action of the drug	Old use	Repurposed use	References
1	Ivermectin	These findings suggest that ivermectin binds to glutamate-activated chloride channels existing in nerve or muscle cells of the nematode with a specific and high affinity, causing hyperpolarization of nerve or muscle cells by increasing the permeability of chloride ions through the cell membrane	Ivermectin was developed as an antiparasitic medication for animals	It has gained attention for its potential antiviral properties and has been studied as a treatment for specific viral infections in humans, including COVID-19. However, its efficacy for this purpose is still a subject of debate and ongoing research	Ikeda (2003), Høeg and Prasad (2023), Kaur et al. (2021)
2	Methylene Blue	Methylthioninium chloride, formally called methylene blue (MB), is an organic thiazine-type compound with a dark blue-green color. MB can be reduced into leucomethylene blue (leucoMB), an antioxidant	Methylene blue has been used as a dye and in various medical procedures	It has been explored for its potential as an antidepressant and as an antimalarial agent	Bužga et al. (2022), Xue et al. (2021), Oz et al. (2011)
3	Colchicine	Colchicine has been known as a treatment for gout for several millennia. A combination of neutrophil and macrophage activation, leukocyte adhesion molecules, inflammasome activation, and IL-1β production mediates inflammation in gout	Colchicine is traditionally used to treat gout and familial Mediterranean fever	It has been investigated for its potential anti-inflammatory effects and is being studied for use in conditions such as cardiovascular disease	Dasgeb et al. (2018), Zhang et al. (2022)

Sirolimus (Rapamycin): Sirolimus is an immunosuppressant in organ transplantation. It has been investigated for its potential in treating various types of cancer.

Metformin: Metformin is a medication used to treat type 2 diabetes. It has been explored for its potential anti-ageing effects and in preventing or treating certain types of cancer.

7.4.4 Newer Therapeutic Approach for Gastrointestinal Disease Conditions

While we do not have access to real-time data, and as our knowledge is limited to information available until January 2022, we can provide some general examples of drugs that have been explored or repurposed for gastrointestinal conditions. Keep in mind that the status and relevance of these examples may have changed since the last update, and we should consult more recent sources for the latest information (Table 7.4):

Thalidomide: Thalidomide was initially developed as a sedative but was withdrawn due to severe teratogenic effects. It has been investigated for its anti-inflammatory properties and has shown promise in treating Crohn's disease.

Mesalamine (5-ASA): Mesalamine is commonly used to treat inflammatory bowel diseases. Research has explored its potential in other gastrointestinal conditions, including diverticulitis and microscopic colitis.

Low-Dose Naltrexone (LDN): Naltrexone is used at higher doses for opioid addiction. Low-dose naltrexone has been studied for its potential anti-inflammatory effects in conditions like Crohn's disease and ulcerative colitis.

Rifaximin: Rifaximin is an antibiotic used to treat bacterial infections in the gut. It has been explored for its potential in irritable bowel syndrome (IBS) and small intestinal bacterial overgrowth (SIBO).

Methotrexate: Methotrexate is used in higher doses as a chemotherapy drug. It has been investigated in lower doses for its immunosuppressive properties in treating inflammatory bowel diseases.

7.5 Conclusion

In conclusion, drug repurposing approach has many significant aspects, such as cost and time efficiency, safety, expanded treatment options, rare diseases, and accelerated drug development. Conversely, there are many challenges to re-evaluating the drugs, such as regulatory challenges, financial issues, clinical trials, intellectual property, and demands for market analysis. It minimizes drug failure, maximizes the drug's therapeutic potential, and enhances the success rate. It also helps to re-evaluate the existing drug for the new therapeutic potential using experimental and computational approaches. This approach helps to reduce the economic burden and the time required to create a novel drug option for a disease condition. New therapeutic opportunities have been created using drug repurposing for some existing disease conditions including CNS, respiratory, and GIT-related disorders. Furthermore, despite its usefulness, there are some challenges regarding financial funding, regulatory and clinical challenges, and intellectual property rights. It is observed that the drug repurposing approach fastens the drug development process for several diseases. Hence, drug repurposing approach could be considered an alternative approach to traditional drug discovery systems.

Table 7.4 Repurposed drugs used in other therapeutic approaches

Sr. no.	Name of the drug	Mechanism of action of the drug	Old use	Repurposed use	References
1	Thalidomide	Thalidomide is also known to inhibit angiogenesis and has several other well-described immunomodulatory properties	Thalidomide was initially developed as a sedative but was withdrawn due to severe teratogenic effects	It has been investigated for its anti-inflammatory properties and has shown promise in treating Crohn's disease	Ginsburg et al. (2001), Hirose et al. (2020)
2	Mesalamine (5-ASA)	The mechanism of action for such chemoprevention has been provided by the agonism of the peroxisome proliferator-activated receptor-gamma by 5-ASA, which unifies its efficacy as an anti-inflammatory and chemopreventive agent	Mesalamine is commonly used for the treatment of inflammatory bowel diseases	Research has explored its potential in other gastrointestinal conditions, including diverticulitis and microscopic colitis	(Ham and Moss (2012), Eronen (2023), Iacucci et al. (2010), Chaudhry et al. (2023)
3	Low-dose naltrexone (LDN)	The proposed mechanism of action of LDN on pain is (1) opioid antagonism	Naltrexone is used at higher doses for opioid addiction	Low-dose naltrexone has been studied for its potential anti-inflammatory effects in conditions like Crohn's disease and ulcerative colitis	Bruun et al. (2021), Kim et al. (2023), Mao and Shen (2023)
4	Rifaximin	Rifaximin is a poorly absorbed bactericidal rifamycin derivative, which inhibits bacterial protein synthesis by irreversibly binding to RpoB, the beta-subunit of the bacterial DNA-dependent RNA polymerase	Rifaximin is an antibiotic used to treat bacterial infections in the gut	It has been explored for its potential in irritable bowel syndrome (IBS) and small intestinal bacterial overgrowth (SIBO)	Koo and DuPont (2010), Pimentel (2016), Piovezani Ramos and Camilleri (2023)
5	Methotrexate	Methotrexate is an FDA-approved folic acid antagonist	Methotrexate is used in higher doses as a chemotherapy drug	In lower doses, it has been investigated for its immunosuppressive properties in the treatment of inflammatory bowel diseases	Wojtczak et al. (2023)

References

Alda M (2015) Lithium in the treatment of bipolar disorder: pharmacology and pharmacogenetics. Mol Psychiatry 20(6):661–670

Arshad U et al (2020) Prioritization of anti-SARS-Cov-2 drug repurposing opportunities based on plasma and target site concentrations derived from their established human pharmacokinetics. Clin Pharmacol Therap 108(4):775–790

Ayache SS, Chalah MA (2017) Fatigue in multiple sclerosis–insights into evaluation and management. Neurophysiol Clin/Clin Neurophysiol 47(2):139–171

Bellingham MC (2011) A review of the neural mechanisms of action and clinical efficiency of riluzole in treating amyotrophic lateral sclerosis: what have we learned in the last decade? CNS Neurosci Ther 17(1):4–31

Bruun KD et al (2021) Low-dose naltrexone for the treatment of fibromyalgia: protocol for a double-blind, randomized, placebo-controlled trial. Trials 22(1):804

Bužga M et al (2022) Methylene blue: a controversial diagnostic acid and medication? Toxicol Res (Camb) 11(5):711–717

Cha Y et al (2018) Drug repurposing from the perspective of pharmaceutical companies. Br J Pharmacol 175(2):168–180

Chatterjee A (2004) Cosmetic neurology: the controversy over enhancing movement, mentation, and mood. Neurology 63(6):968–974

Chaudhry A et al (2023) Mesalamine - a revered drug for inflammatory bowel disease with detrimental effects on the lung and heart. Cureus 15(8):e43567

Cronstein BN, Aune TM (2020) Methotrexate and its mechanisms of action in inflammatory arthritis. Nat Rev Rheumatol 16(3):145–154

Dario A, Tomei G (2004) A benefit-risk assessment of baclofen in severe spinal spasticity. Drug Saf 27(11):799–818

Dasgeb B et al (2018) Colchicine: an ancient drug with novel applications. Br J Dermatol 178(2):350–356

Davies HRH et al (1996) Methotrexate as a steroid sparing agent for asthma in adults. Cochrane Database Syst Rev 2017(9):CD000391

Di Cesare F et al (2023) Gabapentin: clinical use and pharmacokinetics in dogs, cats, and horses. Animals (Basel) 13(12):2045

Dong X et al (2017) Inhibitory effects of thalidomide on bleomycin-induced pulmonary fibrosis in rats via regulation of thioredoxin reductase and inflammations. Am J Transl Res 9(10):4390

Dunn J, Blight A (2011) Dalfampridine: a brief review of its mechanism of action and efficacy as a treatment to improve walking in patients with multiple sclerosis. Curr Med Res Opin 27(7):1415–1423

Elkouzi A et al (2019) Emerging therapies in Parkinson disease—repurposed drugs and new approaches. Nat Rev Neurol 15(4):204–223

Eronen H (2023) Old and novel therapies in inflammatory bowel disease

Eugene AR et al (2021) Head-to-head comparison of sedation and somnolence among 37 antipsychotics in schizophrenia, bipolar disorder, major depression, autism spectrum disorders, delirium, and repurposed in COVID-19, infectious diseases, and oncology from the FAERS, 2004–2020. Front Pharmacol 12:621691

Gatta G et al (2011) Rare cancers are not so rare: the rare cancer burden in Europe. J Eur J Cancer 47(17):2493–2511

Ginsburg PM, Dassopoulos T, Ehrenpreis ED (2001) Thalidomide treatment for refractory Crohn's disease: a review of the history, pharmacological mechanisms and clinical literature. Ann Med 33(8):516–525

Goetz RL, Vijaykumar K, Solomon GM (2022) Mucus clearance strategies in mechanically ventilated patients. Front Physiol 13:834716

Ham M, Moss AC (2012) Mesalamine in the treatment and maintenance of remission of ulcerative colitis. Expert Rev Clin Pharmacol 5(2):113–123

Hinchcliffe M, Smith A (2017) Riluzole: real-world evidence supports significant extension of median survival times in patients with amyotrophic lateral sclerosis. Degener Neurol Neuromuscul Dis 7:61–70
Hirose Y et al (2020) Hypnotic effect of thalidomide is independent of teratogenic ubiquitin/proteasome pathway. Proc Natl Acad Sci 117(37):23106–23112
Høeg TB, Prasad VK (2023) An evidence double standard for pharmacological versus non-pharmacological interventions: lessons from the COVID-19 pandemic. Contemp Clin Trials Commun 33:101108
Iacucci M, de Silva S, Ghosh S (2010) Mesalazine in inflammatory bowel disease: a trendy topic once again? Can J Gastroenterol 24(2):127–133
Ikeda T (2003) Pharmacological effects of ivermectin, an antiparasitic agent for intestinal strongyloidiasis: its mode of action and clinical efficacy. Nihon Yakurigaku Zasshi 122(6):527–538
Jin G, Wong ST (2014) Toward better drug repositioning: prioritizing and integrating existing methods into efficient pipelines. J Drug Discov Today 19(5):637–644
Johnson JW, Kotermanski SE (2006) Mechanism of action of memantine. Curr Opin Pharmacol 6(1):61–67
Ju J, Lee I, Li Y-CE (2022) Stimuli-responsive smart materials for biomedical applications. Front Bioeng Biotechnol 10:1020923
Kamen L, Henney HR III, Runyan JD (2008) A practical overview of tizanidine use for spasticity secondary to multiple sclerosis, stroke, and spinal cord injury. Curr Med Res Opin 24(2):425–439
Kaur H et al (2021) Ivermectin as a potential drug for treatment of COVID-19: an in-sync review with clinical and computational attributes. Pharmacol Rep 73:736–749
Kim JH, Oh C-M, Yoo JH (2023) Obesity and novel management of inflammatory bowel disease. World J Gastroenterol 29(12):1779
Kiriiri GK, Njogu PM, Mwangi AN (2020) Exploring different approaches to improve the success of drug discovery and development projects: a review. Future J Pharm Sci 6(1):1–12
Koo HL, DuPont HL (2010) Rifaximin: a unique gastrointestinal-selective antibiotic for enteric diseases. Curr Opin Gastroenterol 26(1):17–25
Lage OM et al (2018) Current screening methodologies in drug discovery for selected human diseases. J Marine Drugs 16(8):279
Lamore R III et al (2010) Dalfampridine (Ampyra): an aid to walking in patients with multiple sclerosis. P T 35(12):665–669
Langedijk J et al (2015a) Drug repositioning and repurposing: terminology and definitions in literature. J Drug Discov Today 20(8):1027–1034
Langedijk J et al (2015b) Drug repositioning and repurposing: terminology and definitions in literature. Drug Discov Today 20(8):1027–1034
Law GL et al (2013) Drug repurposing: a better approach for infectious disease drug discovery? Curr Opin Immunol 25(5):588–592
Lima LM et al (2006) Thalidomide and analogs as anti-inflammatory and immunomodulator drug candidates. Anti-Inflamm Anti-Allerg Agents Med Chem (Formerly Current Medicinal Chemistry-Anti-Inflammatory and Anti-Allergy Agents) 5(1):79–95
Liu L-W, Hsieh Y-Y, Yang P-M (2020) Bioinformatics data mining repurposes the JAK2 (Janus Kinase 2) inhibitor fedratinib for treating pancreatic ductal adenocarcinoma by reversing the KRAS (Kirsten Rat Sarcoma 2 viral oncogene homolog)-driven gene signature. J Personal Med 10(3):130
Liu Y et al (2021) Drug repurposing for next-generation combination therapies against multidrug-resistant bacteria. Theranostics 11(10):4910
Louie S et al (2013) The asthma–chronic obstructive pulmonary disease overlap syndrome: pharmacotherapeutic considerations. Expert Rev Clin Pharmacol 6(2):197–219
Luo H et al (2011) DRAR-CPI: a server for identifying drug repositioning potential and adverse drug reactions via the chemical–protein interactome. Nucleic Acids Res 39(suppl_2):W492–W498
Lüscher Dias T et al (2020) Drug repositioning for psychiatric and neurological disorders through a network medicine approach. Transl Psychiatry 10(1):141

Mao X, Shen J (2023) Potential roles of enteric glial cells in Crohn's disease: a critical review. Cell Prolif 57:e13536

Marin L et al (2011) Chronic obstructive pulmonary disease: patho-physiology, current methods of treatment and the potential for simvastatin in disease management. Expert Opin Drug Deliv 8(9):1205–1220

McFarland AJ et al (2014) Molecular mechanisms underlying the effects of statins in the central nervous system. Int J Mol Sci 15(11):20607–20637

Mohsen AM et al (2023) Formulation of tizanidine hydrochloride-loaded provesicular system for improved oral delivery and therapeutic activity employing a 2(3) full factorial design. Drug Deliv Transl Res 13(2):580–592

Mower J et al (2022) Improving pharmacovigilance signal detection from clinical notes with locality sensitive neural concept embeddings. In: AMIA annual symposium proceedings. American Medical Informatics Association

Multani I et al (2019) Botulinum toxin in the management of children with cerebral palsy. Pediatr Drugs 21(4):261–281

Munj SA et al (2022) Biomarker-driven drug repurposing on biologically similar cancers with DNA-repair deficiencies. Front Genet 13:1015531

Naylor DM, Kauppi D, Schonfeld J (2015) Therapeutic drug repurposing, repositioning and rescue. J Drug Discov 16:57

Nigam PK, Nigam A (2010) Botulinum toxin. Indian J Dermatol 55(1):8–14

Nishimura Y, Hara H (2018) Drug repositioning: current advances and future perspectives. J Front Pharmacol 9:1068

Novac N (2013) Challenges and opportunities of drug repositioning. J Trends Pharmacol Sci 34(5):267–272

Ovallath S, Sulthana B (2017) Levodopa: history and therapeutic applications. Ann Indian Acad Neurol 20(3):185–189

Oz M et al (2011) Cellular and molecular actions of methylene blue in the nervous system. Med Res Rev 31(1):93–117

Padhy B, Gupta Y (2011) Drug repositioning: re-investigating existing drugs for new therapeutic indications. J Postgrad Med 57(2):153

Park K (2019) A review of computational drug repurposing. Transl Clin Pharmacol 27(2):59–63

Parvathaneni V et al (2019) Drug repurposing: a promising tool to accelerate the drug discovery process. J Drug Discov Today 24(10):2076–2085

Pimentel M (2016) potential mechanisms of action of rifaximin in the management of irritable bowel syndrome with diarrhoea. Aliment Pharmacol Ther 43:37–49

Piovezani Ramos G, Camilleri M (2023) Current and future therapeutic options for irritable bowel syndrome with diarrhea and functional diarrhea. Dig Dis Sci 68(5):1677–1690

Pushpakom S et al (2019a) Drug repurposing: progress, challenges and recommendations. Nat Rev Drug Discov 18(1):41–58

Pushpakom S et al (2019b) Drug repurposing: progress, challenges and recommendations. J Nat Rev Drug Discov 18(1):41–58

Reddy AS, Zhang S (2013) Polypharmacology: drug discovery for the future. J Expert Rev Clin Pharmacol 6(1):41–47

Reisberg B et al (2003) Memantine in moderate-to-severe Alzheimer's disease. N Engl J Med 348(14):1333–1341

Roessler HI et al (2021a) Drug repurposing for rare diseases. J Trends Pharmacol Sci 42(4):255–267

Roessler HI et al (2021b) Drug repurposing for rare diseases. Trends Pharmacol Sci 42(4):255–267

Romito JW et al (2021) Baclofen therapeutics, toxicity, and withdrawal: a narrative review. SAGE Open Med 9:20503121211022197

Sawada R et al (2015) Target-based drug repositioning using large-scale chemical–protein interactome data. J Chem Inf Model 55(12):2717–2730

Shapiro DA et al (2003) Aripiprazole, a novel atypical antipsychotic drug with a unique and robust pharmacology. Neuropsychopharmacology 28(8):1400–1411

Shineman DW et al (2014) Overcoming obstacles to repurposing for neurodegenerative disease. J Ann Clin Transl Neurol 1(7):512–518
Sisignano M, Parnham MJ, Geisslinger G (2016) Drug repurposing for the development of novel analgesics. J Trends Pharmacol Sci 37(3):172–183
Staničová J, Miškovský P, Šutiak V (2001) Amantadine: an antiviral and antiparkinsonian agent. Vet Med Czech 46(9):244–256
Tian H, Cronstein BN (2007) Understanding the mechanisms of action of methotrexate: implications for the treatment of rheumatoid arthritis. Bull NYU Hosp Jt Dis 65(3):168–173
Volkow ND et al (2002) Mechanism of action of methylphenidate: insights from PET imaging studies. J Atten Disord 6(Suppl 1):S31–S43
Wagener JS, Kupfer O (2012) Dornase alfa (Pulmozyme). Curr Opin Pulm Med 18(6):609–614
Wang X, Guan Y (2021) COVID-19 drug repurposing: a review of computational screening methods, clinical trials, and protein interaction assays. Med Res Rev 41(1):5–28
Wermuth HR, Badri T, Takov V (2023) Montelukast. In: StatPearls. StatPearls Publishing Copyright © 2023, StatPearls Publishing LLC, Treasure Island (FL) ineligible companies. Disclosure: Talel Badri declares no relevant financial relationships with ineligible companies. Disclosure: Veronica Takov declares no relevant financial relationships with ineligible companies
Wiffen PJ et al (2017) Gabapentin for chronic neuropathic pain in adults. Cochrane Database Syst Rev 6(6):Cd007938
Wilkinson GF, Pritchard K (2015) In vitro screening for drug repositioning. J Biomol Screen 20(2):167–179
Wojtczak M et al (2023) Can methotrexate be employed as monotherapy for Bullous Pemphigoid? analysis of efficiency and tolerance of methotrexate treatment in patients with Bullous Pemphigoid. J Clin Med 12(4):1638
Xue H, Thaivalappil A, Cao K (2021) The potentials of methylene blue as an anti-aging drug. Cells 10(12):3379
Yang C, Montgomery M (2021) Dornase alfa for cystic fibrosis. Cochrane Database Syst Rev 3:CD001127
Zhang F-S et al (2022) Therapeutic potential of colchicine in cardiovascular medicine: a pharmacological review. Acta Pharmacol Sin 43(9):2173–2190
Zhou S et al (2013) Thalidomide-a notorious sedative to a wonder anticancer drug. Curr Med Chem 20(33):4102–4108

Challenges for Drug Repurposing: Special Emphasis on Regulatory and Patent Issues

8

M. Maheshwari, S. Ghosh, S. Ali, K. Shubham, M. Pal, P. K. Sahoo, and B. Behera

Abstract

Drug repurposing is the method of discovering new therapeutic indications for previously available drugs. It is increasingly appealing to repurpose existing drugs for the treatment of rare and common diseases due to high attrition rates, high costs, and the slower rate of drug discovery and development. The concept of drug repurposing may not be new, but it has gained momentum over the past few years. Several drugs have been effectively repurposed. Over the past decade, numerous pharmaceutical companies have developed new drugs based on the identification of novel biological targets using drug repurposing. This process combines experimental and in silico techniques to rationally develop/identify new applications for pharmacological compounds. However, the prospects of this strategy have not been fully utilized, owing in part to regulatory, technological, and legal challenges. Many promising drugs are shelved because of inadequate efficacy, poor market potential, and consolidation in the industry. In this chapter, we have discussed these challenges. Various innovative strategies that could address and solve these barriers, realizing the full potential of drug repurposing, have also been discussed.

Keywords

Drug repurposing · Intellectual property rights · Patents · Regulatory barriers · Serendipity · In silico technique · Drug discovery

M. Maheshwari and S. Ghosh contributed equally with all other contributors.

M. Maheshwari · S. Ghosh · S. Ali · K. Shubham · M. Pal · P. K. Sahoo · B. Behera (✉)
Delhi Institute of Pharmaceutical Sciences and Research, DPSRU, New Delhi, India
e-mail: beauty.behera@dpsru.edu.in

R. K. Kaundal et al. (eds.), *Drug Repurposing: Exploring Newer Therapeutic Potentials*,
https://doi.org/10.1007/978-981-95-4522-3_8

8.1 Introduction

Drug repurposing, also termed as drug repositioning, re-profiling, switching, or redirecting, is the current state of the art for new indications. Drug repurposing is the technique of finding a novel therapeutic activity for a different ailment for which there exists a clinically approved medication. This also applies to the drugs that were abandoned during the development process and proved to be risky or ineffective in clinical trials (Sudeep Pushpakom et al. 2019). Drug repurposing also involves using medications that have safety concerns in some population. It raises the drug's therapeutic value to its maximum potential, which in turn increases its success rate (Jarada et al. 2020). Unlike traditional drug development strategies, this approach is cost-effective, efficient, and low-risk (Begley et al. 2021). However, when developing such strategies, medications that have not been submitted to clinical testing are excluded (Kirtonia et al. 2021). The idea of repurposing drugs does not include any structural changes made to the medication. Repurposing of a drug allows the use of an approved medication for a new indication by altering its formulation, dose, or mode of administration (Krishnamurthy et al. 2022). Drug repurposing helps us to expand our knowledge without restricting the existing information. Historically, the repurposing of drugs has been opportunistic and unintentional (Parvathaneni et al. 2019). It often involves chance discoveries through serendipity. Serendipitous discoveries are a successful approach to treating various medical conditions (Copeland 2015). Scientists have discovered several repurposed drugs for cancer, cardiovascular diseases, and neurological disorders, among others, by chance. Repurposed medications proved their effectiveness in treating diseases in addition to the illnesses for which they were developed initially (Charlton et al. 2018). For example, sildenafil was initially intended for treating angina but was later found to be effective in the treatment of erectile dysfunction (Goldstein et al. 2019). Similarly, minoxidil has been developed for managing hypertension but was subsequently discovered to stimulate hair growth and became a popular treatment for baldness (Villani et al. 2021). Some drugs were also repurposed based on prospective clinical experience (Cavalla 2019). Thalidomide, which was initially licensed for morning sickness in pregnant women in the 1950s, was later repurposed for treating erythema nodosum leprosum (ENL) in 1964 and multiple myeloma in 1999 (S. Pushpakom 2022). Repurposing old drugs is a topic that can be explored for the discovery of new applications for any medication (S. Pushpakom 2022). Several examples of repurposed drugs, including the older ones, that have left their impact on medical history are described in Table 8.1.

The unwanted side effects of any treatment might also serve as an indication to investigate if the drug would be beneficial for other ailments (Karaman and Sippl 2019). This can be employed for the treatment of both common and rare diseases, and it is becoming an increasingly appealing possibility due to the utilization of low-risk drugs (Sudeep Pushpakom et al. 2019). This technique has become popular because of its shorter time frame for therapeutic development. This approach lowers

Table 8.1 Drugs that have been repurposed and approved

Original drug	Original use	Repurposed use	Approval year	Reference
Aspirin	Pain and fever	Cardiovascular disease prevention	1899	Jourdan et al. (2020)
Chlorpromazine	Antihistamine	Sedative and anti-emetic, antitumor (CRx-026)	1952	Jourdan et al. (2020)
Thalidomide	Nausea and insomnia in pregnancy	Treatment of ENL and multiple myeloma	1964 and 1999	Sudeep Pushpakom et al. (2019)
Finasteride	Prostate enlargement	Cure of hair loss (Propecia)	1998	C. Li and He (2023)
Sildenafil	Angina	Erectile dysfunction (Viagra)	1998	Jourdan et al. (2020)
Atomoxetine	Parkinson's disease	Attention-deficit/ hyperactivity disorder (ADHD)	2002	C. Li and He (2023)
Rituximab	Various cancers	Rheumatoid arthritis	2006	C. Li and He (2023)
Raloxifene	Osteoporosis	Breast cancer	2007	C. Li and He (2023)
Dapoxetine	Analgesia and depression	Premature ejaculation	2012	C. Li and He (2023)
Dimethyl fumarate	Psoriasis	Multiple sclerosis	2013	Jourdan et al. (2020)
Ketoconazole	Fungal infections	Cushing's syndrome	2014	C. Li and He (2023)

the risks and costs by directly moving to the preclinical and clinical trials and skipping the drug development process. The concept of repurposing drugs is now becoming necessary due to the global epidemic of coronavirus disease (COVID-19), which started in China. There is a pressing need to develop new drugs to combat this disease because of the epidemic's rapid onset and high potential for infection (Ng et al. 2021). Many intriguing therapeutic candidates have been uncovered through drug repurposing; some of them are already in clinical trials (Rudrapal et al. 2020). Despite the benefits associated with drug repurposing, this technique can sometimes be expensive, time-consuming, and dangerous than the de novo development method. Risk is the same as it is for any new medicine; late-stage development failures can occur, and certain specialized reformulations might be just as expensive as de novo development (Jarada et al. 2020). Its success in treating rare diseases is not always guaranteed, despite having strong evidence (Dhir et al. 2020). There are also significant patent, regulatory, and technical barriers that prevent drug repurposing from progressing (Bloom and Cadarette 2019). In this chapter, the above barriers and the strategies to overcome these challenges are discussed in detail.

8.2 Approaches for Drug Repurposing

The extraction of raw data linked to disease-drug targets is the first step, which is then followed by the construction of a drug-target-disease association. Computational, experimental, and mixed methods are the primary approaches to drug repurposing (Berger et al. 2014). Hypothesis generation in drug repurposing plays a crucial role in identifying the most suitable drug for a specific indication with a high level of confidence. Once the hypothesis is developed with comprehensive support from possible evidence, a computational or experimental approach may then be chosen (Jourdan et al. 2020). Furthermore, the shortlisted drugs undergo clinical trials for the new indication before being released to the market (Fig. 8.1). The repurposing of drugs based on clinical data is part of the broader fields of precision medicine and drug discovery. The intersection of precision medicine and drug discovery is a promising area of study with huge potential to enhance patient outcomes (Mehta 2022). For instance, electronic health records (EHRs) can be analyzed to obtain meaningful insights for drug repurposing (T. Sharma et al. 2023b). The following section describes in depth the approaches used in drug repurposing.

8.2.1 In Silico Approaches

The in silico approach for drug repurposing involves using several methods that heavily rely on computational data (Akhoon et al. 2019). This process usually involves ligand-based and structure-based drug design complementing artificial intelligence. These methods play a vital role in the repurposing of drugs and use different computational algorithms to estimate the therapeutic efficacy of a drug (Karaman and Sippl 2019). Computational approaches, either alone or combined, can systematically analyze large-scale data to identify potential repurposing opportunities (Akhoon et al. 2019). The in silico techniques involve a combination of machine learning, data mining, and ligand-based and structure-based approaches to understand the complex relationships between diseases, drugs, and targets (Agamah et al. 2020). Diseases can then be identified based on their molecular profile, like genes, environmental variables, signaling pathways, biomarkers, and other parameters that contribute to disease pathogenicity. Data mining and other computer approaches are also used to estimate the degree of similarity between diseases having several genetic features in common. For instance, Parkinson's and Alzheimer's diseases share 48 genes and 4 signaling pathways (Ashburn and Thor 2004). The existence of protein targets shared by these diseases implies that a particular drug may be efficacious against both conditions. Computational techniques for predictive repurposing are a set of powerful tools that can rapidly identify possible applications of pre-existing drugs for new indications (Roy et al. 2021). By using data-driven methods, such as machine learning and artificial intelligence, new therapeutic options can be identified (Jourdan et al. 2020). This approach can be particularly useful in rare conditions for which there is often limited knowledge about disease pathophysiology. In such cases, traditional drug development approaches

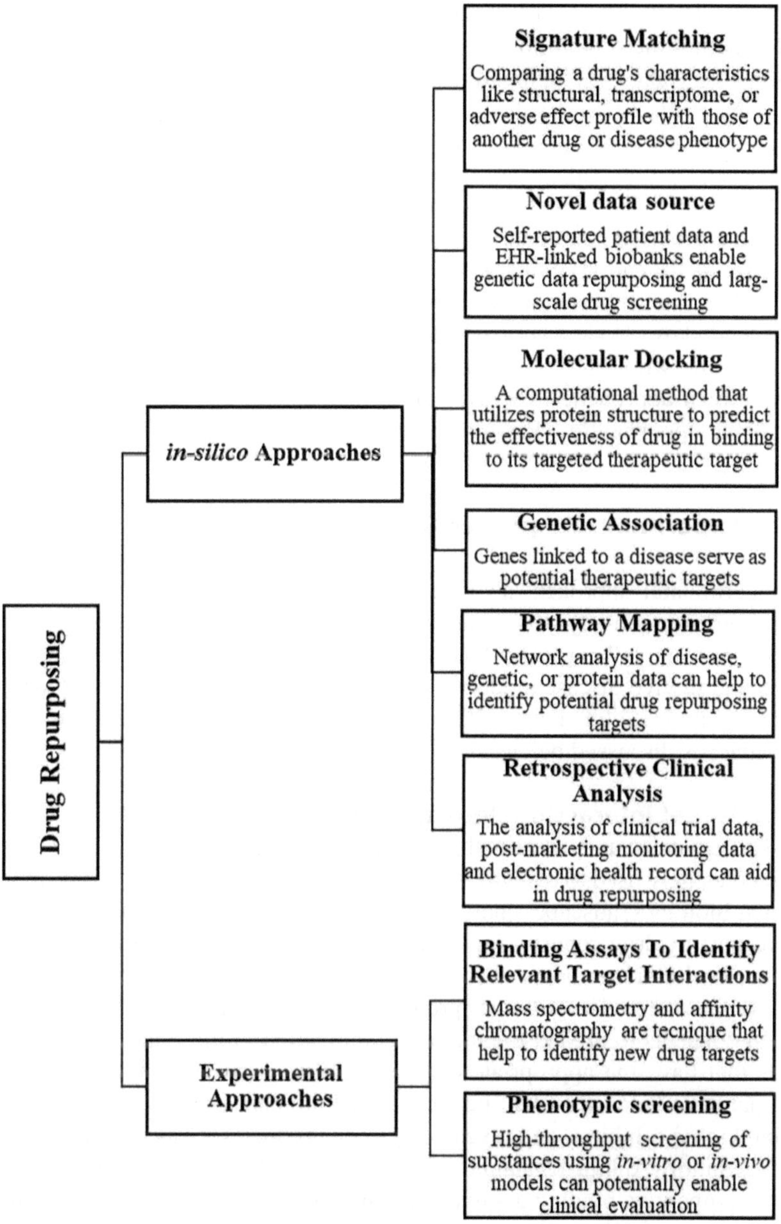

Fig. 8.1 Pathway of a successful drug repurposing

may not be feasible, as these techniques require a deep understanding of the disease mechanism, which may not be available (Rudrapal et al. 2020). Furthermore, with the increasing availability of large-scale genome sequencing initiatives, it is becoming more likely that data will indicate a specific genetic variation responsible for a rare disease (Park 2019). This discovery could lead to fast repurposing of existing drugs targeting the protein in question. This approach could potentially accelerate drug discovery and development, thus offering new treatment options for rare diseases (Cha et al. 2018). Evaluation/validation models may vary from the recommended computational models in certain contexts. Thus, understanding and selecting appropriate validation models is crucial for the effectiveness of the suggested computational models (Rudrapal et al. 2020). Furthermore, identifying the appropriate group of candidate drugs for validation during computational methods is critical. Identification of drug candidates is based on their cost, high toxicity, low bioavailability, and the fact that many treatments have been abandoned or are not recommended by physicians (Roy et al. 2021). These factors might make it difficult to analyze the whole process of in silico–*based* drug repurposing. In addition to computational approaches, experimental methods can also be utilized to identify repurposing opportunities (Parvathaneni et al. 2019).

8.2.2 Experimental Approaches

The experimental approach, also known as activity-based repurposing, involves using experimental assays to screen existing drugs for new therapeutic applications. In this approach, cell/organism-based and protein target-based screens are used in in vitro and in vivo disease models, and no structural information about target proteins is required (Rudrapal et al. 2020). There are several techniques for experimental approaches discussed below.

8.2.2.1 Target Interaction Detection Using Binding Assays

Mass spectrometry and affinity chromatography are proteomic methods that have been increasingly used to recognize the binding partners of several drugs. In today's chemical biology landscape, analyzing drug targets and off-targets has become an integral part of drug repurposing and target validation (Bellera et al. 2015). For instance, a cellular thermostability assay is being introduced as an effective method to assess the interaction of target proteins in cells. This is done by applying biophysical concepts to predict the thermal stabilization of these target proteins using ligands that have the appropriate cellular affinity. Early accomplishments of this technique have led to the identification of cell targets associated with crizotinib, a tyrosine kinase inhibitor, and revealed Quinone reductase-2 as a cellular off-target of paracetamol (Ankita Khataniar et al. 2022). Earlier, researchers employed an in vitro competitive binding assay to examine the efficiency of 38 kinase inhibitors against a group of 317 distinctive human protein kinases. Their investigation uncovered 3175 binding interactions (Cichonska et al. 2017). Additionally, protein kinase inhibitors are well known for their ability to complement genetic approaches, offer unexpected biological insights, and be crucial in preclinical target validation studies. It is anticipated that in the twenty-first century, protein kinases will be the main target during drug repurposing (McCarthy et al. 2013). The initial stages of unbiased affinity techniques are especially helpful in knowing the potential effects of chemicals on cells, such as paradoxical kinase activation through inhibitors, which underlies mechanical off-target tumor development in patients. The relationship between binding and effectiveness in the biological context may also be better understood with the assistance of chemical genetics (Akhoon et al. 2019). These studies often result from industry-driven high-throughput immediate binding or catalytic assays. These involve using different in vitro and organism-based assays to study how small molecules bind to kinases across the kinome. The outcome is a heat map showing biologically meaningful interactions (A. Khataniar et al. 2022).

8.2.2.2 Phenotypic Screening

Phenotypic screening refers to the process of identifying novel medications with disease-relevant properties in models without prior knowledge of the impacted target(s). This screening includes assays that replicate the biology of diseases in human or experimental cells or tissues, like in vitro or ex vivo. These types of assays provide incomprehensible evidence that distinguishes them from recombinant cell lines. By testing the activity of compounds at relevant concentrations, these assays can reveal novel mechanisms of action. In vitro assays are useful for repurposing activity determination. They provide direct knowledge of novel disease settings. These techniques also offer data-driven options for further assessment, permit serendipitous compound evaluation, and allow testing over the entire drug concentration-effect range (Kuzikov et al. 2021). Cell-based assays are another widely used in vitro procedure in various disease areas with different outcomes and throughputs, typically in a 96-well configuration (Iljin et al. 2009). This approach has potential for drug repurposing, particularly when the evaluated compounds are already licensed or under study (Lipinski and Reaume 2020). For instance, high-throughput cell-based screening techniques have been utilized to discover 4910 small molecules for four different types of prostate cancer. During this study, it was identified that Disulfiram, a medication commonly used to treat alcoholism, is an exceptionally selective antineoplastic agent (Iljin et al. 2009). This drug, when combined with copper at clinically relevant levels, has the potential to overcome anticancer treatment resistance in vitro, in vivo, and even in patients. Whole-organism phenotypic testing, including the use of zebrafish models, can also aid in drug repurposing. Previously, in a study, the efficacy of 39 medications approved by the US Food and Drug Administration (USFDA) for treating tobacco dependency was evaluated. In this approach, they discovered that certain chemicals, such as topiramate and apomorphine, altered ethanol and nicotine-induced behavior (Cousin et al. 2014). Drug repurposing effectively requires a thorough literature review to understand the profile of the drug as well as the mechanisms of disease. Up-to-date information on biomarkers and previous studies on gene manipulation should be included during this process. A suitable approach and associated technologies should then be identified for new applications (Sudeep Pushpakom et al. 2019). The successful proposals then advance to the appropriate level of the drug development pipeline.

8.3 Advantages of Drug Repurposing

The expensive and time-consuming method of discovering drugs and their development has slowed progress in the field. Due to the high risk, high costs, and slow pace related to the process of drug discovery, there is a need for a more efficient approach that meets market demands. Drug repurposing is one such technique that can bypass the unexpected failures that may occur during primary preclinical studies. This can significantly reduce the amount of time and money required for drug discovery, saving up to 40% of costs from molecule to market through a shorter pathway (Low et al. 2020). This is a viable method for several reasons (K. Sharma et al. 2023a). For starters, it minimizes the duration and cost of clinical trials by eliminating the need for phase I and II evaluations, which saves a significant amount of money on preclinical, phase I, and phase II expenditures. It provides distribution access to existing pharmaceutical supply networks. The strategy of drug repurposing enables the exploration of new mechanisms of action for existing drugs (Oprea et al. 2011). Lastly, the risk of the repurposed drug failing is very low due to its safety in preclinical models and clinical trials. If the early-stage trials have been completed, this indicates its potential for success. It is worth mentioning that roughly 30% of new market entrants are generated from pre-existing medications, indicating that this technique is effective. Drug repurposing is a less risky investment option that more often provides quicker results, making it a better choice from this perspective (Parvathaneni et al. 2019). These techniques can identify mechanisms directly linked to pharmacological treatments for specific diseases and also reveal the mechanisms associated with pharmaceuticals or diseases (Ankita Khataniar et al. 2022). A survey of 30 pharmaceutical companies revealed that reintroducing a repurposed drug costs $8.4 million, while relaunching a novel formulation costs $41.3 million due to the extensive information available. This leads to a higher success rate for repurposed drugs compared to the original medications, thus lowering their attrition rate (Shruthi et al. 2023). The process of repurposing drugs depends on previous research and development efforts. Any new drug candidate can be quickly established for clinical trials, shortening its review process by many regulatory bodies. If approved, its implementation into healthcare will shorten its overall processing time (Doan et al. 2011). Repurposing drugs avoids the trial-and-error stage involved in the drug development process. The process of turning a potential therapeutic candidate into an authorized medication takes 10–15 years (Parvathaneni et al. 2019). Therefore, drug repurposing procedures are essential to minimize drug processing expenses and time while boosting the efficacy of treatment. Another advantage of this process is accessible bioinformatics databases for drug candidates (Hernández-Lemus and Martínez-García 2021). This technique has become increasingly useful in expanding bioinformatics and chemo-informatics databases. The biomolecular database helps with pathway information such as drug-target interactions. Other databases also provide important information on gene expression and chemical structures and can be used to screen suitable candidate drugs for repurposing (Fotis et al. 2018).

8.4 Phases of Approval of Repurposed Drug

Earlier, Phase II trials assessing the safety as well as efficacy of new treatments encountered high failure rates. To combat this issue, various tactics, including partnerships, acquisitions, and advanced technology, have been employed but with limited success. Drug repurposing has emerged as a feasible method to overcome challenges in drug development and research (Polamreddy and Gattu 2019). Initially, drug repurposing was discovered by chance, but pharmaceutical companies have become more interested in it. The cheminformatics and bioinformatics techniques have been discovered; they have evolved into a data-driven and innovative approach (Cha et al. 2018). The regulations for the approval of repurposed drugs vary across countries globally. Each regulatory agency (Europe, Canada, Australia, the USA, and Japan) has set its timeline for approving repurposed drugs. However, the stages for the approval of drugs remain the same for all countries (Markossian et al. 2018). The major steps involved in the approval of repurposed drugs are depicted in Fig. 8.2 (Kulkarni et al. 2023).

The initial stage involves compound identification. During this stage, the candidate drug is chosen for a certain target in the individuals. Drug development can be sped up and made less risky by identifying chemical probes' on- and off-targets across the system (Kulkarni et al. 2023). The results of previous computational models provide a systematic way to guide mapping efforts, as experimental profiling of the entire target space of drug-like molecules is costly. This improves knowledge of the probes' therapeutic potential and possible adverse effects. In other methods, targets such as receptors, enzymes, proteins, and cells are identified through phenotype-based and target-based drug discovery applications (Xue et al. 2018). The target-ligand interactions are mapped using network pharmacology. To identify the best compound, three major factors—binding, functional, and phenotypic properties of the ligand and target—are taken into account. The following factors are utilized to complement experimentally mapped compound-target networks for drug repurposing applications (Tanoli et al. 2020). Additionally, these network pharmacology approaches may propose entirely unanticipated and unique investigational probes for drug development (Cichonska et al. 2015). Compound acquisition after the identification of drug candidates involves acquiring a license from the regulatory agency for a candidate drug. The drug candidate is authorized

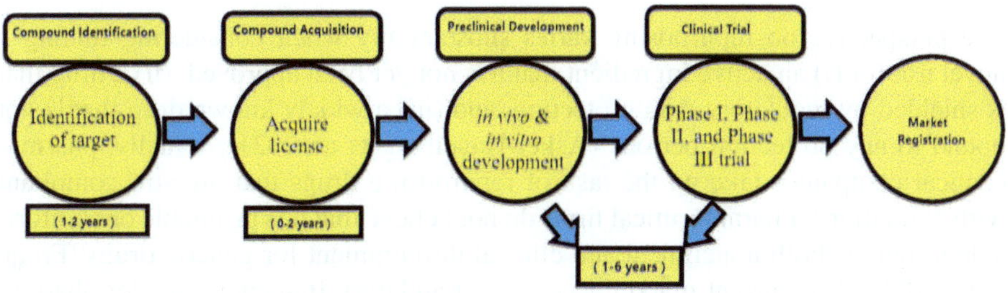

Fig. 8.2 Stages for approval of drug repurposing

to proceed with investigations to determine the medication's safety and effectiveness after completion of the acquisition. Researchers need 0–2 years to license new candidate drugs (Xue et al. 2018).

The preclinical development phase follows this stage where clinical trials are necessary to verify the efficacy and safety of medications. This approach is very similar to the process of discovering new drugs from scratch (Begley et al. 2021). Although the rejection rate is low in comparison to new drug research, the expense of bringing a medicine to market amounts to millions of dollars, even for pharmaceuticals in late-phase clinical trials. Furthermore, the medicine could fail during clinical trials or after its release to the market. Many medications have entered clinical trials for new ailments around the world (Polamreddy and Gattu 2019). In 2017, there were 170 repurposed medications in the development stage, with 72% in clinical development, 10% approved, and the rest in various stages. Corporate sponsors funded 30% of the Phase I and II trials, while academic sponsors funded nearly 70%. Industrial collaboration is necessary for Phase III studies, regulatory approval, manufacture, and commercialization (Inamoto et al. 2019). However, the success rate of drug repurposing has accounted for around 30% of all new FDA-approved medications and immunizations in recent years (Pillaiyar et al. 2020). Once the drug is released into the market, it is observed and monitored for safety purposes. The interest in repurposing drugs has grown in academia and industry. This has resulted in a rise in the quantity of drugs that have the potential to be repurposed. However, many of these medications may not pass validation but go through clinical trials, get regulatory clearance, and be successfully brought to market. Early trials increase the chances of success and save money and time. Previous research can also help reduce the risk of failure in drug repurposing projects (Parvathaneni et al. 2019). Repurposed drugs targeting the same molecular pathway are studied in new patient populations at comparable or lower doses. Clinical data from Phase III and IV trials have provided comprehensive information on the drug's profile, reducing the need for additional research and shortening clinical trial times. Late-stage clinical trials pose challenges for obtaining regulatory approval and marketing drug products (Begley et al. 2021). Drug repurposing often lacks official regulatory approval due to various legal, regulatory, and market constraints.

8.5 Regulatory Barriers

The perspective on repurposing varies substantially when considering finding a novel use for (i) an active ingredient that has not yet been approved, (ii) a drug that is shielded by regulatory data protection, and (iii) a widely known drug that is not bound to any protection periods (Asker-Hagelberg et al. 2021). Usually, pharmaceutical companies take up the task of repurposing drugs that are still compliant with regulations. Pharmaceutical firms do not believe that it is profitable or comfortable to pursue both nonclinical and clinical development for generic drugs (Freije et al. 2020). The current market access and regulatory framework hinder pharmaceutical companies from looking into the potential of repurposing existing drugs.

On the other hand, stakeholders such as learned organizations and academic institutions are more inclined to consider repurposing options once medications have passed their protection periods (Breckenridge and Jacob 2019). Although there are numerous success stories and strong reasons for the drug repurposing of approved drugs, it is not as prevalent as predicted (Parvathaneni et al. 2019).

One of the regulatory barriers is that pharmaceutical companies usually struggle to perceive a satisfactory return on investment (ROI) in research and development to get regulatory licenses and approvals for repurposed drugs (Krishnamurthy et al. 2022). Another barrier is to figure out how to include on-label novel and potential indications that will help patients/practitioners meet an unmet medical need. While off-label use may provide patients with urgent medical needs with timely access to treatments, it also carries significant financial, safety, and liability risks for patients and registered practitioners (Verbaanderd et al. 2020). First of all, the regulatory documents like the patient information leaflet and summary of product characteristics do not include this application. The patients are thus not provided with the information in the patient information booklet regarding the proper circumstances of the specific use and warnings (Asker-Hagelberg et al. 2022). Patients may also face challenges in accessing potentially helpful treatments if the regulatory body does not formally label a new indication that has been created for a medication. This may develop a sense of distrust or uncertainty within the patients regarding their treatment (Persidis 2015). Patients may be unable to pay for off-label therapies because not all nations support the financing of this use. Regulatory agencies often reject the "off-label" use of a drug due to the lack of well-established safety profiles and drug effectiveness for the off-label indication (Dias 2020). Another challenge is the risk of suddenly losing a medication widely used off-label for a specific purpose if its manufacturer chooses to withdraw it from the market (Verbaanderd et al. 2020). The approval procedure for novel drugs and repurposed medications is the same in terms of quality, safety, and efficacy. The risk of repurposing a drug is similar to that of developing a new drug. In the case of drugs where the safety profile has been well understood in the market for a long time, the variations in the benefits of a new indication may alter the overall benefits and harms balance (Breckenridge and Jacob 2019). It may thus become necessary to repeat prior tests if standards in regulatory science continue to rise. The project becomes less appealing because of these two elements, which in turn raise the total cost of development (Verbaanderd et al. 2020). Market access for any new indication is another challenge for repurposed medication. This is because new indication accessibility relies on clinical efficacy, cost-effectiveness, health technology assessment, and regulatory evidence of quality, efficacy, and safety (Persidis 2015).

Biopharmaceutical companies interested in pursuing a drug repurposing campaign must make a significant investment in the research and clinical trial programs for the repurposed drug (Ng et al. 2021). Clinical trials may delay drug repurposing attempts. Large pharmaceutical companies may find it less challenging, while non-profit organizations, smaller biotechnology companies, and academic institutions may find it difficult to afford. Furthermore, even when drug repurposing projects may help find treatment agents for rare diseases, typical clinical trials may not be feasible in such cases. Regulatory bodies thus find it difficult to approve drug repurposing projects (Hernandez et al. 2017). Repurposing can save millions of dollars by potentially reducing preclinical and early-stage research by 6–7 years. Repurposing medications could expedite the preclinical and Phase I trials of the drug discovery process, but Phase II clinical trials may present challenges. This is because, for a larger sample size in subsequent phases, more resources will be required for further validation. Clinical research organizations may also have little expertise in holding trials to meet the necessary regulatory requirements. This leads to the loss of regulatory interest and awareness that in turn may hamper drug repurposing (Mittal and Mittal 2021). A company must have both financial resources and human capital with the appropriate expertise on the medicinal product and investigated indications for drug repurposing (Breckenridge and Jacob 2019). Recent studies have repurposed antiviral and immune-modulating drugs for the treatment of Ebola. The World Health Organization however has rejected these FDA-approved medications because of potential toxicity and insufficient experimental research (J. Li et al. 2016). The procedure of out-licensing offers another company (the licensee) permission to use its intellectual property, technology, and product. In this instance, the licensor usually collects upfront fees, milestone payments, and royalties from the licensee (Gassmann et al. 2016). Although out-licensing has several advantages, there may be significant expenses associated with remanufacturing the active ingredient during this process. These expenses include regulatory paperwork, finalizing research studies, practicing pharmacovigilance, and keeping track of patient safety profiles (Sleigh and Barton 2010). Large drug company management is reluctant to out-license abandoned compounds due to liability concerns. Potential liability issues could arise if a medicine they developed causes several adverse patient actions (e.g., morbidity or fatality). While licensees often give reimbursements, their impact may be restricted. A huge pharmaceutical company is a significantly more attractive candidate for legal action if an unfavorable event occurs, especially if the licensee is a small or conventional company (Chesbrough and Chen 2013). It may be possible to find unexpected toxicities or new adverse events by testing compounds for new populations and indications (Krishnamurthy et al. 2022). The investigator is required to provide safety data to the parent company for studies that are sponsored externally. Pharmaceutical companies do not choose less profitable products because of the high expense of legal costs (Rudrapal et al. 2020).

Creating an application outside a company's specialized area is risky and costly. Obtaining marketing authorization for new indications presents challenges including high administrative costs, product labels, and pharmacovigilance systems (Sultana et al. 2019). There are several challenges in perceiving secondary indication patents. First, the new indication may not be novel or inventive under patent law, as evidence is already published. Patentees must include sufficient evidence of efficacy during application. Before the examiner receives sufficient information to grant the patent, there may be a risk of public exposure (Sudeep Pushpakom et al. 2019). Second, enforcing patents for secondary medical use can be difficult due to the disconnection between stakeholders in the pharmaceutical supply chain. If a generic drug of new indication is available in the market, then the developing company may not recoup their investment. This is because, in most countries, physicians prescribe cheap and generic drugs so that patients can afford the medication (Begley et al. 2021). Finally, patent claims for additional uses of drugs sometimes offer less protection compared to the primary drug patent. This is because the current drug patent system does not effectively promote the development of new applications for drugs that are no longer under patent protection. As a result, there is a higher likelihood that other companies may exploit these repurposed uses. Additionally, obtaining secondary patents can also be expensive (Breckenridge and Jacob 2019). Repurposing minimizes the risks associated with drug development, even though many compounds are already in place and well understood (Sudeep Pushpakom et al. 2019).

8.6 Challenges During Patent Filing

Intellectual property protection is the most frequent obstacle to the drug repurposing process. It has been reported that pharmaceutical corporations patent a large quantity of drugs and prohibit others from producing these substances without a license (Krishnamurthy et al. 2022). Intellectual property terms are regarded as tough and time-exhausting to negotiate because pharmaceutical firms usually wish to safeguard their right to use their compounds. After all, owners desire to keep possession of inventions, be paid, and ensure the availability of medications to patients. Pharmaceutical companies are not able to recover their investments due to the ineffectiveness of the available regimens, particularly for medications for which generic equivalents are already available (Verbaanderd et al. 2020). Repurposing drugs may have a lower ROI if the compound that failed in subsequent phases of development has a short amount of patent term remaining. Depending on the size of the pharmaceutical organization, several thresholds are used to decide if an ROI is worthwhile. A higher ROI could be necessary for larger businesses than for smaller ones (Chesbrough and Chen 2013). The ROI from the small business is not sufficient to develop new medications for patients. Finding a new indication will usually provide a low ROI if the rights to the intellectual property of drugs or formulations are no longer valid (Krishnamurthy et al. 2022). Composition of matter (COM) claims is one of the strongest forms of intellectual property protection for recently synthesized compounds. The COM claims for repurposed drugs are difficult to obtain because the owner of the patent must show that the medication is a strong candidate for new use and must differentiate their claims from previously available information (Fetro and Scherman 2020). According to the literature, it is usual for researchers to patent several compounds during development, protecting not just the candidate drug but also the final drug. It is also necessary to register for a patent for the repurposed drug to safeguard the ROI from the sale of the repurposed medication. This will prevent third parties from manufacturing and commercializing it under the new indication while holding the patent (Krishnamurthy et al. 2022). The patent period for shelved compounds is a major challenge as the companies might not get sufficient time to promote and develop them (Rose 2013). Another difficulty is the involvement with core intellectual property. A license to access patents that protect the shelved/failed compound of the same family is required for any other party to develop these compounds (Chesbrough and Chen 2013). Pharmaceutical companies must develop new intellectual property to capitalize on market exclusivity. Such intellectual property may result from the drug repurposing, an innovative formulation, or any of several other strategies such as dose, route of administration, and drug combination that can enhance the suitability of the drug for a new patient population. Acquiring new intellectual property is a challenge for these companies. This restricts the kinds of chemicals they can investigate, particularly when it comes to drugs for which substantial data is available (Cha et al. 2018). The additional risks for new intellectual property are consistent losses in business due to political, environmental, internal, and external economic factors. Repurposing drugs by the initial organization necessitates an initial investment and convincing management to put money into the development of drugs that were not lucky the first time. Based on this kind of organizational mindset, thalidomide was repurposed by a different company to treat multiple myeloma, instead of its originally intended use for nausea and insomnia in pregnancy (Novac 2013). Intellectual property is a tough obstacle to overcome because it safeguards pharmaceutical investments and prevents rivals from expanding on initial research and repurposing compounds. Material transfer agreements (MTAs) are another difficult and time-consuming obstacle. This is a type of legal document that companies can use to protect their sensitive intellectual property. When companies send samples out to third parties for extra measurement or treatments, as a contractual agreement, MTA is an important document. This document helps the company claim ownership of the samples/drugs and make sure that the third party does not breach their data. The MTA negotiations are particularly contentious when it comes to topics like postponing academic publications to safeguard confidential information, intellectual property provisions, limiting company liability, and restricting compound application to noncommercial research (Marusina et al. 2011). Some legal aspects can hamper the patenting of new medicinal uses and protect patent rights, thereby reducing incentives for the process of drug repurposing. Initially, several national laws make it difficult to get a patent for additional or second medical purposes. Second, numerous unregistered and off-label repurposing indications have been reported in specialized literature or are currently being used in therapeutic practice. This knowledge is already available to the public, even though controlled clinical trials have not yet approved such usage. It may impact patentability and innovation (Talevi and Bellera 2020).

Enforcing a patent for off-patent drugs is challenging when the new indication uses existing strengths and dosage forms. An ideal situation for drug repurposing is when the new indication requires nonmarketed strengths or a unique, new formulation. Changing the drug molecule will deviate from the current repurposing strategy and potentially diminish the effectiveness of the approach (Witkowski 2011). This strategy will alter the drug's composition, thus abandoning the repurposing method. Another obstacle while filing a patent is the process of skinny labeling. Skinny labeling is the practice of generic pharmaceutical manufacturers seeking approval to use the drug label for nonpatented indications of drug products.

An Abbreviated New Drug Application (ANDA) typically necessitates the entire innovator label replication. When a drug product is approved for multiple purposes but only one is patent-protected, ANDAs can be approved with labelling for the non-patent-protected indication. This technique called "skinny labeling," is accomplished by including a statement in the ANDA stating that it does not cover the patent-protected indication (Sudeep Pushpakom et al. 2019). In the USA, a pharmaceutical company can file two types of patents, one of which is a compound patent that stops other firms from manufacturing the chemical. The other indication patent is that restricts other companies from using the patent indication on the label. Sometimes the chemical patent may run out earlier than the indication patent. In that case, other pharmaceutical firms can lawfully make the chemical but cannot distribute it under the claimed indication because the patent has not expired. Instead, they may offer it for another authorized indication that has not yet been patent-protected (via skinny labeling). The inventor firm may incur a loss as a result of this, particularly when the nonpatentable usage may raise the revenue for the product, outweighing the business because of the patented indication by a great deal (Kumar and Roy 2023). Enforcing the patent for a marketed drug that has been patented for the new indication might provide a challenge, particularly if the drug has not been widely utilized for a new indication in practice, Furthermore, because of the practice of significant skinny labeling (labeling done by generic manufacturers for the non-patented indications of the products), it may be difficult to enforce patent rights in the case of generic pharmaceuticals that are manufactured by several pharmaceutical companies (Bell 2021). As long as the medicine does not promote using the medication for the new indication while it is still effective for the stated indication, firms may avoid violating patent rights by this method. Clinicians on the other hand might prescribe the previously mentioned medication for the new indication. Therefore, stopping this off-label use can be challenging, causing a huge revenue loss to the inventors. Thus, big pharmaceutical companies might not prioritize drug repurposing initiatives (Sudeep Pushpakom et al. 2019). Given all these obstacles, pharmaceutical companies must be very persistent and come up with innovative ways to repurpose drugs.

8.7 Strategies to Overcome These Regulatory and Intellectual Property Barriers

Apart from regulatory and patent barriers, another obstacle to drug repurposing is obtaining access to data on dropped/shelved compounds. Although open access to data is gradually gaining popularity in the community of drug discovery, access to some types of trial, experimental, and laboratory data is still restricted. In cases where accessibility is not a concern, certain imaging data is difficult to obtain and integrate (Krishnamurthy et al. 2022). Companies are usually reluctant to share the details of failed/shelved compounds with other firms. Some pharmaceutical companies fail to reveal their chemical libraries of unsuccessful drugs to broaden the applications related to their compound collections which can create a fundamental obstacle to drug repurposing opportunities (Cha et al. 2018). Shelved pharmaceuticals disappear due to a lack of transparent archives for registering the drug for new indications. Another roadblock is reluctance among corporations to release drugs as a shared resource (Parvathaneni et al. 2019). Furthermore, it is logistically challenging to mine enormous datasets and integrate various data types in a user-friendly application (Ng et al. 2021). Identifying another therapeutic use for an established drug still necessitates a significant amount of high-risk capital and investments until the opportunity to earn back along with the profits (Krishnamurthy et al. 2022). Academic researchers might have the essential skills to analyze substances despite lacking access to an array of deprioritized pharmaceuticals (Wasim et al. 2023).

Thus, to overcome these barriers, the following section discusses in depth the various strategies that have been adopted or are currently in progress (Weth et al. 2023).

8.7.1 Development of Business Model

The drug innovation pipeline consists of finding potential drugs for certain conditions, doing preclinical research to simplify the biological mechanisms of action, and conducting clinical trials to ensure safety and efficacy. As a result, drug repurposing requires a significant amount of theoretical, experimental, and clinical understanding (Juárez-López and Schcolnik-Cabrera 2021). Pharmaceutical firms seek to recover their investments used during the drug development process; therefore, an appropriate business model is required. There are three different pharmaceutical business models for drug repurposing: in-house, out-licensing, and extended profiling. In-house models have been used by companies like GlaxoSmithKline, Bayer Healthcare Pharmaceuticals, and Novartis. This approach had limited success because companies failed to fulfill client demands, adapt, or create sufficient ROI and experienced significant downsizing or even bankruptcy (Naylor et al. 2015). Out-licensing allows pharmaceutical firms to provide access to their drugs, which is relatively easy for firms to adopt since it limits exposure to additional costs and risks. Extended profiling evaluates a candidate drug for new application after a successful first-in-human study. This allows developers to learn about efficacy in alternative indications and possible side effects associated with certain comorbidities (Novac 2013). Preclinical and clinical data from already-existing dossiers would enable quick regulatory clearance for a new application. This may maximize the therapeutic value of the repurposed drug while lowering costs for drug development, increasing rates of the drug approval procedure, and decreasing attrition rates (Lodha et al. 2022). This model is more cost-effective than conventional models and has been adopted by several pharmaceutical firms, including Celgene and Allergen (Naylor et al. 2015). Pharmaceutical companies should individually invest in evaluating prospective revenue models for repurposed medications. Thus, few repurposed medications may have a broad target audience, resulting in a potentially big yearly manufacturing volume, making them an attractive business model for generic pharmaceutical manufacturers (Moon et al. 2022).

8.7.2 Potential Solutions to Ease Market Authorization

Obtaining marketing authorization is a complex process for academic researchers, who often lack detailed knowledge and resources to obtain approval. Marketing authorization holders such as pharmaceutical companies often lack interest in seeking authorization. Presently, regulatory standards for extending a label are less risky for generic drugs that have already been on the market for an extended period (Dunne et al. 2013). The dosage definition for efficacy and safety in new indications is critical, considering publicly available data. The European Medicines Agency (EMA) can now request original toxicology reports for registration, enabling negotiation between the originator and small-to-medium enterprises or academic institutions after funding clinical trials (Asker-Hagelberg et al. 2022). The original marketing authorization holder may be reluctant to disclose information that was regulatory compliance 40 years ago but may include deficits if evaluated under contemporary regulations. Retrieving all relevant documentation takes significant effort, especially if the pharmaceutical company that developed the compound is unwilling to release data (Salbu 1999). The EMA has initiated a pilot project to assist academia and nonprofit organizations in generating evidence on repurposing generic drugs for formally authorized indications. However, this initiative fails to alleviate the funding issue, as academics still need to seek funds and meet strict regulatory requirements (van der Pol et al. 2023). Meanwhile, the United Kingdom National Health Service (UK NHS) is developing a comprehensive strategy to support the repurposing of generic drugs, highlighting the need for more funding and support for this innovative approach (Asker-Hagelberg et al. 2022). Thus, the marketing authorization procedure for repurposed generic pharmaceuticals is made simpler by the regulatory bodies. When considering a new use for a generic drug, one possible approach is to empower regulatory agencies to adjust the regulations originally intended for the drug's initial approval (Krishnamurthy et al. 2022). However, this can only occur if the generic drug's safety profile has been demonstrated in clinical practice (through exposure to thousands of patients), at the same dose as suggested for the new indication, and with no (or very minor) changes made during manufacturing. This would lessen the time and financial strain associated with the marketing authorization procedure. There will be additional financial incentives as a result for businesses (or other stakeholders) to seek the marketing authorization procedure (Murteira et al. 2014). In the United Kingdom, an NHS-owned pharmaceutical-contracted manufacturer holds marketing authority for certain medications. This strategy could help, at least in some countries, to obtain repurposed drugs and would help to become market authorization holders (van der Pol et al. 2023).

8.7.3 Introducing New Incentives

The regulatory systems of many nations now have exclusivity policies that incentivize companies to spend money on expanding the usage of already-approved medications through research and development (Verbaanderd et al. 2019). In Europe, the data protection period is for 8 years from the grant of the marketing authorization. The additional 2 years of market exclusivity allow the generic company to rely only on data to support their application, but they cannot launch a product based on that data. If the originator developed a repurposed indication in the protection period, then they have 1 year more exclusivity. In the USA, the regulatory data protection period is 5 years and an additional 3 years for repurposed drugs (Sudeep Pushpakom et al. 2019). Another impetus for a successful drug repurposing is providing tax incentives. Tax incentives can be a beneficial tool for drug owners and researchers, as they can deduct residual product value when donating to a third party, such as a nonprofit organization or governmental nonprofit foundation (Pulley et al. 2018). These incentives may provide benefits in considering the expense of manufacturing shelved drugs. Some government organizations have expanded these incentives to income from intellectual property generated by R&D (Köhler et al. 2012). They can be tailored to specific sorts of R&D activities that vary by firm size, region, or sector and apply differently to different R&D resources. Carry-forward or cash refund opportunities can make these incentives beneficial for firms with no payable taxes in a single fiscal year. The tax incentives can be designed as allowances or credits (Köhler et al. 2012).

8.7.4 Increased Public Research Funding

There is a lack of research funding for larger drug repurposing trials and marketing authorization processes. The primary reasons are that funding initiatives focus on the early-stage development of drugs or may lack funds for larger-scale clinical trials (van der Pol et al. 2023). This issue is particularly prevalent in disease areas where well-financed trials compete with moderately funded trials, leading to hospitals being more reluctant to participate due to budgetary constraints. The government ought to bear the financial burden in drug repurposing research and implementation, as it stands to benefit the most from public health. Governmental organizations must expand the financing opportunity for repurposed pharmaceuticals, which are cost-effective and have demonstrated benefits in clinical trials (Juárez-López and Schcolnik-Cabrera 2021). Insurance providers, whether public or private, are important stakeholders in the healthcare sector. These insurance providers help to reduce the financial burden of health care expenditure. Insurance companies should be encouraged to sponsor or boost financing for research into promising repurposed medications for diseases that are expensive to the healthcare sector. The patients will thus utilize affordable and cost-efficient medication (van der Pol et al. 2023).

8.7.5 Collaborative Models

The conversion from established medications to new, sometimes more expensive, ones can lead to conflicts within health systems. Market access mechanisms often do not acknowledge this effort, resulting in prices remaining the same for comparable drugs without the repurposed indication (Krishnamurthy et al. 2022). To address this, a framework for collaboration between nonprofit organizations and academic, health technology assessment bodies, payers, patient organizations, regulators, and the pharmaceutical industry is recommended. This will facilitate the introduction of new uses for medicines (Verbaanderd et al. 2019). Pharmaceutical companies and academics are increasingly collaborating to discover novel repurposing opportunities. Pharmaceutical firms have experience in translational research and clinical development and access to examining technologies that academic institutions struggle to acquire. Biotechnology companies and academics also possess knowledge of emerging disease biology areas, which could lead to innovative medications (Krishnamurthy et al. 2022). Collaboration can result in human capacity development and exchange. This will also expand the intellectual property options like patent pools and open licensing of existing drugs for neglected or rare diseases. The new collaboration creates funding structures that embrace entrepreneurial capital, the participation of nonprofit organizations, and public funding, which could significantly impact the medical field, where drug repurposing plays an essential role (Asker-Hagelberg et al. 2021).

8.8 Conclusion

Drug repurposing accelerates the accessibility of safe and inexpensive therapeutic alternatives, particularly for patients with rare or neglected illnesses. Drug repurposing offers economic benefits over traditional drug development, which is costly and time-consuming. There is an abundance of information accessible regarding the safety and quality of the medicine that has already been approved and in use. The intention behind drug repurposing is to complement de novo medication discovery rather than to replace it. Furthermore, there is a lower probability of unexpected side effects. Recycling processes are more efficient when used systematically. Data mining combining phenotypic screening and computational methodologies, such as target-based approaches, will generate high-quality data and evaluate medication targeting simultaneously. These models can be used to find new therapeutic interactions. Each approach and strategy of drug repurposing has its benefits as well as drawbacks. Combining different approaches and strategies can often result in a higher success rate. The problems associated with the drug repurposing technique are not only restricted to the initial stages of identification of targets. Limited resources and challenges in securing intellectual property rights and accessibility to data are also key barriers. There is minimal intellectual property protection for repurposed medications. Certain repurposed medications are not allowed to be sold because of the intellectual property issue. This results in a waste of time and money, hence abandoning the repurposing initiatives. The exclusivity policy by various regulatory bodies is another roadblock for drug repurposing. Another regulatory barrier to drug repurposing is obtaining access to the documents of dropped compounds and their trial data. Although the open-source model is gradually gaining popularity in the drug discovery community, public access to some types of data is still restricted. Collaboration between the private and public sectors can be achieved for successful medication repurposing. Government organizations have the authority to develop intricate strategies for dealing with some commercial and legal barriers encountered by medication repurposing initiatives. The confidential, vital data from clinical trials is also accessible to government organizations, sponsors of pharmaceutical companies, and other stakeholders. If published in the public realm, this data might represent a significant advancement in the field of drug repurposing. An open-ended collaboration framework connecting academia, industry, and nonprofit organizations will offer profitable and cost-effective chances to capitalize on abandoned drug items and optimize pharmaceutical companies' portfolios. For instance, the Medical Research Council and the National Center for Advancing Translational Sciences in the United Kingdom and the United States have started extensive funding programs in this field. This kind of collaboration will thus expand the range of new indications for compounds that have previously undergone extensive research and development by the pharmaceutical sector.

Acknowledgments The authors thank the Department of Pharmaceutics and Delhi Pharmaceutical Sciences and Research University (DPSRU) for providing the required facilities.

Author Contributions MM and SG have written and finalised the chapter. SA, MP, and PKS have read and contributed to the final manuscript. B.B. contributed to the conceptualization, writing, and editing of the chapter. All authors have read and edited the final version of the chapter.

Conflict of Interest The authors declare no conflicts of interest among themselves.

References

Agamah FE, Mazandu GK, Hassan R, Bope CD, Thomford NE, Ghansah A, Chimusa ER (2020) Computational/in silico methods in drug target and lead prediction. Brief Bioinform 21(5):1663–1675

Akhoon BA, Tiwari H, Nargotra A (2019) In silico drug design methods for drug repurposing. In: In silico drug design. Elsevier, pp 47–84

Ashburn TT, Thor KB (2004) Drug repositioning: identifying and developing new uses for existing drugs. Nat Rev Drug Discov 3(8):673–683. https://doi.org/10.1038/nrd1468

Asker-Hagelberg C, Boran T, Bouygues C, Eskola SM, Helmle L, Hernández C, Houýez F et al (2021) Repurposing of medicines in the EU: launch of a pilot framework. Front Med 8:817663. https://doi.org/10.3389/fmed.2021.817663

Asker-Hagelberg C, Boran T, Bouygues C, Eskola SM, Helmle L, Hernández C, Houýez F, Lee H, Lingri DD, Louette L (2022) Repurposing of medicines in the EU: launch of a pilot framework. Front Med 8:817663

Begley CG, Ashton M, Baell J, Bettess M, Brown MP, Carter B, Charman WN, Davis C, Fisher S, Frazer I (2021) Drug repurposing: misconceptions, challenges, and opportunities for academic researchers. Sci Transl Med 13(612):eabd5524

Bell JA (2021) Generic drugs and the future of "skinny labels". Harv J Law Technol 35:659

Bellera CL, Di Ianni ME, Sbaraglini ML, Castro EA, Bruno-Blanch LE, Talevi A (2015) Knowledge-based drug repurposing: a rational approach towards the identification of novel

medical applications of known drugs. In: Frontiers in computational chemistry. Bentham Science Publishers Ltd., Shāriqah, pp 44–81
Berger AC, Olson S, Johnson SG, Beachy SH (2014) Drug repurposing and repositioning: workshop summary. National Academies Press
Bloom DE, Cadarette D (2019) Infectious disease threats in the twenty-first century: strengthening the global response. Front Immunol 10:549
Breckenridge A, Jacob R (2019) Overcoming the legal and regulatory barriers to drug repurposing. Nat Rev Drug Discov 18(1):1–2
Cavalla D (2019) Using human experience to identify drug repurposing opportunities: theory and practice. Br J Clin Pharmacol 85(4):680–689
Cha Y, Erez T, Reynolds IJ, Kumar D, Ross J, Koytiger G, Kusko R, Zeskind B, Risso S, Kagan E (2018) Drug repurposing from the perspective of pharmaceutical companies. Br J Pharmacol 175(2):168–180
Charlton RL, Rossi-Bergmann B, Denny PW, Steel PG (2018) Repurposing as a strategy for the discovery of new anti-leishmanials: the-state-of-the-art. Parasitology 145(2):219–236
Chesbrough H, Chen EL (2013) Recovering abandoned compounds through expanded external IP licensing. Calif Manag Rev 55(4):83–101
Cichonska A, Rousu J, Aittokallio T (2015) Identification of drug candidates and repurposing opportunities through compound–target interaction networks. Expert Opin Drug Discov 10(12):1333–1345
Cichonska A, Ravikumar B, Parri E, Timonen S, Pahikkala T, Airola A, Wennerberg K, Rousu J, Aittokallio T (2017) Computational-experimental approach to drug-target interaction mapping: a case study on kinase inhibitors. PLoS Comput Biol 13(8):e1005678
Copeland SM (2015) The case of the triggered memory: serendipitous discovery and the ethics of clinical research. PhD thesis, Dalhousie University, Halifax, Nova Scotia
Cousin MA, Ebbert JO, Wiinamaki AR, Urban MD, Argue DP, Ekker SC, Klee EW (2014) Larval zebrafish model for FDA-approved drug repositioning for tobacco dependence treatment. PLoS One 9(3):e90467
Dhir N, Jain A, Mahendru D, Prakash A, Medhi B (2020) Drug repurposing and orphan disease therapeutics. In: Drug repurposing: hypothesis, molecular aspects and therapeutic applications. IntechOpen
Dias MLCNB (2020) Challenges and opportunities of off-label prescription
Doan TL, Pollastri M, Walters MA, Georg GI (2011) Chapter 23 – The future of drug repositioning: old drugs, new opportunities. In: Macor JE (ed) Annual reports in medicinal chemistry, vol 46. Academic Press, pp 385–401. https://doi.org/10.1016/B978-0-12-386009-5.00004-7
Dunne S, Shannon B, Dunne C, Cullen W (2013) A review of the differences and similarities between generic drugs and their originator counterparts, including economic benefits associated with usage of generic medicines, using Ireland as a case study. BMC Pharmacol Toxicol 14:1–19
Fetro C, Scherman D (2020) Drug repurposing in rare diseases: myths and reality. Therapies 75(2):157–160
Fotis C, Antoranz A, Hatziavramidis D, Sakellaropoulos T, Alexopoulos LG (2018) Network-based technologies for early drug discovery. Drug Discov Today 23(3):626–635. https://doi.org/10.1016/j.drudis.2017.12.001
Freije I, Lamouche S, Tanguay M (2020) Review of drugs approved via the 505 (b)(2) pathway: uncovering drug development trends and regulatory requirements. Ther Innov Regul Sci 54(1):128–138
Gassmann O, Krech CA, Bader MA, Reepmeyer G (2016) Out-licensing in pharmaceutical research and development. In: Value creation in the pharmaceutical industry: the critical path to innovation. Wiley-VCH, Weinheim, pp 362–381
Goldstein I, Burnett AL, Rosen RC, Park PW, Stecher VJ (2019) The serendipitous story of sildenafil: an unexpected oral therapy for erectile dysfunction. Sex Med Rev 7(1):115–128

Hernandez JJ, Pryszlak M, Smith L, Yanchus C, Kurji N, Shahani VM, Molinski SV (2017) Giving drugs a second chance: overcoming regulatory and financial hurdles in repurposing approved drugs as cancer therapeutics. Front Oncol 7:273

Hernández-Lemus E, Martínez-García M (2021) Pathway-based drug-repurposing schemes in cancer: the role of translational bioinformatics. Front Oncol 10:605680. https://doi.org/10.3389/fonc.2020.605680

Iljin K, Ketola K, Vainio P, Halonen P, Kohonen P, Fey V, Grafström RC, Perälä M, Kallioniemi O (2009) High-throughput cell-based screening of 4910 known drugs and drug-like small molecules identifies disulfiram as an inhibitor of prostate cancer cell growth. Clin Cancer Res 15(19):6070–6078. https://doi.org/10.1158/1078-0432.CCR-09-1035

Inamoto Y, Valdés-Sanz N, Ogawa Y, Alves M, Berchicci L, Galvin J, Greinix H, Hale GA, Horn B, Kelly D (2019) Ocular graft-versus-host disease after hematopoietic cell transplantation: expert review from the late effects and quality of life working committee of the CIBMTR and transplant complications working party of the EBMT. Bone Marrow Transplant 54(5):662–673

Jarada TN, Rokne JG, Alhajj R (2020) A review of computational drug repositioning: strategies, approaches, opportunities, challenges, and directions. J Chem 12(1):1–23

Jourdan J-P, Bureau R, Rochais C, Dallemagne P (2020) Drug repositioning: a brief overview. J Pharm Pharmacol 72(9):1145–1151

Juárez-López D, Schcolnik-Cabrera A (2021) Drug repurposing: considerations to surpass while re-directing old compounds for new treatments. Arch Med Res 52(3):243–251. https://doi.org/10.1016/j.arcmed.2020.10.021

Karaman B, Sippl W (2019) Computational drug repurposing: current trends. Curr Med Chem 26(28):5389–5409

Khataniar A, Pathak U, Rajkhowa S, Jha AN (2022) A comprehensive review of drug repurposing strategies against known drug targets of COVID-19. Covid 2:148–167. s Note: MDPI stays neutral with regard to jurisdictional claims in published …

Kirtonia A, Gala K, Fernandes SG, Pandya G, Pandey AK, Sethi G, Khattar E, Garg M (2021) Repurposing of drugs: an attractive pharmacological strategy for cancer therapeutics. Semin Cancer Biol 68:258–278. Elsevier

Köhler C, Laredo P, Rammer C (2012) The impact and effectiveness of fiscal incentives for R&D. Nesta Working Paper No. 12/01

Krishnamurthy N, Grimshaw AA, Axson SA, Choe SH, Miller JE (2022) Drug repurposing: a systematic review on root causes, barriers and facilitators. BMC Health Serv Res 22(1):1–17

Kulkarni VS, Alagarsamy V, Solomon VR, Jose PA, Murugesan S (2023) Drug repurposing: an effective tool in modern drug discovery. Russ J Bioorg Chem 49(2):157–166. https://doi.org/10.1134/S1068162023020139

Kumar S, Roy V (2023) Repurposing drugs: an empowering approach to drug discovery and development. Drug Res 73(9):481–490

Kuzikov M, Costanzi E, Reinshagen J, Esposito F, Vangeel L, Wolf M, Ellinger B, Claussen C, Geisslinger G, Corona A (2021) Identification of inhibitors of SARS-CoV-2 3CL-pro enzymatic activity using a small molecule in vitro repurposing screen. ACS Pharmacol Transl Sci 4(3):1096–1110

Li C, He W-Q (2023) Global prediction of primary liver cancer incidences and mortality in 2040. J Hepatol 78(4):e144–e146

Li J, Zheng S, Chen B, Butte AJ, Swamidass SJ, Lu Z (2016) A survey of current trends in computational drug repositioning. Brief Bioinform 17(1):2–12

Lipinski CA, Reaume AG (2020) High throughput in vivo phenotypic screening for drug repurposing: discovery of MLR-1023 a novel insulin sensitizer and novel Lyn kinase activator with clinical proof of concept. Bioorg Med Chem 28(9):115425

Lodha S, Patel H, Joshi S, Kalyankar G, Mishra A (2022) Regulatory requirements of regulated market. In: Regulatory affairs in the pharmaceutical industry. Elsevier, pp 113–161

Low ZY, Farouk IA, Lal SK (2020) Drug repositioning: new approaches and future prospects for life-debilitating diseases and the COVID-19 pandemic outbreak. Viruses 12(9):1058. https://doi.org/10.3390/v12091058

Markossian S, Ang KK, Wilson CG, Arkin MR (2018) Small-molecule screening for genetic diseases. Annu Rev Genomics Hum Genet 19(1):263–288. https://doi.org/10.1146/annurev-genom-083117-021452

Marusina K, Welsch DJ, Rose L, Brock D, Bahr N (2011) The CTSA pharmaceutical assets portal–a public–private partnership model for drug repositioning. Drug Discov Today 8(3–4):77–83

McCarthy JJ, McLeod HL, Ginsburg GS (2013) Genomic medicine: a decade of successes, challenges, and opportunities. Sci Transl Med 5(189):189sr4

Mehta SS (2022) Commercializing successful biomedical technologies. Cambridge University Press

Mittal N, Mittal R (2021) Repurposing old molecules for new indications: defining pillars of success from lessons in the past. Eur J Pharmacol 912:174569

Moon S, Vieira MCF, Ruiz AA, Navarro DM (2022) New business models for pharmaceutical research and development as a global public good: considerations for the WHO European Region. WHO

Murteira S, Millier A, Ghezaiel Z, Lamure M (2014) Drug reformulations and repositioning in the pharmaceutical industry and their impact on market access: regulatory implications. J Mark Access Health Policy 2(1):22813

Naylor S, Kauppi M, Schonfeld JM (2015) Therapeutic drug repurposing, repositioning and rescue: part II: business review. Drug Discov World 16:57–72

Ng YL, Salim CK, Chu JJH (2021) Drug repurposing for COVID-19: approaches, challenges and promising candidates. Pharmacol Ther 228:107930

Novac N (2013) Challenges and opportunities of drug repositioning. Trends Pharmacol Sci 34(5):267–272

Oprea TI, Bauman JE, Bologa CG, Buranda T, Chigaev A, Edwards BS, Jarvik JW, Gresham HD, Haynes MK, Hjelle B (2011) Drug repurposing from an academic perspective. Drug Discov Today 8(3–4):61–69

Park K (2019) A review of computational drug repurposing. Transl Clin Pharmacol 27(2):59–63

Parvathaneni V, Kulkarni NS, Muth A, Gupta V (2019) Drug repurposing: a promising tool to accelerate the drug discovery process. Drug Discov Today 24(10):2076–2085

Persidis A (2015) Drug repositioning and off-label use—finding the balance and understanding the differences: interview with David Cavalla, MA, PhD, Founder, Numedicus. Drug Repurp Rescue Reposition 1(1):5–7

Pillaiyar T, Meenakshisundaram S, Manickam M, Sankaranarayanan M (2020) A medicinal chemistry perspective of drug repositioning: recent advances and challenges in drug discovery. Eur J Med Chem 195:112275

Polamreddy P, Gattu N (2019) The drug repurposing landscape from 2012 to 2017: evolution, challenges, and possible solutions. Drug Discov Today 24(3):789–795. https://doi.org/10.1016/j.drudis.2018.11.022

Pulley JM, Jerome RN, Shirey-Rice JK, Zaleski NM, Naylor HM, Pruijssers AJ, Jackson JC, Bernard GR, Holroyd KJ (2018) Advocating for mutually beneficial access to shelved compounds. Future Med Chem 10:1395–1398. Future Science

Pushpakom S (2022) Introduction and historical overview of drug repurposing opportunities. In: Drug repurposing. The Royal Society of Chemistry

Pushpakom S, Iorio F, Eyers PA, Escott KJ, Hopper S, Wells A, Doig A, Guilliams T, Latimer J, McNamee C (2019) Drug repurposing: progress, challenges and recommendations. Nat Rev Drug Discov 18(1):41–58

Rose S (2013) Second chances for shelved compounds. Cancer Discov 3(6):598. https://doi.org/10.1158/2159-8290.CD-ND2013-012

Roy S, Dhaneshwar S, Bhasin B (2021) Drug repurposing: an emerging tool for drug reuse, recycling and discovery. Curr Drug Res Rev 13(2):101–119

Rudrapal M, Khairnar SJ, Jadhav AG (2020) Drug repurposing (DR): an emerging approach in drug discovery. In: Badria FA (ed) Drug repurposing. IntechOpen, Rijeka. https://doi.org/10.5772/intechopen.93193

Salbu SR (1999) Off-label use, prescription, and marketing of FDA-approved drugs: an assessment of legislative and regulatory policy. Fla L Rev 51:181

Sharma K, Shrivastava H, Kumar A (2023a) 13 Advantages, opportunities, and challenges to drug repurposing. In: Drug repurposing. De Gruyter, Berlin, Boston, pp 283–292
Sharma T, Padhy I, Sahoo CR (2023b) Approaches, strategies, and advances in computational drug discovery and drug repurposing. In: Drug repurposing and computational drug discovery. Apple Academic Press, pp 27–58
Shruthi NR, Jain MS, Ganesan H, Banerjee A, Zhang H, Sun X-F, Pathak S (2023) Drug repurposing in cancer. In: Drug repurposing for emerging infectious diseases and cancer. Springer, pp 159–179
Sleigh SH, Barton CL (2010) Repurposing strategies for therapeutics. Pharm Med 24(3):151–159
Sultana J, Zaccaria C, De Lisa R, Rossi F, Capuano A, Ferrajolo C (2019) Good pharmacovigilance practice in paediatrics: an overview of the updated European Medicines Agency guidelines. Pediatr Drugs 21(5):317–321
Talevi A, Bellera CL (2020) Challenges and opportunities with drug repurposing: finding strategies to find alternative uses of therapeutics. Expert Opin Drug Discov 15(4):397–401. https://doi.org/10.1080/17460441.2020.1704729
Tanoli Z, Alam Z, Ianevski A, Wennerberg K, Vähä-Koskela M, Aittokallio T (2020) Interactive visual analysis of drug–target interaction networks using drug target profiler, with applications to precision medicine and drug repurposing. Brief Bioinform 21(1):211–220
van der Pol KH, Aljofan M, Blin O, Cornel JH, Rongen GA, Woestelandt A-G, Spedding M (2023) Drug repurposing of generic drugs: challenges and the potential role for government. Appl Health Econ Health Policy 21(6):831–840
Verbaanderd C, Rooman I, Meheus L, Huys I (2019) On-label or off-label? Overcoming regulatory and financial barriers to bring repurposed medicines to cancer patients. Front Pharmacol 10:1664. https://doi.org/10.3389/fphar.2019.01664
Verbaanderd C, Rooman I, Meheus L, Huys I (2020) On-label or off-label? Overcoming regulatory and financial barriers to bring repurposed medicines to cancer patients. Front Pharmacol 10:512464
Villani A, Fabbrocini G, Ocampo-Candiani J, Ruggiero A, Ocampo-Garza SS (2021) Review of oral minoxidil as treatment of hair disorders: in search of the perfect dose. J Eur Acad Dermatol Venereol 35(7):1485–1492
Wasim R, Ansari TM, Siddiqui MH, Ahsan F, Shamim A, Singh A, Shariq M, Anwar A, Siddiqui AR, Parveen S (2023) Repurposing of drugs for cardiometabolic disorders: an out and out cumulation. Horm Metab Res 55(1):7–24
Weth FR, Hoggarth GB, Weth AF, Paterson E, White MPJ, Tan ST, Peng L, Gray C (2023) Unlocking hidden potential: advancements, approaches, and obstacles in repurposing drugs for cancer therapy. Br J Cancer. 130:703–715. https://doi.org/10.1038/s41416-023-02502-9
Witkowski TX (2011) Intellectual property and other legal aspects of drug repurposing. Drug Discov Today 8(3–4):139–143
Xue H, Li J, Xie H, Wang Y (2018) Review of drug repositioning approaches and resources. Int J Biol Sci 14(10):1232

3 Rs: Repurposing and Reformulation of Therapeutics

9

Madhu Kumari, Monika Dwivedi, Anima Pandey, and Shurabhi Chaurasia

Abstract

Drug repurposing and reformulation are current trends in pharmaceutical industry in combating the limitations of slow-paced drug discovery and development process. Drug repurposing utilizes existing clinically approved drug molecules for new therapeutic purpose, thus reviving old drugs by investigating and reformulating them for new pharmacological activity. This therapeutic switching emerged as an efficient mode for developing new pharmacological active drugs, henceforth reducing failures in drug development, timelines of the drug development process, and the cost involved in the process. Moreover, drug repurposing also marks a breakthrough in drug discovery through realizing the 3 Rs principle, as therapeutic switching furnishes a surrogate mode that aligns with replacement, refinement, and reduction of animals for new drug research to fast-track drug discovery. During pandemic, drug repurposing evolved as a resolution for the need of urgent new drug moieties that can offer different mechanisms for preventive and therapeutic efficacy against COVID-19. Emphasizing on this current development as a consequence of trending drug repurposing, this chapter will focus on how drug repurposing is shaping drug research, pharmaceutical industries' trends, and market strategies. This chapter will also offer an insight into how repositioning of drugs came forward to manage COVID-19.

Keywords

Drug repurposing · New targets · COVID-19 · Network pharmacology

M. Kumari · M. Dwivedi (✉) · A. Pandey · S. Chaurasia
Department of Pharmaceutical Sciences, Birla Institute of Technology, Mesra, Ranchi, India
e-mail: monika.dwivedi@bitmesra.ac.in

R. K. Kaundal et al. (eds.), *Drug Repurposing: Exploring Newer Therapeutic Potentials*,
https://doi.org/10.1007/978-981-95-4522-3_9

9.1 Introduction

Drug repurposing, also known as drug repositioning, has evolved as a dynamic and inventive technique for uncovering new therapeutic uses for existing medications in a timely manner. The process of drug repurposing is divided into various steps; the first is discovery of potential molecules. This can be performed using a variety of techniques, such as data mining, bioinformatics, and high-throughput screening. Once a prospective candidate is found, it advances to the preclinical evaluation stage, where its safety and efficacy are tested using in vitro and in vivo models (Fig. 9.1). One of the key benefits of drug repurposing is that it significantly reduces the time and expense required to develop new medicines. Because repurposed medications have previously passed safety testing, the procedure is significantly accelerated, making it especially beneficial during public health crises like the COVID-19 outbreak. Following preclinical testing, repurposed medications are subjected to clinical studies, which include Phase I, Phase II, and Phase III trials, to determine their safety and efficacy in treating the new indication.

Drug repurposing (DR) is aiding in drug discovery and development by identifying novel medical applications for existing drugs, encompassing those that are approved, discontinued, or still in experimental phases (Talevi and Bellera 2020). The process of approving a novel medicine is costly and can take up to 10–15 years. Because of the lengthy discovery process, drug repurposing (repositioning) came

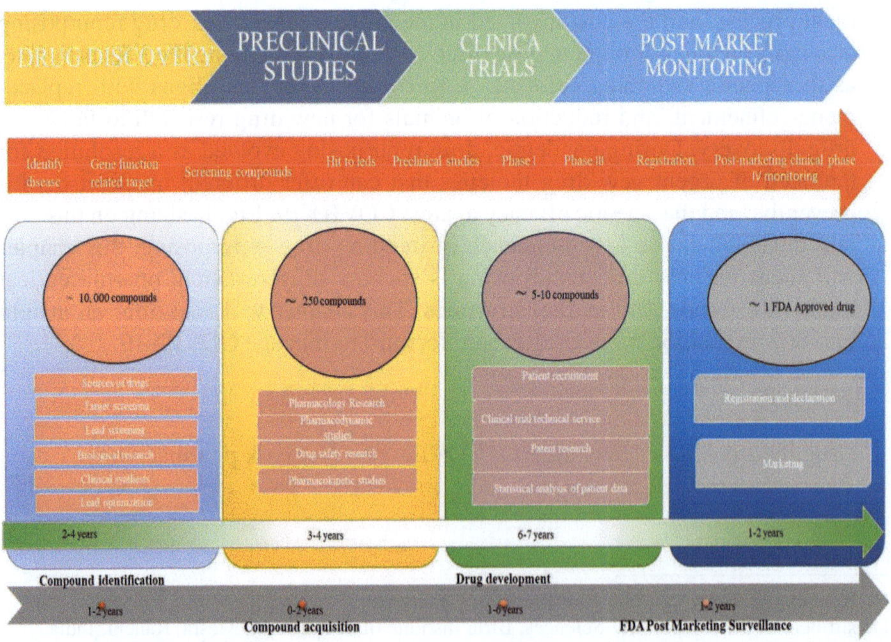

Fig. 9.1 Drug identification and development pathway can take more than a decade in traditional de novo drug discovery (Hua et al. 2022)

forth as an alternate way for reducing the time required to produce a drug. The use of pharmaceuticals licensed by regulatory agencies such as the Food and Drug Administration (FDA), the European Medicines Agency (EMA), and the Medicines and Healthcare Products Regulatory Agency (MHRA), among others, for a new indication is referred to as repurposing. Because the tremendous promise of a shorter development cycle is compelling many pharmaceutical companies, medication repurposing is currently being used to reconstruct some of their FDA-approved and previously unsuccessful pipeline compounds into novel medications for a range of diseases (Parvathaneni et al. 2019).

Since its inception, the idea of "drug repositioning" has been broadened to include active ingredients that were removed from the market due to safety concerns as well as drugs that were toxic or ineffective in clinical trials. It should not, however, include substances that have not yet undergone clinical testing. This excludes chemicals stored in chemical libraries by academic and industrial research groups from being screened for new biological qualities that differ from the features for which they were originally intended and prepared. In order to get the desired novel therapeutic effect, such molecules typically necessitate hit-to-lead chemistry. Drug repurposing leverages the pleiotropic effects of existing drugs; furthermore, genomic evidence indicates that diseases such as Parkinson's and Alzheimer's share common molecular targets, including 48 genes and 4 signaling pathways (Jourdan et al. 2020). Drug repositioning, which identifies alternate therapeutic uses for already-approved pharmaceuticals, is a promising topic in the research and discovery of therapeutic medicines. Drug repurposing lessens restrictions on the introduction of a drug to the market because the drug development process often requires billions of dollars and numerous years of labor. It decreased the probability of failure, compared to the usual drug discovery program's failure rate of about 45%, which is linked to toxicity or safety issues, and had the added benefit of shortening the typical drug discovery period by up to 5–7 years. By accurately forecasting therapeutic utility, repurposing research ultimately aims to cut down the duration and cost of medication discovery and development (Schuler et al. 2022) . The use of repurposed medications that have already undergone human testing provides an understanding of the pharmacokinetics, pharmacodynamics, dosage, metabolic profiles, molecular pathways, the mechanism of actions of the medication, various target interactions, etc. Thus, in repurposed drugs, examining medication pharmacokinetic traits and toxicity profiles will therefore require less investigation (Turabi et al. 2022). The safety evaluation and formulation development methods have already been established; thus, this technique has progressed toward drug development much more quickly. For previously abandoned, outmoded drugs, repositioning also implies significant investment possibilities, especially for long-term safety considerations. Drug repurposing necessitates an in-depth understanding of the biological and molecular pathways that a drug can influence, as well as its interactions with endogenous biomolecules. This hypothesis risk can be considerably reduced with full knowledge of the drug and its impacted pathways, resulting in successful drug repurposing (Parvathaneni et al. 2019).

The drugs that could be candidates for repurposing fall into one of three categories:

1. Generic medications, which are prescribed medications not protected by patents.
2. Failed pharmaceuticals refer to drugs that have progressed through various stages of clinical development but have not yet reached the market.
3. Patented drugs—those that have been authorized or are nearing the end of the clinical development process.

Major contributors in drug reprofiling including academia, biotech, and pharma have a common resource, i.e., generic medications. These drugs are readily available for both preclinical research and clinical development, and they have been on the market for an extended duration. Furthermore, repurposed drugs boast well-established and thoroughly documented safety records. Nevertheless, when it comes to drugs falling into the other two categories, namely, those that were unsuccessful or are still protected by patents, pharmaceutical companies have a distinct advantage in terms of possessing comprehensive data related to their specific compounds. As data accessibility can sometimes exhibit biases favoring pharmaceutical companies, these companies play a pivotal role as essential collaborators in systematic efforts to identify new applications for existing medications (Cha et al. 2018).

There are some important elements in drug repurposing that determine the ideal drug/disease combination by examining the signaling and biochemical pathways and further assessing through human trials to confirm the new assertion (Mehta and Dhapte-Pawar 2021).

9.2 Opportunities in Drug Repurposing

9.2.1 Systems Medicine

Network pharmacology and systems medicine offer an integrated viewpoint that reconciles two seemingly disparate paradigms in drug discovery that are the phenotypic and target-oriented approaches. By leveraging network and metabolic control analysis, these approaches can be valuable tools for not only identifying multiple potential drug targets but also for selecting synergistic pharmacological combinations that exhibit enhanced therapeutic efficacy.

9.2.2 Collaborative Models

The pharmaceutical firms and academics can make complementary contributions in the identification of innovative repurposing potential of existing drug molecules. Pharmaceutical research explored immensely valuable chemical libraries of unsuccessful or shelved drugs; however, these drug candidates are far from translating as successful drug candidates. They also have first-hand knowledge of translational research and clinical development (Talevi and Bellera 2020).

9.2.3 Rare and Untreated Ailments

Since for uncommon and neglected illnesses, the economics of medication development is not very attractive, therefore, nonprofit and academic organizations predominate in the drug development events for those diseases, and also, certain public policies and regulatory measures promote research into these kinds of illnesses. Drug reprofiling is a particularly attractive approach for these conditions. Owing to drug reprofiling, certain courses of drug development economy are let off, such as tax exemptions, expedited approval procedures, grants, and waivers of regulatory fees. Notably, a sizable portion of DNDi's pharmacological portfolio currently conducting clinical trials relates to repurposed medications, such as miltefosine, fexinidazole, and fosravuconazole. In fact, fexinidazole is also rediscovered as human African trypanosomiasis, and it is the first medication approved in 30 years with the ability to treat advanced stages of sleeping sickness (Talevi and Bellera 2020).

Drug reprofiling is a frequent strategy for discovering new therapeutic uses for already approved medications opposite to looking for brand-new therapeutic agents. The method is increasingly being utilized to get beyond traditional drug discovery's typical resource, timeframe, and cost constraints. The quick development of novel computational algorithms has also been fostered by the availability of advanced information and genetic datasets (Berdigaliyev and Aljofan 2020).

9.3 History of Drug Repurposing

Between 2007 and 2009, approximately 30–40% of the new medicines and biologics that earned US FDA approval were identified as repositioned or repurposed products. Similarly, a separate study covering the period from 1984 to 2009 revealed that 35% of transformative products approved by the FDA, commonly referred to as drugs, were actually repurposed products (Krishnamurthy et al. 2022).

Drug reprofiling or redirection is a very appealing, cost-effective, and time-efficient method since it involves adding novel indications to previously existing medications. According to one recent study, more than 30% of US FDA-approved medications and vaccines have undergone drug reprofiling. Reprofiling of drugs offers a distinct advantage over traditional strategies by harnessing new technologies that effectively address significant challenges in drug discovery. In a survey conducted in 2007, it was found that the average cost of reprofiling amounted to $8.4 million. The well-established profiles of these compounds contributed to a notably higher success rate in the case of reprofiled medications compared to the conventional approach. The survey from 2007 indicated that only 25% of drugs from phase II clinical trials and 65% of those from phase III clinical trials eventually reached the market, in contrast to 10% and 50% of new molecular entities, respectively (Latif et al. 2022).

The earliest example of drug repurposing can be traced back to acetylsalicylic acid, commonly known as aspirin. Originally marketed as an analgesic by Bayer in 1899, Aspirin saw its first repositioning in the 1980s when it was repurposed as an

Table 9.1 List of repurposed medicines classified by biological action

Sl. NO	Drug name	Original indication	Novel indication	References
1.	Chlorpromazine	Antiemetic, antihistamine	Nonsedating tranquillizer	Kumar et al. (2019)
2.	Valproate	Epilepsy	Epilepsy	Jourdan et al. (2020)
3.	Aspirin	Analgesic, anti-inflammatory, antiplatelet aggregation	Cardiovascular events, colorectal cancer	Jourdan et al. (2020)
4.	Metformin	Type 2 diabetes therapy, oral hypoglycemic agent	IPF therapy with AMPK pathway regulation	Mehta and Dhapte-Pawar (2021)
5.	Niclosamide	Oral anthelmintic agent	Agent that reduces virulence in the treatment of CF	Mehta and Dhapte-Pawar (2021)
6.	Raloxifene	Osteoporosis	Breast cancer	Latif et al. (2022)
7.	Sildenafil	Hypertension	Erectile dysfunction	Berdigaliyev and Aljofan (2020)

antiplatelet aggregation medication, particularly at low doses. Presently, aspirin may undergo yet another transition, this time with a new role in oncology, where it continues to be frequently employed to prevent cardiovascular events as a secondary indication. Research has demonstrated that taking aspirin every day for a minimum of 5 years can avert the onset of numerous malignancies, with colorectal cancer being the most susceptible. Pfizer first researched sildenafil as a potential antihypertensive medication in 1985. This physiological consequence prompted Pfizer to sell sildenafil for erectile dysfunction in 1998 (Jourdan et al. 2020). Ramosetron is another such drug that was previously used as an antiemetic. Because of its adverse effect, constipation, it was later reprofiled for irritable bowel syndrome. As a result, drug repurposing encompasses the scientific reproduction of present medications' pharmacological properties (Latif et al. 2022). Chloroquine was a well-known antimalarial chemical that was produced in 1934 and was then focused toward many other disorders such as parasite diseases (before 1960), fever, and lupus skin rashes (Kumar et al. 2019). According to a recent study, remdesivir is a possible medication for the treatment of COVID-19. Remdesivir works by blocking RNA polymerase and incorporating it into nascent viral RNA. It has two important consequences: it terminates the viral RNA chain and, as a result, suppresses viral genome replication (Kifle et al. 2021) (Table 9.1).

9.4 Need for Drug Repurposing

Finding new therapeutic uses for existing medications that were previously approved for some other indication or that failed in clinical trials is drug repurposing. The pharmaceutical industry is taking notice of this technique as it speeds up medication development, lowers costs, and lowers the likelihood of failure. The new drug discovery process is a lengthy process that yields outcomes, requiring extensive financial support and continuous effort from researchers and development teams to develop new treatments. This is mainly because of bottlenecks in the therapeutic development process. As a result, it frequently takes more than 14 years to convert a promising chemical into an approved drug. Repurposing of drugs is one tactic that can help people get through these difficulties (Ahmad et al. 2021). It is also known as therapeutic switching or pharmacological repositioning. Drug repurposing is a crucial method in drug development, and according to some statistics, between 2007 and 2009, the US Food and Drug Administration (FDA) authorized 30–40% of novel pharmaceuticals and biologics that can be categorized as repurposed or repositioned products (Krishnamurthy et al. 2022). Collaboration projects, chemical libraries, and regulatory incentives were the most often mentioned drug repurposing facilitators. Moreover, drug repurposing also reduces the burden on animal and human subjects for experimentation for research. Drug repurposing tactics are multiplying, particularly for the treatment of unusual and unmet medical needs (Sahoo et al. 2021).

There are several compelling reasons for pursuing drug repurposing, including (Ahmad et al. 2021; Sahoo et al. 2021):

- *Cost-Efficiency*: Creating a new medication from scratch is a time-consuming, costly, and dangerous procedure that can take years and cost billions of dollars. In contrast, because a large amount of safety and pharmacokinetic data is already accessible, repurposing existing medications can drastically cut the time and expense associated with drug development.
- *Safety Profile*: Since repurposed medications frequently go through comprehensive safety research in people, a lot is known about their adverse effects, doses, and potential drug interactions. This can speed up the regulatory approval procedure and lessen the possibility of unanticipated adverse effects.
- *Quicker Development*: Repurposed medications can reach patients more quickly than entirely novel molecules because they have previously completed several clinical trial stages. This is particularly significant for treating illnesses for which no viable treatments are available or in emergency situations requiring quick action, such as pandemics.
- *Investigation of New Mechanisms*: Through repurposing, researchers can investigate if a medication created for one ailment can also be effective for another condition with a different underlying mechanism. This promotes understanding the mechanisms of disease and creating novel treatment approaches.
- *Treatment for Orphan Diseases*: Repurposing can aid in finding medicines for orphan diseases, which are rare ailments that may not have drawn considerable research and development funding for the creation of novel drugs.
- *Combination Therapies*: To increase the effectiveness of repurposed medications, they might be coupled with other medicines or treatments. Synergistic effects and more effective treatment alternatives may result from this strategy.

- *Sustainability*: Creating new medications from scratch uses a lot of resources and could harm the environment. As it reduces the need for extra drug synthesis and production, repurposing current medications can be a more environmentally friendly strategy.
- *Drug Resistance*: Current therapies can pose severe problems for infectious diseases and cancer. Alternative therapy options that might not have the same resistance problems can be provided through repurposing.
- *Investigating New Targets*: Although certain medications were created with a specific ailment in mind, they can have favorable effects when used to treat other conditions. This may result in the identification of fresh therapeutic targets.

Drug repurposing is a valuable method in drug research since it increases the effectiveness of already prescribed pharmaceuticals, hastens the creation of novel treatments, and may provide answers to various medical problems. It represents a potentially fruitful path for innovation in the medical and pharmaceutical fields. It totally encourages 3 Rs—Replacement, Reduction, and Refinement in animal experimentation for research.

9.5 Different Stages of Drug Repurposing

A medicine can be repurposed at any point in its development; for instance, sildenafil was initially studied for the treatment of hypertension but was later approved for the treatment of erectile dysfunction. Other examples include the diabetes medication metformin, which is used off-label to treat polycystic ovarian syndrome (PCOS), the antiviral activity of calcium channel blockers, gliotoxin, and the anticancer potential rediscovered in antifungal itraconazole. Moreover, the antifungal meltefosine shows anti-leishmania action in balb/c mice. Even herbal products have the potential for repurposing as well in different therapies (Berdigaliyev and Aljofan 2020).

By mixing old or existing pharmaceuticals and FDA-approved clinically utilized drugs, new medicinally active compounds are generated. To speed up the medication repurposing process, in silico techniques, as well as artificial intelligence (AI) technologies, together with structure-based and ligand-based drug design, have been employed recently. In comparison to traditional approaches to drug discovery, drug repurposing offers various advantages, including reduced time spent on research, reduced complexity, and lower process costs. The typical drug discovery strategy is predicted to take 10–12 years to develop new medication compounds. In the case of drug repositioning, the predicted period is between 1 and 3 years. By using the typical drug development procedure, the average cost of bringing a new pharmacologically active medicine to market is USD 1.24 billion. In contrast, the medication repurposing procedure costs around 60% of the cost of typical drug development methods. Drug repurposing shortens the time required for drug development at a lower cost and with a lower risk of failure or a high success rate. Pre-collected data pertaining to structural optimization, pharmacokinetics, toxicology, clinical efficaciousness, and safety profile of drugs related to the traditional drug discovery approach are available. Antiviral medications such as favipiravir,

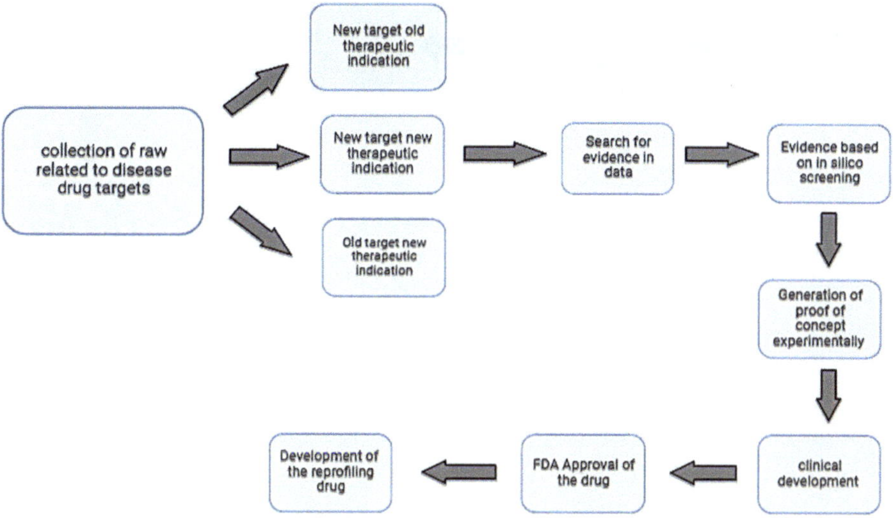

Fig. 9.2 Various stages of drug repurposing (DR) (Sharma et al. 2023)

remdesivir, and lopinavir have previously been tried in clinical trials to treat AIDS, SARS, and MERS. A drug repurposing study is currently being conducted to explore the efficacy of these medicines against COVID-19 (Sahoo et al. 2021).

Likewise, failure of exploratory medications has implications for new therapeutic fields. Saracatinib, an anticancer medication produced by AstraZeneca, has shown significant symptom reversal in Alzheimer's disease (AD) animal models and successfully completed Phase 1 (NCT01864655) studies against AD. Medication like Sildenafil was a licensed medicine with a specific adverse effect; however, it was rediscovered by chance to have superior therapeutic potential in other indications. Sildenafil that was originally created to treat hypertension and angina pectoris has now amassed a sizable market for the treatment of erectile dysfunction. Moreover, screening for the off-label use of pharmaceuticals that have received FDA approval is the existing trend for accelerating the drug discovery path. Itraconazole has been found to have an extra anti-angiogenesis function while being approved primarily as an antifungal agent. Examining the aforementioned characteristic guided the compound's clearance through Phase 2 trials to discover how effective it is as a second-line therapy agent for treating prostate and lung cancer (NCT00887458 and NCT02357836, respectively) (Fig. 9.2).

9.6 Methods and Approaches Employed for Drug Repurposing

The methods for repurposing drugs can be classified into three categories, taking into account the information accessible about the physicochemical, pharmacokinetic, toxicological, and biological characteristics of drug compounds. These categories include approaches that are drug-focused, target-focused, and those centred

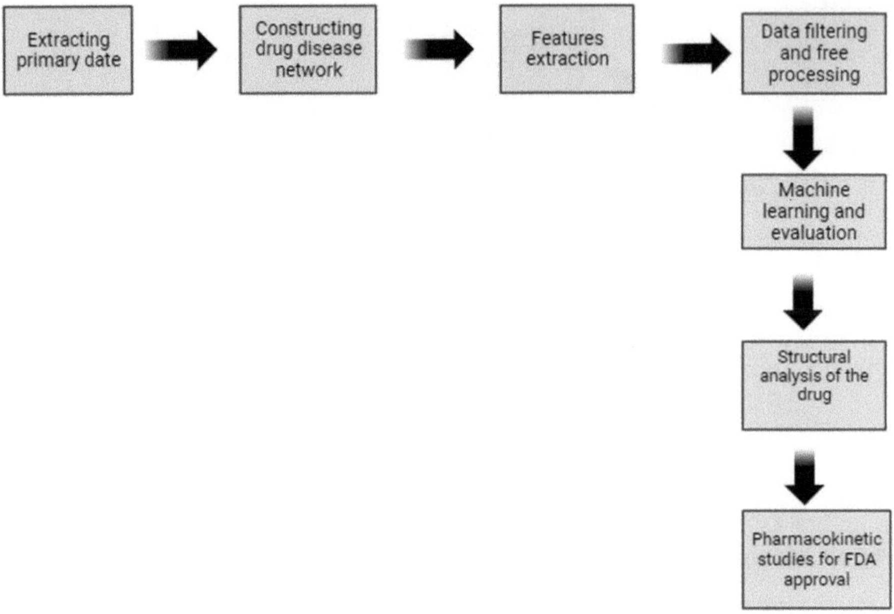

Fig. 9.3 Flowchart of drug repurposing (Niranjan et al. 2022)

around disease- or therapy-related factors (Sahoo et al. 2021). Repurposing is done in two steps: the initial step is to screen marketed medications or screen licensed drugs against a specific therapeutic target in silico, and the shortlisted compounds are then investigated by specific pathophysiological mechanisms of the disease of interest utilizing in vitro and in vivo techniques (Fig. 9.3).

The start of clinical trials for suitable indications is the second step in the medication repurposing process (Gns et al. 2019). There are several techniques for addressing the problem, including biological approaches, in silico techniques, experimental approaches, knowledge-based approaches, and mixed techniques (Kumar et al. 2019). The two main categories of systematic repurposing approaches are experimental screening procedures and in silico approaches that use available data to find possible new drug–disease connections (Cha et al. 2018).

9.6.1 Experimental Screening Approaches

Experimental screening procedures are employed to find hits for both drug discovery and drug repurposing, with significant differences in their application and outcomes. In drug development programs, searches are often conducted for de novo candidate hits, which are fueled by high-throughput screening campaign, which necessitates highly specialized screening equipment and compound libraries holding millions of compounds. Repurposing programs focus on advanced known compounds that have been approved or failed and have some knowledge of their safety or mechanism of action available, are led by in-depth screening, and have smaller compound libraries. A notable differentiation between drug discovery and repurposing lies in the outcome of identified compounds. In the context of repurposing, a compelling hit denotes a therapeutic molecule ready for advanced development. This contrasts with the role of a high-throughput screening (HTS) hit in drug discovery, which marks the initial phase of a newly discovered medicinal chemistry program that undergoes reiterative refinement. Additionally, the efficiency of an assay in drug development screens hinges on its simplicity and speed, enabling the assessment of a large number of potential compounds. In contrast, repurposing screens operate on a smaller scale, affording a broader spectrum of assay complexity (Cha et al. 2018).

9.6.2 In Silico Approaches

This strategy involves utilizing several public databases, sourcing information from research, clinical trials, label use reports, and various published data sources. Bioinformatics technologies and artificial intelligence are then employed to identify interaction networks between drug targets and medications. Notable databases like ChemBank, DrugBank, KEGG, Pubmed, OMIA, as well as disease–drug knowledge repositories, and genomic databases such as PDB, GenBank, GEO, and MIPS, play a crucial role in this process. These data contribute to the efficient development of cost-effective computing algorithms with fewer barriers. In silico techniques are rooted in the integration and analysis of extensive data related to drug–disease interactions, with some methods relying on gene expression studies in pertinent cell lines post-pharmacological treatment. There are numerous points of view, such as repurposing based on biological networks, and data-driven repurposing. Along with this, text mining, cluster approaches, and semantic techniques can all be used for repurposing, depending on the methodology (Kumar and Roy 2023).

9.6.2.1 Text Mining Approach

A substantial dataset containing linkages between biological concepts can be harnessed from the published literature, alongside a wealth of valuable insights into drug repurposing. Many of these insights are unique, and text mining serves as a valuable tool to extract and unveil these linkages. Text mining involves the extraction of novel information from a diverse range of published resources using computational methods. In the field of biology, text mining encompasses activities like retrieving pertinent information from filtered and structured literature, recognizing biological entities through controlled vocabularies, extracting and discovering biological insights from these concepts, and identifying relationships between drugs and their targets or diseases. The application of natural language processing techniques accelerates the development of text mining tools, thereby facilitating the discovery of pharmaceuticals suitable for repurposing (Dash et al. 2023).

9.6.2.2 Semantic Approaches

These are often used to extract information and images that can be used in drug repurposing. Biological entity relationships are discovered using data from medical databases, and a semantic network is established on top of an existing ontology network, with algorithms designed to uncover linkages in the network. The semantic network comprises drug–drug, target–target, and drug–target relationships built by the authors. The suggested method was assessed on a network by combining edge-partitioning techniques with semantic link prediction techniques. The suggested approach produced precise predictions about the relationships between drugs and targets because it made use of extensive semantic knowledge. The semantic data-driven algorithm ranked potential drug–disease relationships using a Bayesian framework that incorporates prior biological knowledge. These ranked associations were then integrated with connections to additional biological entities to construct a comprehensive semantic drug discovery network. A semantic drug discovery network was then created by integrating ranked relationships with the association of other biological entities. Pharmacological databases are essential tools for repurposing medications. These databases gather information on drug properties as well as interactions between pharmaceuticals and other living things. Many computational approaches are based on pharmacological data. DrugBank is a special cheminformatics and bioinformatics resource that combines extensive drug target information with detailed drug data. An additional kind of data resource for drug repositioning is proteomics databases. These approaches for drug repositioning are based on networks utilizing the most significant protein–protein interaction (PPI) networks found in proteomics databases. Proteomics databases are also valuable tools for creating heterogeneous networks, such as those between drugs, proteins, and diseases. MIPS is a well-known proteomics database that contains manually annotated, high-quality PPI data from academic publications. For certain network approaches, PPI information can be obtained via MIPS. It is also a useful tool for assessing the experimental outcomes of some computational techniques.

9.6.2.3 Cluster Approaches

Cluster techniques are useful for finding new links between drugs and diseases or targets. These are based on the idea that biological things (drugs, diseases, proteins, etc.) and networks within the same module have the same characters. Using cluster algorithms, different modules, groups, or subnetworks are found that display drug–drug, drug–target, or drug–disease interactions. Using clustering algorithms, distinct modules or subnetworks can be identified that reveal meaningful patterns in drug–drug, drug–target, and drug–disease interactions. Drug–drug, target–target, and drug–target relationships are just a few of the relationships that have been covered in these modules. Overlapping clusters are invisible to the most popular network-based cluster approaches, such as DBSCAN, CLIQUE, STING, and OPTICS. Researchers investigated the therapeutic repositioning of drugs used for SCLC (small-cell lung cancer) to resolve overlapping using a k-means-based network cluster algorithm. Candidate drug compounds with significant similarities to approved lung cancer drugs and genes related to lung cancer were chosen through chemical–chemical and chemical–protein interactions (Kumar et al. 2019). Furthermore, another network cluster approach called ClusterONE was developed on two kernel components: a greedy growth process and a cohesiveness score. This method's versatility is its strength; it can reliably forecast protein interactions in protein–protein networks as well as disease–drug relationships in disease–drug networks. This method is used for building disease–protein–drug networks using a symmetrical conditional probability (Kumar et al. 2019) (Table 9.2).

Table 9.2 Various data resources utilized by researchers and pharmaceutical companies for strategizing repurposing of drugs (Kumar and Roy 2023)

Drug database searches in use		
S.No.	Database name	Application
1	DrugBank	An extensive database that offers details on medications, their targets, and their modes of action (Wishart et al. 2008).
2	ChEMBL	A database of bioactive small molecules that resemble drugs, offering details on their targets, uses, and activities (Gaulton et al. 2012).
3	Gene Expression Omnibus (GEO)	A gene expression data repository that can be utilized to find candidates for repurposing and possible therapeutic targets (Clough et al. 2023).
4	The TCGA (Cancer Genome Atlas)	Offers genetic information on a range of cancer types, which can be utilized to find novel applications for currently approved medications (Tomczak et al. 2015).
5	Cheminformatics software	Chemical structures and properties are analyzed using programs such as ChemDraw, Cheminformatics Workbench, and others (Walker et al. 2010).
6	Computational Modeling and Virtual Screening	Drug–target–protein interactions can be predicted through molecular docking and molecular dynamics simulations.
7	Analysis of Biological Pathways	A resource for biological pathways and possible targets for medication repurposing is KEGG (Kyoto Encyclopedia of Genes and Genomes) (Aghajani et al. 2022).
8	Literary Exploration	Searching patent databases and scientific literature for details on possible drug candidates and their potential uses (Verbaanderd et al. 2021).
9	Resources for Clinical Trials	Finding medications that have demonstrated promise in treating other conditions by looking through data from clinical trials (Verbaanderd et al. 2021).
10	Databases for Phenotype and Disease	Disease Ontology and OMIM (Online Mendelian Inheritance in Man): These resources offer insights into phenotypic and genetic associations that may be useful in identifying prospective therapeutics for drug repurposing (Amberger et al. 2014).
11	Cooperation Projects	The goal of collaborative initiatives like the NIH National Center for Advancing Translational Sciences (NCATS) and the Drug Repurposing Hub is to advance drug repurposing efforts (Jonker et al. 2024).
12	AI and Machine Learning	In order to find possible opportunities for drug repurposing, machine learning and artificial intelligence techniques are being used more and more to analyze and integrate data from various sources (Herráiz-Gil et al. 2025).

(continued)

Table 9.2 (continued)

Drug database searches in use		
S.No.	Database name	Application
13	Clinical and Practical Information	Real-world data sources, like claims databases and electronic health records, can shed light on the applications of currently available medications and their potential efficacy for novel ailments (Girman et al. 2022).
14	Online Pharmacy Network	This approach considers the interactions between drugs, targets, and diseases as a network, allowing for the identification of potential repurposing opportunities based on network analysis (Truong et al. 2022).
15	Drug Screening Libraries	Screening compound libraries for new therapeutic uses by testing existing drugs against different targets or diseases. These resources and strategies are used in combination to identify potential candidates for drug repurposing, ultimately accelerating the development of new treatments and expanding the applications of existing drugs (Wall and Lopez-Ribot 2020).

9.6.2.4 Network Propagation Approach

Propagation approaches based on networks represent yet another significant category of network-based approaches. With these methods, previous data spreads from the source node to every network node as well as a few subnetwork nodes. These approaches can be categorized into two groups based on the various methods of propagation: local approaches and global approaches. The effectiveness of these techniques in identifying disease–gene, disease–target, and disease–drug interactions has been demonstrated by numerous studies. Local propagation techniques may not always produce accurate predictions because they only consider a small portion of the network's data. On the other hand, global methods that utilize data from the entire network outperform local methods.

In order to attain exceptional performance, the majority of researchers today focus on global approaches. To find novel disease–gene interactions, Kohle et al., for instance, constructed a network propagation approach based on a network's global information. The method was divided into three stages: (i) building a disease–gene network and extracting drug–disease relationships; (ii) using a random walk propagation algorithm within the network to obtain the global information of the network; and (iii) defining global metrics to predict novel disease–gene relationships. Compared to other approaches, the suggested method performed better.

Researchers frequently have to choose between different approaches depending upon their investigation aim. There are two types of networks used in these methods: homogeneous, i.e., protein–protein interaction (PPI) network and heterogeneous network. A homogeneous network is frequently used in some approaches to identify drug targets involved with multiple pathways because disease pathways can be constructed from PPI network analysis. PPI networks have become more accurate due to the establishment and updating of multiple PPI databases through experiments, in addition to the deeper research related to the network.

Concluding from data resource imputs for drug screening, it is well recognized that drug repurposing can save money throughout the preclinical phases I and II of the drug development process (Niranjan et al. 2022).

9.7 Drug Repurposing in Various Diseases

Drug repurposing has been investigated in several disorders across numerous medical specialties. Repurposing existing drugs is paramount in the ongoing battle against rapidly spreading infectious diseases like HIC, influenza, hepatitis C, Ebola, dengue, and numerous other life-threatening illnesses (Sakurai et al. 2015). Thalidomide, initially developed for sedation in the 1950s but withdrawn due to fetal risks, found new life in treating multiple myeloma and tumors through drug repositioning. Likewise, minoxidil, initially developed for hypertension, was repositioned and reformulated into a topical treatment for hair loss. Thus, reformulation improved the profile of drugs like amiodarone with intravenous use for better myocardial distribution. Combination therapy enhances drug effectiveness, as seen in cardiovascular drugs like acetazolamide and loop diuretics for heart failure, statins with ezetimibe for atherosclerosis, and angiotensin receptor blockers with β-blockers for Marfan syndrome. In the past few decades, numerous drugs have been repurposed to treat various diseases such as cancer, neurological disorders, severe infections, cardiovascular disease, metabolic disorders, viral diseases, and rare diseases (Austin and Gadhia 2017; Novack 2021; Deotarse 2015; Okuyama 2023).

9.7.1 Cancer

Existing medications that were initially created for other uses have been investigated for their potential to treat cancer. The medications with existing pharmacokinetic, pharmacodynamic, and toxicity profiles can hasten the development process and thus contribute to the benefits of drug repurposing in cancer therapy. For instance, the ability of the diabetic medication metformin to limit tumor growth has made it a promising candidate for treatment of cancer. Drug repurposing for cancer treatment is a promising approach to discover new therapeutic options for cancer patients. In the past few decades, various drugs were explored for the repositioning of drugs for cancer therapy, as presented in Table 9.3.

Drug repurposing for cancer treatment offers the advantage of faster development and reduced costs, as these drugs have already undergone extensive safety and toxicity testing. It also provides new treatment options for patients, especially in cases where standard therapies have limited effectiveness. Continuous efforts are underway to explore additional drug candidates for repurposing in the fight against cancer.

Table 9.3 Drug repurposed for cancer treatment (Dash et al. 2023)

S.No.	Drugs	Applications
1	Thalidomide and Lenalidomide	Treat multiple myeloma and myelodysplastic syndromes. They are immunomodulatory drugs that have anti-angiogenic and anti-inflammatory properties, making them effective against certain cancers.
2	Imatinib (Gleevec)	Imatinib, initially developed to treat chronic myeloid leukemia (CML), was later found to be effective against other cancers, including gastrointestinal stromal tumors (GISTs). It targets the BCR–ABL fusion protein, which is also present in some GISTs.
3	Metformin	Used to manage type 2 diabetes; has been studied for its potential anticancer effects. It may help reduce the risk of certain cancers and improve outcomes in cancer patients by affecting cellular metabolism and growth.
4	Bisphosphonates	Bisphosphonates like zoledronic acid, primarily used to treat osteoporosis and bone-related conditions, have been explored for their potential to prevent cancer-related bone complications, such as bone metastases in breast and prostate cancer.
5	Vorinostat and Romidepsin	Treat cutaneous T-cell lymphoma. They target the epigenetic regulation of gene expression in cancer cells.
6	Tamoxifen	Tamoxifen, originally developed as a breast cancer treatment, has also been explored for its potential in preventing breast cancer in high-risk individuals.
7	Eflornithine (DFMO)	Treat sleeping sickness; eflornithine has shown promise in inhibiting polyamine synthesis and has been repurposed for the treatment of neuroblastoma, a childhood cancer.
8	Propranolol	Propranolol, a beta-blocker used to treat high blood pressure and anxiety, has shown potential for repurposing in the treatment of hemangiomas, a type of benign tumor, as well as for reducing the growth of certain solid tumors.
9	Disulfiram	Originally used to treat alcohol addiction, disulfiram has been investigated for its potential anticancer effects. It appears to inhibit key signaling pathways in cancer cells.
10	Sirolimus (Rapamycin)	Sirolimus, an immunosuppressant used in organ transplantation, has been repurposed for the treatment of lymphangioleiomyomatosis (LAM), a rare lung disease associated with tumors.

9.7.2 Neurological Disorders

Drug repurposing has been pursued in neurological disorders and successfully added existing therapeutics for neurological disorders. Millions of people around the world are affected by neurological illnesses, but only a select few medications are completely effective in treating them. In order to identify improved treatments for neurological illnesses, medication repurposing has attracted attention over the past 10 years as a method that is underutilized. For instance, cimetidine, an antihistamine, has been investigated for its ability to help Alzheimer's sufferer's cognitive function. This approach can significantly reduce the time and cost of drug development for neurological conditions. Table 9.4 details about drugs repurposed in the context of neurological diseases.

Table 9.4 Drugs repurposed for neurological disorders (Hua et al. 2022)

S.No.	Drugs	Applications
1	L-DOPA (Levodopa)	Originally developed as a treatment for Parkinson's disease, L-DOPA is still a primary therapy for managing the motor symptoms of this condition. It is converted to dopamine in the brain, compensating for dopamine deficiency in Parkinson's.
2	Gabapentin and Pregabalin	Originally developed as antiepileptic drugs, gabapentin and pregabalin have been repurposed for the management of neuropathic pain, such as diabetic neuropathy and postherpetic neuralgia.
3	Baclofen	Baclofen, initially used as a muscle relaxant to treat spasticity, has been explored for its potential in managing alcohol use disorder and reducing alcohol cravings.
4	Lamotrigine	Originally developed for epilepsy, lamotrigine is used as a mood stabilizer in the treatment of bipolar disorder.
5	Memantine	Memantine, approved for Alzheimer's disease, has been considered for the treatment of other neurological conditions, such as frontotemporal dementia and Huntington's disease.
6	Riluzole	Developed for amyotrophic lateral sclerosis (ALS), riluzole has been studied for its potential in the treatment of other neurodegenerative diseases, including Huntington's disease and Alzheimer's disease.
7	Valproic Acid	Valproic acid, used primarily as an antiepileptic medication, has been repurposed as a mood stabilizer for bipolar disorder.
8	Aripiprazole	Aripiprazole, an atypical antipsychotic, has been repurposed as an augmentation therapy in the treatment of major depressive disorder and as a treatment for irritability associated with autism.
9	Minocycline	Originally an antibiotic, minocycline has shown promise as a potential neuroprotective agent in neurodegenerative diseases like Huntington's and amyotrophic lateral sclerosis (ALS).
10	Naltrexone	Naltrexone, originally developed for opioid addiction treatment, has been explored for its potential in the management of multiple sclerosis and other autoimmune neurological conditions.
11	Cannabinoids	Cannabinoid-based drugs, such as nabiximols (Sativex), have been developed for spasticity in multiple sclerosis and have potential applications in pain management and epilepsy.
12	Thalidomide	Thalidomide, repurposed from its original use, has been investigated for its potential in the treatment of neuroinflammatory diseases like multiple sclerosis.
13	Niacin (Vitamin B3)	Niacin has been studied for its potential neuroprotective effects in Alzheimer's disease and other neurodegenerative conditions.

9.7.3 Infectious Diseases

The upsurging need of new therapeutics for managing severe infections during the pandemic highlighted the benefits of medicinal repurposing in infectious diseases. During this period, current antiviral and antimalarial drugs were tested against the novel corona virus. Pharmaceuticals created to address past viral infections have also been investigated for their potential in treating emerging infectious diseases. Repurposing is gained attention and paced off during the COVID-19 pandemic owing to the FDA approval to many therapeutic molecules for repurposing in

Table 9.5 Drug repurposed for managing infectious diseases, antifungals, and antiviral drugs (Hua et al. 2022)

S.No.	Drugs	Applications
1	Hydroxychloroquine and Chloroquine	Conventional use for treatment of malaria and autoimmune diseases. They were explored as potential treatments for COVID-19 due to their anti-inflammatory and antiviral properties.
2	Remdesivir	Initially developed for the treatment of Ebola, remdesivir was repurposed to treat COVID-19 as it is an antiviral drug that inhibits viral RNA synthesis.
3	Ivermectin	Ivermectin, commonly used to treat parasitic infections in humans and animals, has been considered as a potential treatment for various viral infections, including COVID-19, based on its antiviral and anti-inflammatory properties.
4	Azithromycin	Azithromycin, an antibiotic commonly used to treat bacterial infections, was investigated for its potential in combination with other drugs to treat respiratory infections, including COVID-19.
5	Amantadine	Originally used as an antiviral medication for influenza, amantadine has been explored as a potential treatment for Parkinson's disease, as it has properties that affect dopamine release and reuptake.
6	Zinc	Zinc supplements have been examined for their potential to reduce the severity and duration of viral infections, such as the common cold and COVID-19.
7	Ribavirin	Ribavirin was originally developed for hepatitis C and repurposed for the treatment of various viral infections, including respiratory syncytial virus (RSV) and Lassa fever.
8	Favipiravir	Favipiravir, developed as an antiviral agent for influenza, has been repurposed for the treatment of emerging viral infections like Ebola and COVID-19.
9	Interferons	Interferon therapies, developed for various indications, have been repurposed to enhance the immune response against viral infections, including hepatitis B and C, HIV, and respiratory viruses.

COVID management. Remdesivir, sold as Veklury, was developed as an antiviral for RNA-based viruses and tested against Ebola during pandemic; it was licenced as a COVID-19 medication within 6 months of the pandemic. Drug repurposing for infectious disease treatment is a valuable strategy for rapidly identifying new treatment options for emerging or neglected infectious diseases. This approach involves repurposing existing drugs, compounds, or therapies developed for other indications to combat infectious pathogens. Drugs repurposed for managing infectious diseases are listed in Table 9.5.

9.7.4 Repurposed Antifungals and Antiviral Drugs

Recently, researchers have investigated some antifungal drugs, such as itraconazole and posaconazole, for antiviral activity and had studied them as potential treatments for viral infections, including respiratory syncytial virus (RSV) and others (Table 9.5). Researchers explore existing antiviral drugs, such as ribavirin, as potential treatments for a range of viral infections, particularly when no specific antiviral drugs are available. Drug repurposing for such infectious disease treatment offers low cost, fast process of new drug development, and drug approval for clinical use in case. This approach is particularly important in responding to emerging infectious diseases and addressing the need for effective treatments when no specific antiviral drugs are available.

9.7.5 Cardiovascular Diseases

Currently, medications in use for cardiovascular disease are being investigated for their potential to treat some other CVS disorders, for instance, medications for high blood pressure may be used to treat heart failure. Drugs like statins that are initially in use for CVS disorders like heart failure and stroke have proven successful in managing hyperlipidemia and atherosclerosis. Moreover, asprin that is a pain reliever and anti-inflammatory drug turns to be a success for multiple issues like heart attacks and strokes. In this league, ezetimibe and ranolazine have also emerged as repurposed drugs for reducing atherosclerosis and as an anti-anginal drug, respectively (Table 9.6).

9.7.6 Drug Repurposing for Autoimmune Diseases

Immune-mediated diseases are reported to affect nearly 20% of global population, and autoimmune diseases also affect the skin. In the past few decades, drugs that were explored for other ailments are repurposed for treatment of autoimmune diseases. Drugs used for rheumatoid arthritis have displayed promising results for treatment of inflammatory bowel disorders like Crohn's disease and ulcerative colitis. Table 9.7 represents a list of repurposed drugs recently exploited for autoimmune diseases.

9.7.7 Drug Repurposing for Metabolic Disorders

Re-evaluation and repurposing of existing drugs or compounds for other indications have demonstrated relevant activity for metabolic conditions. Reprofiled medications for metabolic illnesses can generally be divided into three categories. Group A includes those drugs for which metabolic problems are primary and secondary indications. Group B included drugs that were initially licensed for conditions other than metabolic disorders but have been repurposed to treat metabolic disorders. Drugs in Group C are those that were first approved for metabolic illnesses but are now being used to treat nonmetabolic ailments. Drug repurposing in metabolic diseases listed in Table 9.8, such as diabetes, can result in creative therapies.

Table 9.6 List of repurposed drugs for cardiovascular diseases

S.No.	Drugs	Applications
1	Statins	Originally developed as cholesterol-lowering drugs, statins like atorvastatin and simvastatin have been found to reduce the risk of cardiovascular events, including heart attacks and strokes. They are widely used to manage hyperlipidemia and atherosclerosis (Zhou and Liao 2009).
2	Aspirin (Acetylsalicylic Acid)	Aspirin, primarily known as a pain reliever and anti-inflammatory agent, is widely used to prevent heart attacks and strokes. It functions as an antiplatelet agent to reduce the risk of blood clot formation (Dai and Ge 2012).
3	Ezetimibe	Ezetimibe, used to lower cholesterol levels, has been repurposed as a cholesterol absorption inhibitor, helping to reduce the risk of atherosclerotic cardiovascular disease (Shaya et al. 2020).
4	Ranolazine	Ranolazine was initially developed as an anti-anginal medication. It has also been explored for its potential to treat arrhythmias and heart failure by improving myocardial energy metabolism (Vadnais and Wenger 2010).
5	Minoxidil	Originally developed as an antihypertensive medication, minoxidil has been used in topical formulations to stimulate hair growth in androgenetic alopecia (Suchonwanit et al. 2019).
6	Ivabradine	Ivabradine, developed as a treatment for angina, has been repurposed to reduce heart rate and improve outcomes in heart failure patients (Kaski et al. 2018).
7	Sacubitril/Valsartan	This combination medication, originally approved for heart failure, is used to reduce hospitalization and cardiovascular mortality in patients with heart failure with reduced ejection fraction (Chang et al. 2019).
8	Aliskiren	It is developed as an antihypertensive drug and has been studied for its potential to reduce cardiovascular events and diabetic kidney disease in hypertensive patients (Pimenta and Oparil 2009).
9	Hydralazine and Isosorbide Dinitrate	This combination of medications, initially used to treat high blood pressure, has been repurposed for the management of heart failure in African American patients (Golwala et al. 2013).
10	Omega-3 Fatty Acids:	Omega-3 fatty acids found in fish oil supplements have been studied for their potential in reducing cardiovascular events, particularly in individuals with elevated triglycerides (Jump et al. 2012).
11	Colchicine	Colchicine, traditionally used to treat gout, has been investigated for its anti-inflammatory effects in reducing the risk of cardiovascular events in patients with a history of heart disease (Siddiqui et al. 2023).
12	SGLT2 Inhibitors	Sodium–glucose cotransporter 2 (SGLT2) inhibitors, initially developed for diabetes management, have been repurposed to reduce the risk of heart failure and cardiovascular events in diabetic and nondiabetic patients (Marfella et al. 2022).
13	Beta-Blockers	While beta-blockers were initially developed to treat hypertension; they have found broad applications in the management of various cardiovascular conditions, including arrhythmias, heart failure, and angina (Larochelle et al. 2014).

Table 9.7 List of repurposed drugs for autoimmune diseases (Li et al. 2017)

S.No.	Drugs	Applications
1	Corticosteroids	Corticosteroids like prednisone and prednisolone, originally developed for their anti-inflammatory and immunosuppressive properties, are commonly used to manage various autoimmune conditions, such as rheumatoid arthritis, lupus, and multiple sclerosis.
2	Methotrexate	Methotrexate, originally developed as a chemotherapy drug for cancer, is widely used to treat autoimmune diseases, including rheumatoid arthritis and psoriasis, due to its immunosuppressive properties.
3	Cyclophosphamide	Cyclophosphamide, used as a chemotherapy agent, is also employed in the treatment of severe autoimmune disorders like systemic lupus erythematosus (SLE) and vasculitis.
4	Azathioprine	Azathioprine, developed as an immunosuppressant for organ transplantation, has been repurposed to treat autoimmune diseases, including inflammatory bowel disease (IBD) and autoimmune hepatitis.
5	Hydroxychloroquine	Originally developed as an antimalarial drug, hydroxychloroquine is commonly used to manage autoimmune diseases like systemic lupus erythematosus (SLE) and rheumatoid arthritis due to its anti-inflammatory and immunomodulatory properties.
6	Cyclosporine	Cyclosporine, developed as an immunosuppressant for organ transplantation, is used to treat autoimmune conditions, including psoriasis and certain types of arthritis.
7	Tocilizumab and Sarilumab:	These drugs, originally developed for rheumatoid arthritis, are repurposed for the treatment of cytokine release syndrome associated with CAR-T cell therapy and other autoimmune diseases.
8	Rituximab	Rituximab, developed for B-cell lymphoma, is used to treat autoimmune diseases characterized by abnormal B-cell activity, such as rheumatoid arthritis and systemic lupus erythematosus.
9	Alemtuzumab	Alemtuzumab, originally developed for chronic lymphocytic leukemia, is repurposed to treat multiple sclerosis by targeting abnormal immune responses.
10	Abatacept	Abatacept, initially approved for rheumatoid arthritis, is used to treat autoimmune conditions like psoriatic arthritis and juvenile idiopathic arthritis by inhibiting T-cell activation.
11	Apremilast	Apremilast, developed for psoriatic arthritis and psoriasis, is repurposed to treat autoimmune diseases like Behcet's disease and oral ulcers associated with Behcet's.
12	Vitamin D Analogues	Vitamin D analogues like calcitriol have been explored for their immunomodulatory effects and potential in the treatment of autoimmune diseases such as psoriasis and multiple sclerosis.

Table 9.8 List of drugs repurposed for metabolic disorders (Finsterer and Frank 2013)

S.No.	Drugs	Applications
1	Metformin	Metformin, originally developed as an antidiabetic drug, is commonly used to manage type 2 diabetes and polycystic ovary syndrome (PCOS). It improves insulin sensitivity and glucose control.
2	Orlistat	Orlistat, developed for weight loss, has been repurposed to manage obesity and associated metabolic disorders. It works by reducing the absorption of dietary fats.
3	Liraglutide and Other GLP-1 Receptor Agonists	Originally developed for diabetes management, drugs like liraglutide have been repurposed for weight loss and as potential therapies for non-alcoholic fatty liver disease (NAFLD).
4	Empagliflozin and Other SGLT2 Inhibitors	Sodium–glucose cotransporter 2 (SGLT2) inhibitors, used to lower blood sugar in diabetes, have been repurposed to reduce the risk of heart failure and improve cardiovascular outcomes.
5	Fenofibrate	Fenofibrate, initially developed for lipid management, is repurposed to treat high triglycerides and mixed dyslipidemia, particularly in patients with metabolic syndrome.
6	Bempedoic Acid	Bempedoic acid, used to lower LDL cholesterol levels, has been repurposed to manage hypercholesterolemia and lower cardiovascular risk in metabolic disorder patients.
7	Naltrexone and Bupropion Combination	The combination of naltrexone and bupropion, originally developed for opioid addiction and smoking cessation, has been repurposed as a weight loss therapy.
8	Carnitine	Carnitine supplements are explored for their potential in managing fatty acid oxidation disorders and improving energy metabolism.
9	Resveratrol	Resveratrol, a compound found in red wine, is studied for its potential in improving insulin sensitivity and reducing the risk of metabolic diseases.
10	Levothyroxine	Levothyroxine, used to replace or supplement thyroid hormone, is employed to manage thyroid disorders that can impact metabolism.

9.7.8 Drug Repurposing for Rare Diseases

Since rare or orphan diseases affect only a small portion of the population, they typically receive limited research attention and minimal funding, resulting in few new therapeutic options for affected individuals. These conditions often receive limited attention in drug development due to their low prevalence, making drug repurposing a cost-effective and efficient approach for rare diseases.

9.7.9 Drug Repurposing for Genetic Disorders

Exploration of human genetics to connect genes with specific diseases offered new candidate targets for genetic disorders. However, drug repurposing came forth as an effective mode for reducing timelines and for improving target-specific repurposed drugs that can be utilized for genetic disorders.

Patients may find hope in already available drugs that may be effective against the underlying processes of these rare and genetic disorders listed in Table 9.9.

Table 9.9 List of drugs repurposed for rare and genetic disorders (Shah et al. 2021)

S.No.	Drugs	Applications
1	Enzyme Replacement Therapy (ERT)	ERT is a well-established method for treating certain rare lysosomal storage disorders like Gaucher disease, Fabry disease, and Pompe disease. Enzymes that are deficient in these conditions are replaced with recombinant enzymes to alleviate symptoms.
2	Nitisinone	Originally developed to treat tyrosinemia type 1, nitisinone has been repurposed for the treatment of alkaptonuria, a rare genetic metabolic disorder.
3	Dimethyl Fumarate (DMF)	DMF, initially approved for multiple sclerosis, has shown promise in the treatment of Friedreich's ataxia, a rare neurological disorder.
4	Sirolimus and Everolimus	Sirolimus and everolimus, originally developed as immunosuppressants, have been repurposed for the treatment of certain rare diseases like lymphangioleiomyomatosis (LAM) and tuberous sclerosis complex (TSC).
5	Eculizumab	Eculizumab, developed to treat paroxysmal nocturnal hemoglobinuria and atypical hemolytic uremic syndrome, is used to manage rare disorders involving complement system dysregulation.
6	Deferiprone	Deferiprone, initially developed for iron chelation therapy in thalassemia, is repurposed to treat neurodegeneration with brain iron accumulation (NBIA) disorders.
7	Laronidase	Laronidase is an enzyme replacement therapy used in mucopolysaccharidosis type I (MPS I), a rare lysosomal storage disorder.
8	Cyclosporine A	Cyclosporine A, an immunosuppressant used for organ transplantation, has been repurposed to treat certain rare dermatologic diseases such as pemphigus vulgaris.
9	Eliglustat	Eliglustat, developed for Gaucher disease, offers a non-ERT option for treating this rare lysosomal storage disorder.
10	Erdafitinib	Erdafitinib, originally developed for cancer, was approved for the treatment of a rare form of bladder cancer with specific genetic mutations.
11	Cysteamine Bitartrate	Cysteamine bitartrate, used to manage cystinosis, a rare metabolic disorder, helps remove excess cystine from cells.
12	Nitisinone	Nitisinone, used for tyrosinemia, is repurposed to treat alkaptonuria, a rare metabolic disorder.
13	Canakinumab	Canakinumab, initially developed for autoinflammatory conditions, is explored for its potential to treat certain rare periodic fever syndromes.
14	Fampridine:	Fampridine, originally developed for multiple sclerosis, is used to improve walking in patients with Lambert–Eaton myasthenic syndrome (LEMS), a rare neuromuscular disorder.

9.7.10 Drug Repurposing for Respiratory Diseases

The repurposing of drug for respiratory diseases also attempted for prophylaxis and therapy of severe respiratory diseases. Drug repurposing for respiratory diseases has been proven to be an effective approach for managing obstructive pulmonary disease and asthma utilizing inhaled bronchodilators and corticosteroids. Repurposed drugs used for respiratory disorders are presented in Table 9.10.

9.8 Drug Repurposing for Corona Virus Disease 19 (COVID-19)

The COVID-19 pandemic has necessitated extensive research into multiple strategies for managing the illness, with drug repurposing emerging as a crucial component. Numerous medications are currently undergoing clinical trials to explore their potential for repurposing against COVID-19. These drugs include favipiravir, hydroxychloroquine, remdesivir, tocilizumab, kaletra, and chloroquine. These medications can be categorized into four groups based on their mechanisms of action: inhibitors of RNA-dependent RNA polymerase, inhibitors of serine protease TMPRSS2, inhibitors of the primary cysteine protease 3CLpro, and drugs targeting inflammation and the cytokine storm. Chloroquine and hydroxychloroquine, which have been in use for over seven decades, exhibit antiparasitic properties by raising lysosomal pH and also have immunosuppressive effects. Remdesivir, initially developed as an antiviral treatment for Ebola virus infection, functions by inhibiting viral RNA-dependent RNA polymerase. Multiple randomized controlled trials with placebos have been conducted to assess remdesivir's efficacy, and it received approval as an anti-COVID-19 medication in Japan on May 7, 2020. Favipiravir has been authorized for addressing both new and recurrent influenza virus infections. When inside the body, this medication undergoes a metabolic transformation into its active triphosphorylated state, which effectively hinders the RNA-dependent RNA polymerase derived from the virus. Camostat and nafamostat exhibit the capability to inhibit proteolytic enzymes, particularly TMPRSS2, and are known to alleviate symptoms associated with acute pancreatitis. Additionally, these drugs can prevent blood coagulation during extracorporeal circulation, particularly in patients with hemorrhagic lesions. In the realm of antiretroviral therapy for HIV/AIDS, lopinavir and ritonavir function as protease inhibitors. The combination of lopinavir and ritonavir in a fixed-dose formulation is believed to obstruct the primary cysteine protease of SARS-CoV-2. Ritonavir is thought to enhance lopinavir concentrations by inhibiting CYP3A4. Likewise, ivermectin, an anthelmintic medication approved for addressing intestinal illnesses and scabies, is considered to impede the replication of SARS-CoV-2. Moreover, dexamethasone, a steroid initially used to manage severe infections and interstitial pneumonia, is now under consideration as a treatment option for severe COVID-19 cases marked by a systemic inflammatory response. This inflammatory reaction can result in lung injury and multi-organ failure, and the anti-inflammatory properties of steroids like dexamethasone are

Table 9.10 List of drugs repurposed for respiratory disorders (Newman 2018)

S.No.	Drugs	Applications
1	Ivermectin	Originally developed as an antiparasitic drug, ivermectin has been explored for its potential to treat respiratory infections, including COVID-19.
2	Montelukast	Montelukast, initially developed for the treatment of asthma, is repurposed for allergic rhinitis and allergic conjunctivitis.
3	Tiotropium and Other Anticholinergics	Anticholinergic drugs like tiotropium are used for the management of chronic obstructive pulmonary disease (COPD) and have been studied for their potential in asthma treatment.
4	Methylxanthines	Methylxanthines like theophylline, traditionally used to treat asthma, are explored for their bronchodilatory effects in COPD.
5	Ketamine	Ketamine, initially developed as an anesthetic, has been repurposed as a treatment for refractory asthma and chronic pain associated with respiratory diseases.
6	Surfactant Replacement Therapy	Surfactant replacement therapy with drugs like beractant and poractant alfa is used to treat respiratory distress syndrome in premature infants.
7	Iloprost:	Originally developed for pulmonary arterial hypertension, iloprost has been studied for its potential to treat acute respiratory distress syndrome (ARDS) and other severe lung conditions.
8	Lung Surfactants	Natural or synthetic lung surfactants are used to treat respiratory distress syndrome in newborns and premature infants.
9	Nitric Oxide (Inhaled)	Inhaled nitric oxide has been repurposed for the treatment of pulmonary hypertension and as a supportive therapy in ARDS.
10	Budesonide and Other Inhaled Corticosteroids	Inhaled corticosteroids like budesonide are used to manage asthma and have been explored for their potential in other respiratory conditions, such as chronic cough.
11	Pirfenidone and Nintedanib	These drugs, originally developed for idiopathic pulmonary fibrosis, have been investigated for their potential in the treatment of progressive fibrotic interstitial lung diseases.
12	Colchicine	Colchicine, traditionally used for gout, is explored for its anti-inflammatory effects in treating respiratory diseases such as chronic obstructive pulmonary disease (COPD).
13	Amiloride	Originally developed as a diuretic, amiloride is being studied for its potential in treating cystic fibrosis by affecting ion transport in the airways.
14	Fluvoxamine	Fluvoxamine, an antidepressant, has been studied for its potential in reducing the severity of COVID-19 symptoms and preventing respiratory deterioration.
15	Pulmonary Vasodilators	Drugs like sildenafil and tadalafil, developed for erectile dysfunction, have been repurposed to treat pulmonary arterial hypertension.
16	Sildenafil and tadalafil	Developed for erectile dysfunction, these have been repurposed to treat pulmonary arterial hypertension.

Table 9.11 Most extensively researched drugs and molecules being repurposed for the treatment of COVID-19 (Rodrigues et al. 2022)

S.No.	Repurposed drug	Original therapeutic application	COVID-19 probable mechanism of action
1	Umifenovir	Influenza and other respiratory viruses	Blocks virus attachment and entry. Modulates immune response and interferon production (Rodrigues et al. 2022).
2	Dexamethasone	Inflammatory conditions such as bronchial asthma, endocrine, and rheumatic disorders	Binds to the cellular glucocorticoid receptor, modulates production of pro-inflammatory and anti-inflammatory signals (Kino et al. 2021).
3	Chloroquine and Hydroxychloroquine	Malaria and chronic inflammatory diseases	Prevents virus entry and decapsidation. Modulates the host immune system (Lei et al. 2020).
4	Lopinavir–Ritonavir	AIDS/ HIV	Inhibits the virus 3CL protease (Magro et al. 2021).
5	Baricitinib	Rheumatoid arthritis	Modulates cytokine production (Kubo et al. 2018).
6	Masitinib	Asthma, Alzheimer's disease, multiple sclerosis, amyotrophic lateral sclerosis	Inhibits the virus 3CL protease (Drayman et al. 2021).
7	Molnupiravir	Influenza viruses and encephalitic alphaviruses	Inhibits virus RNA synthesis (Ojha et al. 2024).
8	Remdesivir	Ebola virus	Inhibits virus RNA synthesis (Tchesnokov et al. 2019).
9	Tocilizumab	Rheumatoid arthritis, other autoimmune rheumatic diseases	Inhibits IL-6 activity (Emery et al. 2008).

believed to mitigate or prevent this response. Additional anti-inflammatory medications are also being explored to target the cytokine storm that occurs after SARS-CoV-2 infection (Kato et al. 2022) (Table 9.11).

Furthermore, the mechanisms by which these drugs target their respective agents should align or bear resemblance to those of their viral family members. The concept of drug repurposing has been corroborated by previous observations of successful treatment in patients with mild COVID-19 symptoms. For example, an additional study has elucidated and proposed the repurposing of other medications, such as Bruton tyrosine kinase inhibitors (BTK), interleukin inhibitors, interferons, vascular endothelial growth factor (VEGF) inhibitors, immune checkpoint inhibitors, and XPO-1 (a selective inhibitor of nuclear export), as potential alternatives for expediting the treatment of COVID-19 (Niranjan et al. 2022). Drug repositioning, often referred to as drug reprofiling, drug rescuing, drug recycling, or therapeutic flipping, unquestionably entails investigating new therapeutic applications for already-approved medications. Repurposing antiviral treatments like remdesivir and favipiravir, combined with antimalarial medicines like chloroquine and hydroxychloroquine, for treating COVID-19 is a good example. Similarly, therapeutic switching is used to increase R&D productivity by coming up with new uses for stereoisomers and drug metabolites that are already on the market or covered by patents. According to Martorana et al. (2016), this technique successfully creates drug candidates with novel therapeutic properties or pharmacological effects (Sahoo et al. 2021). Many experts contend that, as compared to traditional drug development techniques, repurposing already-approved medications can have benefits in terms of speed, cost-effectiveness, lower risk, and higher success rates. This is mainly because earlier phases of development aimed at proving drug safety have already been finished, allowing researchers to skip them (Ashburn and Thor 2004).

Nevertheless, it is only sometimes apparent how much time, risk, and money may be saved through repurposing, and the evidence for this claim needs to be more consistent. According to some studies, over 30% of repurposing initiatives result in medications that are successfully approved for marketing, compared to a success rate of around 10% for new drug applications in a more general sense (Hauser et al. 2017). Contradictory views contend that efficacy, not safety, frequently determines whether a treatment is safe or effective and that repurposed drugs do not necessarily have a higher success rate than newly produced ones (Hernandez et al. 2017). Despite significant advancements in technology and our understanding of various diseases, the pace of discovering new medications has shown a comparatively slower rate of progress (Mohammad et al. 2021). It has been said that finding and creating brand-new medicines can take between 10 and 17 years. Repurposed drugs, however, frequently receive approval more quickly, in a timeframe of 3–12 years, and for a price that is roughly half as high (Deotarse et al. 2014; Novack 2021). Drug repurposing presents a valuable method for rapidly and effectively identifying novel therapeutic targets in these disease areas. Researchers and medical practitioners can explore unknown medicinal uses and potentially enhance patient outcomes by utilizing currently available medications with well-established safety characteristics. The objective of discovering novel drug–disease correlations has been attained through various techniques, including computational methods, high-throughput screening, and clinical trials (Hua et al. 2022). However, there are obstacles to drug repurposing, including a lack of financial incentives for pharmaceutical corporations, obstructions from regulations, and the requirement for fresh clinical trials. Despite these difficulties, medication repurposing has the potential to be a time- and money-saving drug development method, particularly for uncommon diseases. Scientists may now evaluate potential medications for repurposing with a brand-new database called NICEdrug.ch. (Roessler et al. 2021). Drug repositioning is one of the most practical approaches when a sudden, big global public health event like COVID-19 is considered. In the search for repurposed COVID-19 therapies, various approaches have been used. Some aim to prevent viral multiplication or entrance, while others aim to modify the host immunological response, decreasing the intensity of symptoms. Although several medications use various identification strategies for repositioning, most remain until the phase II clinical trial, which is still far from the markets.

9.9 Conclusion

In conclusion, drug repurposing is a cost-effective, efficient, and versatile strategy for accelerating the development of treatment across various medical conditions. Its well-established safety profiles, ability to explore new mechanisms, and adaptability in addressing rare diseases make it a valuable asset in the pharmaceutical industry. While it faces challenges, including regulatory obstacles and financial incentives, drug repurposing has proven its worth, particularly during global health crisis. Drug repurposing come up as an efficient solution for managing COVID-19. Developing a new drug typically takes an average 10–15 years and costs more than $2 billion (DiMasi et al. 2016). Approximately 90% of drugs that enter clinical evaluation fail at some phase, because of safety concerns or lack of efficacy (Nagaraj et al. 2018), and many drugs are withdrawn or discontinued after being brought to market (Siramshetty et al. 2016). The development process can be made 2–3 times faster by repurposing (repositioning) existing drugs (Hurle et al. 2013), because the adverse effects have already been investigated and manufacturing procedures have been established. Compared with newly developed drugs, repurposed drugs are 3 times more likely to be approved (Hernandez et al. 2017). Larger drug development outcomes have been limited by the present, expensive, and time-consuming paradigm of drug discovery. An alternate strategy is desperately needed to restart medication development to meet market demands because of the high risks, significant costs, and sluggish speed of drug discovery (Low et al. 2020). With an urgent need for efficient medicines, repurposing existing medications with well-established safety profiles provided a practical solution. During the pandemic, repurposed medications such as hydroxychloroquine, remdesivir, and dexamethasone received a lot of attention. It is important to note that not all repurposed medications have proven successful, highlighting the significance of comprehensive clinical evaluation. Drug repositioning emphasizes the possibility for existing knowledge and resources to be used to respond quickly to developing health problems. As we continue to explore this innovative approach, we hold the promise of improving patient care, reducing development costs, and advancing medical science in our ongoing quest for effective treatments.

References

Aghajani J, Farnia P, Farnia P, Ghanavi J, Velayati AA (2022) Molecular dynamic simulations and molecular docking as a potential way for designed new inhibitor drug without resistance. Tanaffos 21(1):1–14

Ahmad S, Qazi S, Raza K (2021) Translational bioinformatics methods for drug discovery and drug repurposing. In: Translational bioinformatics in healthcare and medicine, Academic Press, U.K.; 127–139

Amberger JS, Bocchini CA, Schiettecatte F, Scott AF, Hamosh A (2014) OMIM.org: Online Mendelian Inheritance in Man (OMIM®), an online catalog of human genes and genetic disorders. Nucleic Acids Res 43(D1):D789–D798

Ashburn TT, Thor KB (2004) Drug repositioning: identifying and developing new uses for existing drugs. Nat Rev Drug Discov 3(8):673–683

Austin BA and AD Gadhia (2017) New Therapeutic Uses for Existing Drugs. Rare Diseases Epidemiology: Update and Overview. M. Posada de la Paz, D. Taruscio and S. C. Groft. Cham, Springer International Publishing, Switzerland; 233–247

Berdigaliyev N, Aljofan M (2020) An overview of drug discovery and development. Future Med Chem 12(10):939–947

Cha Y, Erez T, Reynolds I, Kumar D, Ross J, Koytiger G, Kusko R, Zeskind B, Risso S, Kagan E (2018) Drug repurposing from the perspective of pharmaceutical companies. Br J Pharmacol 175(2):168–180

Chang H-Y, Feng A-N, Fong M-C, Hsueh C-W, Lai W-T, Huang K-C, Chong E, Chen C-N, Chang H-C, Yin W-H (2019) Sacubitril/valsartan in heart failure with reduced ejection fraction patients: real world experience on advanced chronic kidney disease, hypotension, and dose escalation. J Cardiol 74(4):372–380

Clough E, Barrett T, Wilhite S, Ledoux P, Evangelista C, Kim I, Tomashevsky M, Marshall K, Phillippy K, Sherman P, Lee H, Zhang N, Serova N, Wagner L, Zalunin V, Kochergin A, Soboleva A (2023) NCBI GEO: archive for gene expression and epigenomics data sets: 23-year update. Nucleic Acids Res 52:D138

Dai Y, Ge J (2012) Clinical use of aspirin in treatment and prevention of cardiovascular disease. Thrombosis 2012:245037

Dash R, M Yadav, J Biswal, S Samanta, T Sharma and S Mohapatra (2023) Drug repurposing a compelling cancer strategy with bottomless opportunities: Recent advancements in computational methods and molecular mechanisms. Indian Journal of Pharmacology 55(5): 322–331

Deotarse P, A Jain, M Baile, N Kolhe and A Kulkarni (2015) Drug repositioning: a review. Int. J. Pharma Res. Rev 4: 51–58

DiMasi JA, HG Grabowski and RW Hansen (2016) Innovation in the pharmaceutical industry: New estimates of R&D costs. J Health Econ 47: 20–33

Drayman N, DeMarco JK, Jones KA, Azizi SA, Froggatt HM, Tan K, Maltseva NI, Chen S, Nicolaescu V, Dvorkin S, Furlong K, Kathayat RS, Firpo MR, Mastrodomenico V, Bruce EA, Schmidt MM, Jedrzejczak R, Muñoz-Alía M, Schuster B, Nair V, Han KY, O'Brien A, Tomatsidou A, Meyer B, Vignuzzi M, Missiakas D, Botten JW, Brooke CB, Lee H, Baker SC, Mounce BC, Heaton NS, Severson WE, Palmer KE, Dickinson BC, Joachimiak A, Randall G, Tay S (2021) Masitinib is a broad coronavirus 3CL inhibitor that blocks replication of SARS-CoV-2. Science 373(6557):931–936

Emery P, Keystone E, Tony HP, Cantagrel A, van Vollenhoven R, Sanchez A, Alecock E, Lee J, Kremer J (2008) IL-6 receptor inhibition with tocilizumab improves treatment outcomes in patients with rheumatoid arthritis refractory to anti-tumour necrosis factor biologicals: results from a 24-week multicentre randomised placebo-controlled trial. Ann Rheum Dis 67(11):1516–1523

Finsterer J, Frank M (2013) Repurposed drugs in metabolic disorders. Curr Top Med Chem 13(18):2386–2394

Gaulton A, Bellis LJ, Bento AP, Chambers J, Davies M, Hersey A, Light Y, McGlinchey S, Michalovich D, Al-Lazikani B, Overington JP (2012) ChEMBL: a large-scale bioactivity database for drug discovery. Nucleic Acids Res 40(Database issue):D1100–D1107

Girman CJ, Ritchey ME, Lo Re V (2022) Real-world data: assessing electronic health records and medical claims data to support regulatory decision-making for drug and biological products. Pharmacoepidemiol Drug Saf 31(7):717–720

Gns HS, Gr S, Murahari M, Krishnamurthy M (2019) An update on drug repurposing: re-written saga of the drug's fate. Biomed Pharmacother 110:700–716

Golwala HB, Thadani U, Liang L, Stavrakis S, Butler J, Yancy CW, Bhatt DL, Hernandez AF, Fonarow GC (2013) Use of hydralazine-isosorbide dinitrate combination in African American and other race/ethnic group patients with heart failure and reduced left ventricular ejection fraction. J Am Heart Assoc 2(4):e000214

Herráiz-Gil S, Nygren-Jiménez E, Acosta-Alonso DN, León C, Guerrero-Aspizua S (2025) Artificial intelligence-based methods for drug repurposing and development in cancer. Appl Sci 15(5):2798

Hua Y, Dai X, Xu Y, Xing G, Liu H, Lu T, Chen Y, Zhang Y (2022) Drug repositioning: Progress and challenges in drug discovery for various diseases. Eur J Med Chem 234:114239

Hauser AS, MM Attwood, M Rask-Andersen, HB Schiöth and DE Gloriam (2017) Trends in GPCR drug discovery: new agents, targets and indications. Nat Rev Drug Discov 16(12): 829–842

Hernandez JJ, M Pryszlak, L Smith, C Yanchus, N Kurji, VM Shahani and SV Molinski (2017) Giving Drugs a Second Chance: Overcoming Regulatory and Financial Hurdles in Repurposing Approved Drugs As Cancer Therapeutics. Front Oncol 7: 273

Hurle MR, L Yang, Q Xie, DK Rajpal, P Sanseau and P Agarwal (2013) Computational drug repositioning: from data to therapeutics. Clinical Pharmacology & Therapeutics 93(4): 335–341

Jonker AH, O'Connor D, Cavaller-Bellaubi M, Fetro C, Gogou M, Hoen PACT, de Kort M, Stone H, Valentine N, Pasmooij AMG (2024) Drug repurposing for rare: progress and opportunities for the rare disease community. Front Med (Lausanne) 11:1352803

Jourdan JP, Bureau R, Rochais C, Dallemagne P (2020) Drug repositioning: a brief overview. J Pharm Pharmacol 72(9):1145–1151

Jump DB, Depner CM, Tripathy S (2012) Omega-3 fatty acid supplementation and cardiovascular disease. J Lipid Res 53(12):2525–2545

Kumar S and V Roy (2023) Repurposing Drugs: An Empowering Approach to Drug Discovery and Development. Drug Res (Stuttg) 73(9): 481–490

Kaski JC, Gloekler S, Ferrari R, Fox K, Lévy BI, Komajda M, Vardas P, Camici PG (2018) Role of ivabradine in management of stable angina in patients with different clinical profiles. Open Heart 5(1):e000725

Kato Y, Nishiyama K, Nishimura A, Noda T, Okabe K, Kusakabe T, Kanda Y, Nishida M (2022) Drug repurposing for the treatment of COVID-19. J Pharmacol Sci 149(3):108–114

Kifle ZD, Ayele AG, Enyew EF (2021) Drug repurposing approach, potential drugs, and novel drug targets for COVID-19 treatment. J Environ Public Health 2021:6631721

Kino T, Burd I, Segars JH (2021) Dexamethasone for severe COVID-19: how does it work at cellular and molecular levels? Int J Mol Sci 22(13):6764

Krishnamurthy N, Grimshaw AA, Axson SA, Choe SH, Miller JE (2022) Drug repurposing: a systematic review on root causes, barriers and facilitators. BMC Health Serv Res 22(1):970

Kubo S, Nakayamada S, Sakata K, Kitanaga Y, Ma X, Lee S, Ishii A, Yamagata K, Nakano K, Tanaka Y (2018) Janus Kinase inhibitor Baricitinib modulates human innate and adaptive immune system. Front Immunol 9:1510

Kumar R, Harilal S, Gupta SV, Jose J, Thomas Parambi DG, Uddin MS, Shah MA, Mathew B (2019) Exploring the new horizons of drug repurposing: a vital tool for turning hard work into smart work. Eur J Med Chem 182:111602

Larochelle P, Tobe SW, Lacourcière Y (2014) β-Blockers in hypertension: studies and meta-analyses over the years. Can J Cardiol 30(5, Supplement):S16–S22

Latif K, Ullah A, Shkodina AD, Boiko DI, Rafique Z, Alghamdi BS, Alfaleh MA, Ashraf GM (2022) Drug reprofiling history and potential therapies against Parkinson's disease. Front Pharmacol 13:1028356

Lei ZN, Wu ZX, Dong S, Yang DH, Zhang L, Ke Z, Zou C, Chen ZS (2020) Chloroquine and hydroxychloroquine in the treatment of malaria and repurposing in treating COVID-19. Pharmacol Ther 216:107672

Li P, Zheng Y, Chen X (2017) Drugs for autoimmune inflammatory diseases: from small molecule compounds to anti-TNF biologics. Front Pharmacol 8:460

Low ZY, Farouk IA, Lal SK (2020) Drug repositioning: new approaches and future prospects for life-debilitating diseases and the COVID-19 pandemic outbreak. Viruses 12(9):1058

Magro P, Zanella I, Pescarolo M, Castelli F, Quiros-Roldan E (2021) Lopinavir/ritonavir: repurposing an old drug for HIV infection in COVID-19 treatment. Biom J 44(1):43–53

Marfella R, D'Onofrio N, Trotta MC, Sardu C, Scisciola L, Amarelli C, Balestrieri ML, Grimaldi V, Mansueto G, Esposito S, D'Amico M, Golino P, Signoriello G, De Feo M, Maiello C, Napoli

C, Paolisso G (2022) Sodium/glucose cotransporter 2 (SGLT2) inhibitors improve cardiac function by reducing JunD expression in human diabetic hearts. Metabolism 127:154936

Mehta PP, Dhapte-Pawar VS (2021) Repurposing drug molecules for new pulmonary therapeutic interventions. Drug Deliv Transl Res 11(5):1829–1848

Mohammad Sadeghi H, I Adeli, T Mousavi, M Daniali, S Nikfar and M Abdollahi (2021) Drug repurposing for the management of depression: where do we stand currently? Life 11(8): 774

Nagaraj AB, QQ Wang, P Joseph, C Zheng, Y Chen, O Kovalenko, S Singh, A Armstrong, K Resnick, K Zanotti, S Waggoner, R Xu and A DiFeo (2018) Using a novel computational drug-repositioning approach (DrugPredict) to rapidly identify potent drug candidates for cancer treatment. Oncogene 37(3): 403–414

Newman SP (2018) Delivering drugs to the lungs: The history of repurposing in the treatment of respiratory diseases. Advanced Drug Delivery Reviews 133: 5–18

Novack GD (2021) Repurposing medications. Ocul Surf 19: 336–340

Niranjan V, Setlur AS, Karunakaran C, Uttarkar A, Kumar KM, Skariyachan S (2022) Scope of repurposed drugs against the potential targets of the latest variants of SARS-CoV-2. Struct Chem 33(5):1585–1608

Ojha D, Hill CS, Zhou S, Evans A, Leung JM, Schneider CA, Amblard F, Woods TA, Schinazi RF, Baric RS, Peterson KE, Swanstrom R (2024) N4-Hydroxycytidine/molnupiravir inhibits RNA virus-induced encephalitis by producing less fit mutated viruses. PLoS Pathog 20(9):e1012574

Okuyama R (2023) Advancements in Drug Repurposing: Examples in Psychiatric Medications. Int J Mol Sci 24(13)

Parvathaneni V, Kulkarni NS, Muth A, Gupta V (2019) Drug repurposing: a promising tool to accelerate the drug discovery process. Drug Discov Today 24(10):2076–2085

Pimenta E, Oparil S (2009) Role of aliskiren in cardio-renal protection and use in hypertensives with multiple risk factors. Ther Clin Risk Manag 5(3):459–464

Rodrigues L, Bento Cunha R, Vassilevskaia T, Viveiros M, Cunha C (2022) Drug repurposing for COVID-19: a review and a novel strategy to identify new targets and potential drug candidates. Molecules 27(9):2723

Roessler HI, NV Knoers, MM van Haelst and G van Haaften (2021) Drug repurposing for rare diseases. Trends in pharmacological sciences 42(4): 255–267

Sakurai Y, AA Kolokoltsov, C-C Chen, MW Tidwell, W E Bauta, N Klugbauer, C Grimm, C Wahl-Schott, M Biel and RA Davey (2015) Two-pore channels control Ebola virus host cell entry and are drug targets for disease treatment. Science 347(6225): 995–998

Siramshetty VB, J Nickel, C Omieczynski, BO Gohlke, MN Drwal and R Preissner (2016) withdrawn-a resource for withdrawn and discontinued drugs. Nucleic Acids Res 44(D1): D1080–1086

Sahoo BM, Ravi Kumar BVV, Sruti J, Mahapatra MK, Banik BK, Borah P (2021) Drug Repurposing Strategy (DRS): emerging approach to identify potential therapeutics for treatment of novel Coronavirus infection. Front Mol Biosci 8:628144

Schuler J, Falls Z, Mangione W, Hudson ML, Bruggemann L, Samudrala R (2022) Evaluating the performance of drug-repurposing technologies. Drug Discov Today 27(1):49–64

Shah S, Dooms MM, Amaral-Garcia S, Igoillo-Esteve M (2021) Current drug repurposing strategies for rare neurodegenerative disorders. Front Pharmacol 12:768023

Sharma K, Ahmed F, Sharma T, Grover A, Agarwal M, Grover S (2023) Potential repurposed drug candidates for tuberculosis treatment: progress and update of drugs identified in over a decade. ACS Omega 8:17362

Shaya FT, Sing K, Milam R, Husain F, Del Aguila MA, Patel MY (2020) Lipid-lowering efficacy of Ezetimibe in patients with atherosclerotic cardiovascular disease: a systematic review and meta-analyses. Am J Cardiovasc Drugs 20(3):239–248

Siddiqui MU, Junarta J, Sathyanarayanan S, Kochar K, Ullah W, Fischman DL (2023) Risk of coronary artery disease in patients with gout on treatment with Colchicine: a systematic review and meta-analysis. IJC Heart Vasc 45:101191

Suchonwanit P, Thammarucha S, Leerunyakul K (2019) Minoxidil and its use in hair disorders: a review. Drug Des Devel Ther 13:2777–2786

Talevi A, Bellera CL (2020) Challenges and opportunities with drug repurposing: finding strategies to find alternative uses of therapeutics. Expert Opin Drug Discov 15(4):397–401

Tchesnokov EP, Feng JY, Porter DP, Götte M (2019) Mechanism of inhibition of Ebola virus RNA-dependent RNA polymerase by Remdesivir. Viruses 11(4):326

Tomczak K, Czerwińska P, Wiznerowicz M (2015) The Cancer Genome Atlas (TCGA): an immeasurable source of knowledge. Contemp Oncol (Pozn) 19(1a):A68–A77

Truong TTT, Panizzutti B, Kim JH, Walder K (2022) Repurposing drugs via network analysis: opportunities for psychiatric disorders. Pharmaceutics 14(7):1464

Turabi KS, Deshmukh A, Paul S, Swami D, Siddiqui S, Kumar U, Naikar S, Devarajan S, Basu S, Paul MK, Aich J (2022) Drug repurposing—an emerging strategy in cancer therapeutics. Naunyn Schmiedeberg's Arch Pharmacol 395(10):1139–1158

Vadnais DS, Wenger NK (2010) Emerging clinical role of ranolazine in the management of angina. Ther Clin Risk Manag 6:517–530

Verbaanderd C, Rooman I, Huys I (2021) Exploring new uses for existing drugs: innovative mechanisms to fund independent clinical research. Trials 22(1):322

Walker T, Grulke CM, Pozefsky D, Tropsha A (2010) Chembench: a cheminformatics workbench. Bioinformatics 26(23):3000–3001

Wall G, Lopez-Ribot JL (2020) Screening repurposing libraries for identification of drugs with novel antifungal activity. Antimicrob Agents Chemother 64(9):e00924

Wishart DS, Knox C, Guo AC, Cheng D, Shrivastava S, Tzur D, Gautam B, Hassanali M (2008) DrugBank: a knowledgebase for drugs, drug actions and drug targets. Nucleic Acids Res 36(Database issue):D901–D906

Zhou Q, Liao JK (2009) Statins and cardiovascular diseases: from cholesterol lowering to pleiotropy. Curr Pharm Des 15(5):467–478

QSAR-Based Studies for Drug Repurposing

10

Gul Naz Fatima, Hera Fatma, and Shailendra K. Saraf

Abstract

The search for new therapeutic agents is a constant endeavor. The synthetic drug pipelines are drying out, and now the clinically established agents are being evaluated for other ailments. Drug repurposing, drug reprofiling, drug repositioning, or therapeutic switching is a method in which drug development for a new indication starts with an already known drug. Repurposing could reduce the 10–17 years' process to 3–12 years. This facilitates the circumvention of several pre-approval assessments necessary for a newly developed medication. Inclusion of novel informatics approaches and systems biology to reveal unknown targets or mechanisms of approved drugs improves drug repurposing methods. A combination of both, computational and experimental, assays is desirable to repurpose drugs for new indications. A computational platform and algorithm, a comprehensive database and toolset, resource for protein sequence and functional information are the essentials for drug repurposing. Various approaches in computer-aided drug design processes are proving effective in the design of new molecules, virtual screening, and lead optimization. The approach could be furthered by sequence alignment, molecular docking, and molecular dynamics simulations. Thus, computational screening for repurposing of drugs has become a vital approach in the development of new therapeutic agents, offering significant reduction in time, investment, and an improved success rate.

Keywords

Drug repurposing · Computational screening · Ligand-based · Target-based · Quantitative structure activity relationship—QSAR

G. N. Fatima · H. Fatma · S. K. Saraf (✉)
Division of Pharmaceutical Chemistry, Faculty of Pharmacy, Babu Banarasi Das Northern India Institute of Technology, Ayodhya Road, Chinhut, Lucknow, Uttar Pradesh, India

R. K. Kaundal et al. (eds.), *Drug Repurposing: Exploring Newer Therapeutic Potentials*,
https://doi.org/10.1007/978-981-95-4522-3_10

10.1 Introduction

The process of discovery of novel pharmaceuticals is both costly and time-intensive. According to the Eastern Research Group (ERG), it takes an average of 8–15 years and incurs a high cost to develop a novel molecular entity. Also, the success rate for this process is just 2.0%. An approach where drug development for a new indication starts with an already known drug is known as drug reprofiling, or drug repositioning (March-Vila et al. 2017), or therapeutic switching (Zapata-Cardona et al. 2023a). It enables bypassing of multiple pre-approval tests required for a newly developed drug (Fig. 10.1). In general, drug repurposing offers several advantages to the drug discovery process, such as decreased risk of failure, less investment (Pan et al. 2022), and a shorter development period (Zapata-Cardona et al. 2023b). Although this strategy is not new, the approach has experienced substantial growth in popularity in the last decade. Approximately one-third of recent drug approvals are attributed to drug repurposing, contributing to over 25% of the pharmaceutical industry's annual revenue assay (Parveen 2020).

The drug-repurposing research using QSAR comprises the application of several techniques, methods, and strategies. All of them are within the realm of computational work. The development of in silico drug repurposing has been made possible by two technologies: one is high-throughput data from various sources, and the other is due to advances in computational and data sciences (Park 2019). Drug repurposing has emerged as a crucial method in the development of new drugs due to a great reduction in time and a higher success rate. Under unforeseen circumstances, such as the outbreak of SARS-Cov-19 or the increasing incidence of

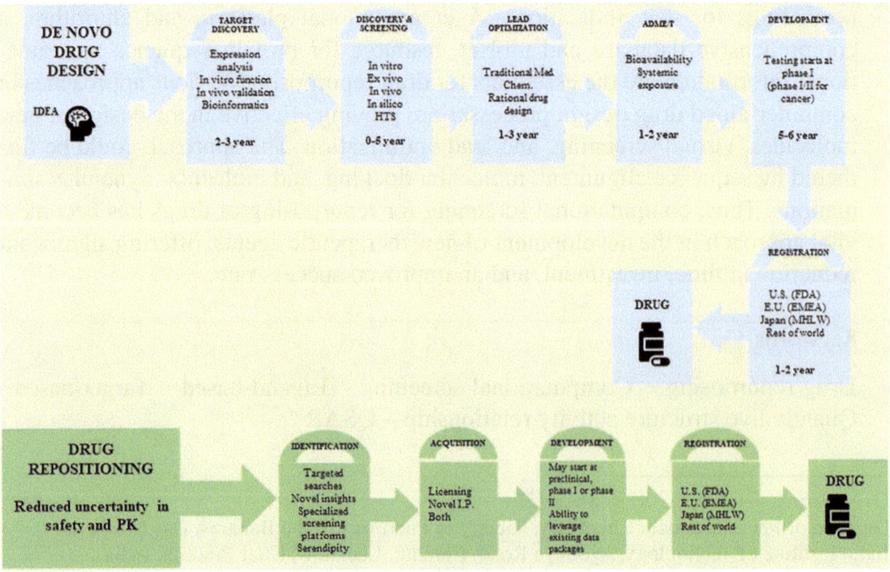

Fig. 10.1 De novo drug discovery and drug repurposing: A comparative representation

cancer, there is a critical requirement for expedited drug research and development. Recently, pharmacophore-based approaches have been found to play a significant role in several computer-aided drug designing processes and have been effectively used for designing new molecules, virtual screening, and lead optimization (Parveen 2020).

Machine learning techniques make use of the manually constructed descriptors for the drug-repurposing process. These may lead to the identification of potential candidates for further clinical trials. However, the majority of machine learning techniques rely substantially on feature engineering and domain expertise in the early stages, since these approaches could only handle fixed-sized inputs. The quantity of chemical and biological data that is currently available for use in drug research and discovery has steadily increased in recent years. The task of efficiently examining the vast domain data becomes crucial in the context of pharmacological repurposing (Pan et al. 2022).

Deep learning methods (Ahmad et al. 2019) have also been used in drug repurposing. Deep learning is superior to previous machine learning techniques because it can understand the intricate correlations between input attributes and outputs from a massive quantity of data. This capacity has been made possible by increasing the computing hardware processing power of the machines. Despite being in the early stages of development, their application in drug repurposing has already demonstrated considerable potential (Pan et al. 2022).

Computational drug screening, also referred to as in silico drug screening, involves the use of computer tools and simulations to predict the interactions between small molecules and biological targets. The process encompasses a range of methodologies and approaches, as discussed below.

10.2 Drug Repurposing: Ligand-Based

The goal of ligand-based drug repurposing is to find new uses for ligands by analyzing their chemical properties. Chemical similarity comparisons, in which the degree of similarity between two ligands typically suggests their potential of sharing comparable features, such as off-targets, side effects, and other bioactivities, are a prevalent theme among such methods. A well-known drug may, thus, be repurposed for a novel indication if it shares structural similarities with compounds from annotated databases like PubChem, ChEMBL, ZINC, or ChemProt. This is known as the chemical similarity principle. For in silico drug repositioning, a variety of ligand-based target prediction techniques, such as ChemMapper, SuperPred, HitPick, PASS, and Similarity Ensemble Approach (SEA), have been developed. In order to compare two ligands, first, they must be transformed into specific mathematical expressions called chemical fingerprints (Yang et al. 2022). These include the well-known "daylight fingerprinting," which uses path information taken from a chemical graph to characterize ligands, and the "Molecular ACCess System (MACCS) fingerprinting," which uses predetermined substructural pieces (Durant et al. 2002). The metrics that identify the common elements between two comparison vectors, like the Tanimoto coefficient (Tc) (Rácz et al. 2018), can then be used to assess how similar two fingerprints are to one another. Regardless of representation, a high Tc value frequently suggests that the two compounds are structurally similar and, according to the chemical similarity principle, may have comparable bioactivities.

Methods, such as SEA, assess chemical similarity in relation to a random background that is similar to the one used in Basic Local Alignment Search Tool (BLAST) (Bal and Hujol 2007) sequence searches in order to calculate a statistical score. The activity cliff in which structural alterations in compounds result in appreciable changes in their bioactivities, and do not follow the linear assumption of a quantitative structure–activity relationship (QSAR), can violate the chemical similarity principle even though it is generally valid. The idea of ligand similarity has led to the development of the concept of network pharmacology, which explores for novel bioactivities of molecules for therapeutic repurposing as well as multibody relationships between the ligands. Recently, chemical similarity network analysis pull-down (CSNAP) (Lo et al. 2015), a unique network-based method for large-scale molecule profiling, has been introduced. Using a preset Tc threshold, CSNAP is an instance-based unsupervised learning technique that concurrently combines drugs and query compounds obtained from the ChEMBL database (Gaulton et al. 2012) into a network similarity graph. Subsequently, every query compound in the network is given a consensus statistics score of targets, which is determined by the observed target frequency from its first-order neighbors. This score is comparable to the one used for functional predictions in a network of protein–protein interactions. CSNAP facilitates a comprehensive study of compounds by leveraging their chemical diversity and possesses the capability to do extensive off-target profiling of vast compound collections. A CSNAP analysis of 200 drug-like compounds from a cell-cycle screening accurately identifies several cancer targets as well as new compounds, which grouped together with known drugs, suggesting their potential for repurposing in other applications.

Drug–target network (Vogt and Mestres 2010) is another network-based method of drug repurposing. It is a bipartite network that specifically takes into account the interaction between the drug and target through their binary associations. This formulation served as the foundation for the development of a supervised learning algorithm that integrates chemical and genomic space with high precision to predict drug–target interactions. Cheng et al. effectively repurposed two preexisting drugs, simvastatin and ketoconazole, and demonstrated their potency to inhibit cell growth in a human MDA-MB-231 breast cancer cell line using such a network-based inference method (Cheng et al. 2012).

One significant drawback of the ligand-based drug-repurposing strategy is its inability to find targets for newly developed drugs that have unique chemical structures different from the drugs already in existence. Consequently, a basic chemical search of the bioactivity database will not yield any results. In the current scenario, a number of novel methods have been suggested to address this difficulty. An effective method involves evaluating the 3D pharmacophore and/or shape similarity of compounds to ascertain whether two ligands being compared are scaffold hopping pairs. This indicates that despite their distinct chemical structures, the two compounds may interact with a shared drug-binding site through consensus functional groups (Armstrong et al. 2010).

Lately, Lo et al. have created the CSNAP3D program (Lo et al. 2016), which is an enhanced version of the original CSNAP framework. This tool incorporates 2D and 3D similarity metrics to automatically identify scaffolds. CSNAP3D successfully enhances target identification for difficult drug classes, including HIV reverse transcriptase (HIV-RT), which had previously shown subpar performance when assessed by numerous 2D similarity-based compound target profiling systems. A study used CSNAP3D analysis on the drug collection of HIV-RT inhibitors obtained from the directory of usable decoys (DUDs) and identified three different inhibitors, namely, Efavirenz, Nevirapine, and Tivirapine, based on the shared chemical characteristics. These were employed in the investigation to identify a new structure that can stimulate the formation of microtubules, similar to the effects of Paclitaxel. An alternative method for identifying chemical scaffold hopping involves conducting 2D substructure searches to discover common elements present in the structures of two drugs. Wu et al. developed the substructure–drug–target network-based inference (SDTNBI) to predict drug–target interactions (DTI) and reposition drugs on a large scale. They used this approach to reposition nonsteroidal anti-inflammatory drugs (NSAIDs) as anticancer agents that target AKR1C3, CA9, CA12, or CDK2 (Wu et al. 2017).

10.3 Drug Repurposing: Target-Based

The increasing number of protein crystal structures in the Protein Data Bank (PDB), the availability of the 3D structure of the target (Jamir et al. 2023), and the progress in the in silico structural bioinformatics analysis (Mullins 2022) make the target-based drug repurposing strategy more favorable. The target-based drug repurposing employs the 3D structure of proteins to uncover other off-targets that may have similar bioactivities to the original ligands. One such approach involves comparing the similarities of binding sites between protein targets that are on and off the protein. This helps to identify new off-targets for the cocrystal ligands that have the potential to interact. In another approach, the compound library is directly docked to a target or a target ligand is docked to a panel of proteins. These approaches offer different ways for predicting new interactions between the proteins and the ligands, which might be useful for drug repositioning (Fig. 10.2).

Just like the ligand-based drug-repurposing method, the binding site of a target protein can be utilized to search a database of drug-binding pockets. This search helps in identifying off-targets that have the potential to interact with the corresponding ligands. Multiple pocket similarity comparison algorithms have been developed, which offer potential prospects for structure-based drug repositioning (Haupt and Schroeder 2011). One simple technique for comparing pockets involves performing a structural alignment by superimposing two sites. ProBis-Ligands is a strategy that utilizes local structure alignment to transfer ligands from one protein structure to another that is homologous. Konc and Janezic stated this method to have

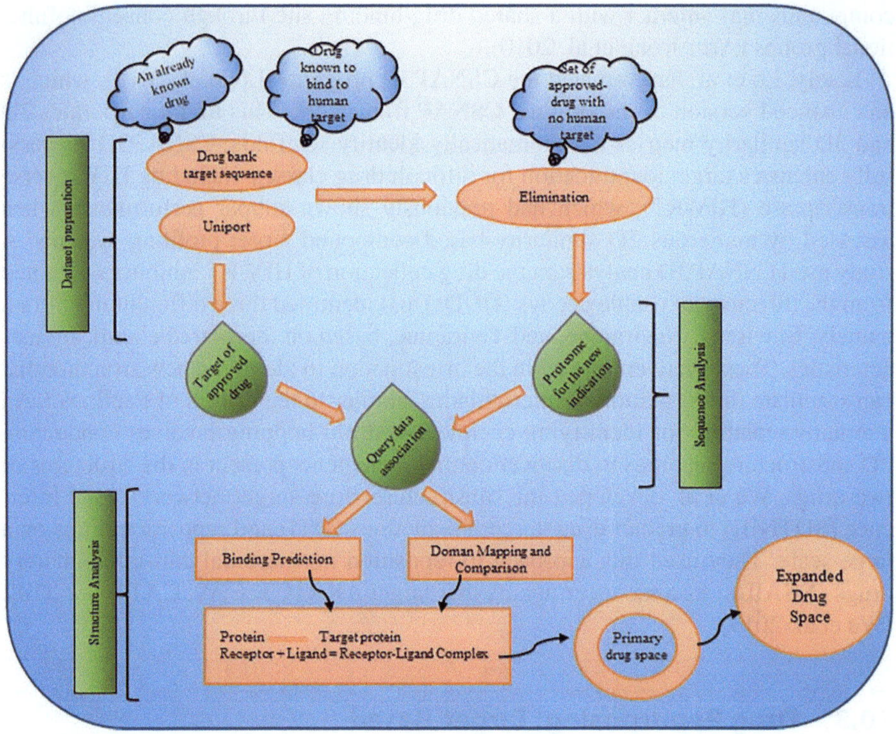

Fig. 10.2 Overview of computational strategy for drug repurposing

the potential to be used in drug repositioning. Franchi et al. conducted a study to examine how protein kinase inhibitors interact with synapsin I, a protein that binds to ATP, and controls the release of neurotransmitters in the synapse. The researchers conducted a methodical comparison of the staurosporine-binding site in the Pim-1 kinase proto-oncogene with 6412 protein–ligand binding sites from the scPDB database. They demonstrated that the ATP-binding site of synapsin I has the potential to recognize the pan-kinase inhibitor staurosporine. A drawback associated with this approach is that it does not take into account the chemical characteristics that are necessary for ligand binding. As a result, there is a higher likelihood of false-positive results (Konc and Janežič 2014).

Unlike experimental high-throughput screening (HTS), molecular docking algorithms enable the simulation of binding modes for many ligands from a vast database. This has led to the development of a new computer-aided drug screening method. The technique is known as virtual screening (VS) (Walters et al. 1998) or structure-based virtual screening (SBVS). In the forward virtual screening (VS) process, a collection of chemical compounds is computationally tested against a target protein. The resulting docking scores can then be utilized to determine the most promising ligands for further synthesis and experimental evaluation. The structure-based drug repurposing involves the use of docking techniques to investigate the secondary actions of approved drugs by binding them to a desired protein target. Bisson et al. utilized several structural models of antagonist-bound androgen receptor (AR) to dock a library of marketed oral medications, in order to identify nonsteroidal antagonists against AR, a therapeutic target for prostate cancer. By evaluating compounds with the highest docking scores, scientists discovered that acetophenazine, fluphenazine, and periciazine, which are phenothiazine derivatives, were the most favorable. Thus, the compounds which have been utilized as antipsychotic medications in clinical settings were repurposed as potential AR antagonists (Bisson et al. 2007).

Reverse virtual screening (VS) docking, commonly referred to as "structure-based target fishing," involves docking a single ligand to many drug-binding pockets in order to detect and prioritize potential off-targets based on docking scores. In such a case, a target promiscuity profile can be created for a specific drug to discover new bioactivities with the objective of drug repositioning. An important obstacle in the technique is the necessity to create an automated procedure for preparing binding sites from PDB data. Recently, the problem has been resolved through the development of various databases that contain preconfigured protein pockets, such as the scPDB database. Additionally, there are servers available, such as the TarFisDock server, that can do reverse target fishing (Lo and Torres 2019).

Thus, the key considerations for drug development through repurposing using in silico methods are the determination of molecular docking pose and the prediction of binding affinity (Arulanandam. 2021). For this purpose, the following tools are majorly employed.

A. *Molecular Docking Tools*: Molecular docking is a computational tool used in the development and discovery of drugs to predict the interaction of the ligand in its preferred orientation, whenever it binds to a target protein receptor. By providing insights into the binding interactions between a drug candidate and its target protein, it rationalizes the designing of drugs. The objective of molecular docking is to predict the binding affinity and the conformations (Trott and Olson 2010). The docking studies have been successfully utilized in drug repurposing. Screening huge libraries of both the ligands and the targets can be efficiently and rapidly accomplished using this convenient method. It is particularly suitable for both drug-based and target-based drug repurposing. Despite its limitations, docking is widely accepted and experimentally validated for predicting interactions between the new drug entities and the targets (March-Vila et al. 2017). The two approaches used are as follows.
   (i) *Computer simulation approach:* This analysis involves energy profiling. Initially, the ligand and the target molecules are separated, and then the ligand is allowed to bind to the target's pocket. This method has the advantage of being more compatible. However, it requires a longer period of time to evaluate the most suitable docked conformer of the ligand (Agarwal and Mehrotra 2016).
   (ii) *Shape complementarity approach:* This uses surface structures of the ligand and the target for carrying out molecular docking. The molecular surface

area is elucidated with respect to its solvent-accessible surface area and estimation of the target protein molecule's hydrophobicity. It scans several ligands for target binding, yielding a fast and reliable result (Agarwal and Mehrotra 2016). Numerous ligands can be scanned in seconds to determine their binding characteristics on the target surface (Kumar and Kumar 2019).

The prediction helps to screen virtual libraries of drug-like molecules to obtain lead for further drug development (Trott and Olson 2010). The different types of molecular methodology can be categorized as (Kumar and Kumar 2019), rigid ligand and rigid receptor docking, flexible ligand and rigid receptor docking, flexible ligand and flexible receptor docking, ensemble docking, or hybrid method. The docking program is selected on the basis of three criteria, which include availability, use of conventional file formats as input, and easy application to virtual screening. There are different tools that can be used for molecular docking such as DOCK, FRED, GLIDE, GOLD, SLIDE, SURFLEX, QXP, AUTODOCK VINA, SWISS MODEL, AUTODOCK TOOL 4.2, iGEMDOCK software, and Dock-Thor (Kellenberger et al. 2004).

AutoDock Tool 4.2 In silico study was employed to repurpose the antifungal drug itraconazole for anticancer treatment. The compound was subjected to molecular docking, using AutoDock tool 4.2, on the selected target protein (PDB ID 4JKV) (Dhorje et al. 2020). Recently, it has been found that itraconazole exhibited an inhibitory effect on the Hedgehog (Hh) pathway. The binding affinity of itraconazole to a smoothened receptor (SMO) was predicted by molecular docking, allowing for the quantification of its usefulness as an anticancer medication. The study aimed to determine the potential efficacy of itraconazole as a therapeutic agent for triple negative breast cancer (TNBC). Moreover, the drug's toxicity on malignant cells was demonstrated by using various dosages of itraconazole on the TNBC cell line through repeated cell viability experiments. Itraconazole exhibited superior interactions at two binding sites of SMO with minimal binding energy. Thus, it can be inferred that itraconazole exerted inhibitory effects on SMO, making it a potential candidate for use as an anticancer medication. In 2022, Sakshi Kamboj and her colleagues conducted a study on the targeting of nonstructural proteins of Hepatitis C virus in order to anticipate the repurposing of drugs. Naftifine, Butalbital (NS3), Vinorelbine, Epicriptine (NS3/4A), Pipecuronium, Trimethaphan (NS5A), Olodaterol, and Vemurafenib (NS5B) are drugs that specifically target the HCV NS protein. The selection of a highly efficient drug could be made possible by molecular docking. The visualization was conducted using PyMOL Version 2.5.2 (Kamboj et al. 2022).

AutoDockVina One of the most commonly used molecular docking software is AutoDockVina. Vina carries out grid-based energy estimation along with the Lamarckian genetic algorithm to identify drugs exhibiting affinity with the target protein (Dhorje et al. 2020; Parveen 2020). It was developed by Dr. Oleg Trott. It has enhanced accuracy of the binding energy prediction in comparison to AutoDock

4. It is accessible for both corporate and private use, following the terms of a permissive Apache license. Utilizing the software, a study was undertaken by Zapata-Cardona et al. to predict the binding energy for few known antiretroviral agents against the crucial proteins, namely, RdRp, ExoN-NSP10, and 3CLpro of SARS-CoV-2, to predict their repurposing potential for the same (Zapata-Cardona et al. 2023a). In another study, Nelfinavir, a quinoline moiety-containing drug, was repurposed as an anticancer agent by targeting Bcl-2, CDK-6, and VEGFR-2. Similarly, Artemisinin, Artenimol, Artemether, Artemotil, and Artesunate have been repurposed by Parveen et al. The docking was carried out to investigate the molecular pattern of Artemisinins 1–5 within the active pocket of the anticancer targets' crystal structures (Parveen 2020).

In another study by Guillem et al., SARS-CoV-2-pro protein, 6LU7, and 6WQF were taken as the targets, and the drug molecules were docked using three distinct docking programs, FRED, Glide, and Autodock Vina (Macip et al. 2022). Alireza Mohebbi et al. conducted a repurposing study employing Ciclopirox, which was based on fragment-based pharmacological discovery for the inhibition of the Hepatitis B virus–core protein (Mohebbi et al. 2023). More than 24 derivatives with $\Delta G_{binding}$ (-16.56 ± 1.46 Kcal.mol^{-1}) greater than that of Ciclopirox were identified. The compounds were docked using the AutodockVina interface with the target protein (PDB ID 6J10). The study found that Ciclopirox derivatives might suppress the HBV assembly and subsequently result in viral replication inhibition by binding directly to the viral core protein's carboxyl-terminal domain. This was predicted by the use of structure-based virtual screening against the SARS-CoV-2 main protease (M^{pro}). The software used was Vina, with grid box dimensions of 15 Å centered at $x = -20.313$; $y = 18.742$; and $z = -28.341$. It considered inter- and intramolecular interactions for the enthalpic term and the rotatable bonds between heavy atoms in the ligand for the entropic term. The study showed that Bedaquiline (a very lipophilic antibiotic used for multidrug-resistant tuberculosis), Glibenclamide (a second-generation sulphonyl urea for diabetes mellitus type 2 treatment), and Miconazole (an azole prescribed for the treatment of fungal infections) were promising M^{pro} inhibitors (Ferraz et al. 2020). In an in silico prediction of potential inhibitors for M^{pro} of SARS-CoV-2 by use of molecular docking and simulation-based drug repurposing, a total of 10 compounds were docked, and the best binding energy was observed for Lopinavir–Ritonavir (−10.6 kcal/mol) (Kumar et al. 2020a). A program called PLANTS, i.e., Protein–Ligand ANT System, is also used for carrying out molecular docking studies for proteins and ligands (Maffucci and Contini 2020).

iGEMDOCK One hundred and thirty drugs licensed by the US FDA were selected that were used for the treatment of conditions such as hypertension, cardiovascular diseases, respiratory tract infections (RTI), or as antibiotics and antiviral drugs. The study aimed to determine the drugs' ability to be repurposed as anti-SARS-CoV-2 agents (Anand et al. 2021). The binding affinity was assessed using iGEMDOCK software (v 2.1), which, through a graphical environment for recognizing pharmacological interactions and virtual screening, predicted 15 drugs including Baloxavir marboxil, Danoprevir, Sofosbuvir, and Fosinopril as potential anti COVID-19 drug candidates.

DockThor The DockThor program is a protein–ligand docking method developed to deal with highly flexible ligands like HIV-1 protease inhibitors. In this, both protein and ligand including cofactor molecules are treated with the same force field in docking. It uses a phenotypic crowding multiple solution steady-state genetic algorithm as the search method (Santos et al. 2020). Ferraz et al. (2020) in a study predicted a good binding affinity of two oral drugs (Bedaquiline and Glibenclamide) and one buccal drug (Miconazole) to be promising inhibitors of the main viral protease and therefore effective against SARS-CoV-2 (Ferraz et al. 2020).

Glide Program Glide is a ligand docking program for predicting protein–ligand binding modes and ranking ligands via high-throughput virtual screening. Glide utilizes two different scoring functions, SP and XP GlideScore, to rank the compounds. The experiments are most conveniently prepared using the Maestro graphical user interface (GUI). It uses an atom-pair similarity scoring algorithm, where the two molecules being compared are first processed to generate sets of atom pairs (Repasky et al. 2007).

B. *Visualization*: To visualize the molecular docking interaction analysis, PyMOL molecular visualization tool, Discovery Studio Visualizer, LIGPLOT$^+$ (Kumar et al. 2020a; Zapata-Cardona et al. 2023a), and CHIMERA are usually used (Parveen 2020).

LIGPLOT Visualizer LIGPLOT generates the 2D structure of protein–ligand complexes from standard Protein Data Bank to generate images that summarize the molecular structure (Wallace et al. 1995a). The interactions shown by it are hydrogen bonds and the hydrophobic contacts. It takes the 3D structure of the ligand and unrolls it, flattening it out onto the 2D page. It shows not only the water molecules with hydrogen bonds to the ligand but also the residues from the protein that have hydrogen bonds with water (Wallace et al. 1995b). Yogesh et al. (2020) predicted in silico potential inhibitors for SARS-CoV-2 M^{pro} (PDB ID 6Y2F) and used LIGPLOT for visualization of the interactions (Kumar et al. 2020a).

Discovery Studio Visualizer Discovery Studio is a comprehensive software suite for molecular modeling, sequences analysis, and other data relevant to researchers. It has a free visualizer that can be used to view data generated by other software. It offers a wide variety of tools for working with and visualizing data. Tools can be accessed from menus, toolbars, and tool panels. An in silico study by Kumar et al. predicted potential inhibitors for SARS-CoV-2 M^{pro} using docking and simulation-based drug repurposing, and the interaction visualization analysis studies were performed using Discovery Studio visualizer (Kumar et al. 2020a).

PyMOL It was created by Warren Lyford DeLano and commercialized initially by DeLano Scientific LLC. It is a molecular visualization system and the most widely used tool for the purpose. It is written in Python programming language. It is used for 3D visualization of the macromolecules and supports 12 different stereo visualization modes. It can edit dihedral angles, construct peptides, mutate residues, etc. Additionally, it possesses another beneficial feature known as real-time energy minimization through sculpting. However, sculpturing is flawed as it lacks the implementation of force fields. The utilities of PyMOL have been extensively enhanced by various plugins, including macromolecular analysis, homology modeling, protein–ligand docking, pharmacophore modeling, VS, and MD simulations (Yuan et al. 2017). Sanika Dhorje et al. (2020) repurposed itraconazole for cancer using molecular docking on the target protein (PDB ID 4JKV), and PyMOL was used for visualization. The in silico study was also confirmed by using in vitro MTT assay (Dhorje et al. 2020). In another study utilizing in silico methods such as docking and simulation-based drug repurposing, the possible inhibitors for the primary protease of SARS-CoV-2 were obtained. PDB ID 6Y2F was obtained from RCSB PDB. The examination of the interaction visualizations was conducted using the PyMOL software for molecular visualization.

C. *Identification of the FDA-Approved Drugs*: PubChem chemical database can be used to identify the potent inhibitors for an indication. A research group made use of the database to download around 75 drug compounds for SARS-CoV-2 (Kumar et al. 2020a). Another algorithm to screen FDA-approved drugs is the FINDSITE. Regardless of evolutionary relationship, the algorithm makes use of the similarities in pocket binding to similar ligands (Snell et al. 2016).
D. *Databases for Drug Inputs*: In order to search a vast set of drug substances, there is a need for some comprehensive resources (Fig. 10.3). DrugBank and UniProt are examples of such resources. A study uses DrugBank version 5.0.10 (Dhorje et al. 2020) to obtain information on the protein–target sequences of just those approved drugs that were not stated to have a recognized human target. This helped to ensure minimization of chances of the off-target effects (Chakraborti et al. 2019). ChEMBL is another programming interface that can be used to identify chemical structural similarity. It works on the simplified molecular-input line-entry system (SMILES) format, which uses American Standard Code for Information Interchange (ASCII) strings to express chemical structures (MacMahon et al. 2023). Also, ZINC is a freely accessible virtual screening database that provides access to over 750 million compounds that can be purchased. ZINC offers over 230 million molecules in 3D dockable configurations. Furthermore, it facilitates the identification of analogs in less than a minute (Parveen 2020). The Developmental Therapeutics Program (DTP) provides various repositories to researchers globally for the procurement, retention, and dissemination of chemical and biological samples and standards (Ambrosio et al. 2023).
E. *Target Information*: The drug targets can be collected from DrugBank, Therapeutic Target Database (TTD), and STITCH. TTD is a database that

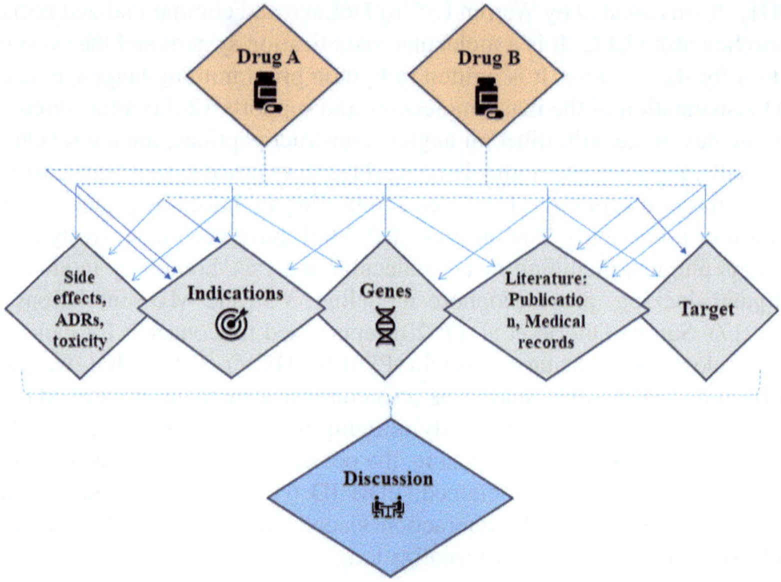

Fig. 10.3 Illustration of information integration from different resources

includes target-specific drug-binding data from published research on recognized therapeutic targets. The database STITCH comprises scores of interactions between small molecules and proteins, supported by evidence from biochemical and experimental data, correlations found in databases, and details found in abstracts of scholarly publications. STITCH defines a score of 0.7 as strong confidence (MacMahon et al. 2023).

F. *Molecular Dynamics (MD) Simulations*: Molecular dynamics simulations help to computationally observe the physical movement of atoms and molecules. This provides structural integrity and conformational changes that might occur in a protein–ligand complex, given in terms of root-mean-square deviation (RMSD) and root-mean-square fluctuation (RMSF). YASARA is a simulation software that involves AMBER14 force field, in a time frame of 10 ns, and numerous snapshots. In a study, drugs such as Lopinavir–Ritonavir, Tipranavir, and Raltegravir were analyzed using MD simulations. It suggested the drugs to be stable and conformationally flexible in the enzyme's active site, when docked into the binding pocket of CoV-2 M^{pro} (PDB ID 6Y2F) (Kumar et al. 2020a; Ambrosio et al. 2023). Another tool, Desmond v. 3.844 Simulation Interaction Diagram tool, is also utilized to examine the behavior of the protein and the ligands. The simulation study also includes RMSD and RMSF computations (Ambrosio et al. 2023).

G. *Protein Structure Modeling Approach*: SWISS-MODEL was developed by Computational Structural Biology Group at the Swiss Institute of Bioinformatics (SIB) and the Biozentrum University of Basel (Schwede et al. 2003). It is the first automated protein homology modeling server and has been continuously enhanced over the past 25 years. Ghavami et al., in a study on drug repurposing, used SWISS-MODEL to predict drugs that could be repurposed for SARS-CoV-2. The SWISS-MODEL service is intended to work with a minimal amount of user input, i.e., only the amino acid sequence of a target protein is the requisite. Some comparison modeling projects may require additional user input due to their changing complexity, such as selecting a new template or adjusting the target–template alignment. As a result, the SWISS-MODEL server provides the user with the basic interaction choices. Its modeling functionality has recently been expanded to include the modeling of homo- and heteromeric complexes, with the amino acid sequences of the interaction partners serving as the starting point. Other recently disclosed advances include the creation of a new modeling engine, ProMod3 (Studer et al. 2021), which improves the model accuracy, as well as an improved local model quality estimation approach (QMEANDisCo) (Studer et al. 2020), which is based on a novel version of QMEAN.

H. *Databases for Sequence Alignments*: The technique of sequence alignment helps to understand the evolutionary relationship between the previously known target and the newer ones (Table 10.1). The database, named MEGA 6.0, can be used to construct the phylogenetic tree and determine the genome sequence alignment. A group of researchers led by Y. Kumar carried out a study for the repurposing of a drug for SARS-CoV-2 using the software (Kumar et al. 2020a). In addition, EMBOSSWater is an algorithm that carries out the local alignment of a sequence to one or more additional sequences, which can be used for pairwise sequence alignments, and the Multiple Sequence Comparison by Log-Expectation (MUSCLE), a technique that accomplishes the alignment of multiple sequences that can be used to seek numerous sequence alignments between a query protein and a collection of identified target proteins. The study by Chakraborti et al. makes use of these program suites to obtain potential target proteins for *C. albicans*. Nitrofural, Fusidic acid, Novobiocin, and Cetrimonium are examples of few drugs shortlisted as potentially repurposable for *C. albicans* (Chakraborti et al. 2019).

I. *Databases to Genomic Sequence*: The analysis of the genome sequence alignments requires the use of the complete reference genome sequence. This can be downloaded with the help of NCBI GenBank database. According to a study on SARS-CoV-2, SL-CoVZXC45 was found to be the most closely related virus, with a query coverage of 95% and a sequence identity of 89.11%. This also helps to conclude the origin of the current infectious form of the agent; e.g., the SARS-CoV-2, responsible for the pandemic, is found to originate from Bat-SL-CoV-2 (Kumar et al. 2020a). A protein pair's percentage sequence identity provides a rough indication of how close they are structurally and functionally. It is highly probable that related proteins that have sequence identities more than 35–40% are structurally similar. Comparing the structures of two pairs with a series sequence identity of 20–35% is known as the "twilight zone" and is much less common. A study used the HMMER 3.0 suite of jackhammer for sensitive profile-based iterative sequence search information on *C. albicans* proteins. The

Table 10.1 Few examples of drugs belonging to different pharmacological classes, repurposed for new indication

Already known drug	Indication	Target	New indication	Target	References
Ciclopirox	Antifungal	Inhibition of metal-dependent enzyme	Hepatitis B virus	Viral replication inhibition	Mohebbi et al. (2023)
Betrixaban	Anticoagulant	Inhibition of factor Xa	SARS-corona virus-2	SARS-cov-2 M^{pro}	Ambrosio et al. (2023)
Paromomycin	Anti-amoebiasis	Inhibition of protein synthesis by binding to 16S ribosomal RNA	SARS-corona virus-2	SARS-cov-2 M^{pro}, SARS-cov-2 nsp10 nsp16 hetero-oligomeric complex	Ghavami et al. (2023)
Binodenoson	Stress agent	Adenosine receptor	SARS-corona virus-2	SARS-cov-2 nsp4, SARS-cov-2 nsp10 nsp16 hetero-oligomeric complex	Ghavami et al. (2023)
Lamivudine, Emtricitabine	Antiretroviral	Inhibition of viral DNA synthesis	SARS-corona virus-2	RNA-dependent RNA polymerase, exon-NSP10, 3clpro	Zapata-Cardona et al. (2023a)
5-fluorouracil	Anticancer	Thymidylate synthase inhibitor	SARS-corona virus-2	SARS-cov-2 M^{pro}	Matondo et al. (2022)
Remdesivir	Antiviral		SARS-corona virus-2	Rdrp enzyme inhibition	Parihar et al. (2022)
Zuclopenthixol, Flupentixol	Antipsychotic	Antagonism of D1 and D2 dopamine receptors	Anticancer	VEGFR-2, COX-2	Parveen et al. (2022)
Artemisinin	Antimalarial	Inhibition of flavivirus replication	Chemotherapeutics	Bcl-2, CDK-6, VEGFR-2	Parveen. (2020)
Ritonavir, Darunavir	Antiviral	Protease inhibitor	SARS-corona virus-2	SARS-cov-2 M^{pro}	Djokovic et al. (2021)
Rutin	Antiviral		SARS-corona virus-2	SARS-cov-2 M^{pro}	Agrawal et al. (2021)
Cimetidine	Gastric ulcer	Inhibition of histamine	SARS-corona virus-2	SARS-cov-2 M^{pro}	Macip et al. (2022)

Brilacidin	Antibiotic	Disruption of bacterial cell membrane	SARS-corona virus-2	SARS-cov-2 M^{pro}	Cavasotto and Filippo (2021)
Lumefantrine	Antimalarial	Binding to hemin, preventing detoxification	HIV	HIV-1 PR inhibitor	Arulanandam (2021)
Baricitinib, Chloroquine	Antiviral	Inhibition of JAK protein	SARS-corona virus-2	SARS-cov-2 M^{pro}	Anand et al. (2021)
Doxorubicin, Epirubicin	Antibiotics	Interruption of DNA structure	TNBC	Inhibition of DNA, synthesis forming an anthracycline–DNA–topoisomerase II ternary complex	Ávalos-Moreno et al. (2020)
Gemcitabine	Antiviral		TNBC	Inhibition of DNA replication	Ávalos-Moreno et al. (2020)
Lopinavir, Ritonavir, Tipranavir	Antiviral	Inhibition of CYP3A enzymes	SARS-corona virus-2	SARS-cov-2 M^{pro}	Kumar et al. (2020a)
Nelfinavir	Antiviral	Binding to protease active site	SARS-corona virus-2	SARS-cov-2 M^{pro}	Kumar et al. (2020b)
Indinavir	Antiviral	Binding to protease active site	SARS-corona virus-2	SARS-cov-2 M^{pro}	Maffucci and Contini (2020)
Itraconazole	Antifungal	Inhibition of cytochrome P-450-dependent enzyme	TNBC	TNBC cell line (MDA-MB-231)	Dhorje et al. (2020)
Tadalafil	Erectile dysfunction	PDE-5 inhibitor	Anti-aging	Transient receptor potential gene 7, ribosomal protein S6 polypeptide 2 gene, forkhead box C gene	Snell et al. (2016)
Sildenafil	Erectile dysfunction	Cgmp PDE-5 inhibitor	Anti-aging		
Dexamethasone, baricitinib	Anti-inflammation	Suppression of migration of neutrophils	COVID-19	Inhibition of SARS-cov-2 replication	MacMahon et al. (2023)

elimination of the brief alignment stretches would lead to capture the important target sequence and structural data in the context of its binding pockets included in the functional domains. Such a prediction can increase the credibility of the assessment (Chakraborti et al. 2019).

J. *Tools for Conversion, Analysis, and Storage of Chemical Data*: OPENBABEL is a computational tool that can be used for format interconversion of compounds. One may search, transform, analyze, and store chemical data using this open, collaborative chemical toolbox. It is a versatile tool that can convert between more than 110 different formats (Onyango et al. 2022).

K. *In Silico Pharmacokinetics*: The compounds under investigation can be determined for ADME properties using the SwissADME tool (Anand et al. 2021; Mohebbi et al. 2023). The solubility of the drug, pharmacokinetic characteristics, drug likelihood, and Lipinski's rule of five are majorly taken into consideration as the main elements for ADME-associated attributes (Ghavami et al. 2023; Parveen 2020). For the prediction of toxicity, ProTox-II can be used (Parveen 2020). The toxicity of the chemicals was also assessed by the toxicity estimation software tool (T.E.S.T) (Anand et al. 2021). A software, namely, admetSAR version 2.0, is also used for in silico prediction of the drugs. The initial version of this product was available in the year 2012 and included a total of 27 predictive models. The main objective was to enhance and optimize the existing models by significantly improving both the quantity and quality of the training data (Matondo et al. 2022). About 40 models were implemented for new chemical ADMET properties which are trained by state-of-the-art machine learning methods including neural network and k-nearest neighbor (KNN) algorithms. In the year 2021, a study on repurposing was carried out to analyze 130 drugs against SARS-CoV-2, from which 15 were found to be potent. For the predicted potent drugs, the lethal dose and toxicity were determined using admetSAR 2.0 (Anand et al. 2021).

L. *Determination of Druggable Sites*: For identification of the active site, protein–ligand identifier profiler (PLIP) BIOTEC Du Dresden is one of the software used. Chandel et al. predicted and identified, through this method, the FDA-approved drugs to be repurposed against novel coronavirus (Ghavami et al. 2023).

10.4 Conclusion

This chapter illustrates the importance of in silico methods in drug repurposing with the help of various examples. It is evident that a combination of computational and experimental techniques, along with the integration of diverse data from different sources, is crucial for the success of future drug repositioning initiatives. The techniques are used to understand the relationship between the mechanisms of action of drugs by evaluating phenotypic data, which includes both indications and side effects. This contributes to the development of quantitative structure–activity relationships. It turns out that integrating information from many resources is always preferable for strong in silico predictions. This is because none of the existing methods, when used alone, can guarantee a 100% success in finding new indications for drugs. In conclusion, this chapter emphasizes that the purpose of computational approaches is not to anticipate experimentally validated outcomes but rather to stimulate experimental investigations that generate novel concepts for drug repositioning studies, which must eventually be confirmed through experimentations.

References

Agarwal, Mehrotra (2016) An overview of molecular docking. JSM Chem 4(2):1024–1028

Agrawal, Agrawal, Blunden (2021) Rutin: a potential antiviral for repurposing as a SARS-CoV-2 main protease (Mpro) inhibitor. Nat Prod Commun 16(4):1934578X21991723

Ahmad, Farman, Jan (2019) Deep learning methods and applications. In: Khan M, Jan B, Farman H (eds) Deep learning: convergence to Big Data analytics. Springer Singapore, Singapore, pp 31–42

Ambrosio, Costa, Romeo, Esposito, Alkhatib, Salpini et al (2023) Targeting SARS-CoV-2 main protease: a successful story guided by an in silico drug repurposing approach. J Chem Inf Model 63(11):3601–3613

Anand, Ghildiyal, Madhwal, Bhatt, Verma, Rai (2021) Computational guided approach for drug repurposing against SARS-CoV-2. Future Virol 16(3):211–243

Armstrong, Morris, Finn, Sharma, Moretti, Cooper et al (2010) ElectroShape: fast molecular similarity calculations incorporating shape, chirality and electrostatics. J Comput Aided Mol Des 24:789–801

Arulanandam (2021) In silico approach on drug repurposing-Antimalarial drugs against HIV-1 protease. bioRxiv. 2021:2021.01. 12.426148

Ávalos-Moreno, López-Tejada, Blaya-Cánovas, Cara-Lupiañez, González-González, Lorente et al (2020) Drug repurposing for triple-negative breast cancer. J Personal Med 10(4):200

Bal H, Hujol J (2007) Introduction to basic local alignment search tool. In: Bal H, Hujol J (eds) Java for bioinformatics and biomedical applications. Springer US, Boston, pp 25–82

Bisson, Cheltsov, Bruey-Sedano, Lin, Chen, Goldberger et al (2007) Discovery of antiandrogen activity of nonsteroidal scaffolds of marketed drugs. Proc Natl Acad Sci USA 104(29):11927–11932

Cavasotto, Filippo D (2021) In Silico drug repurposing for COVID-19: targeting SARS-CoV-2 proteins through docking and consensus ranking. Mol Inform 40(1):2000115

Chakraborti, Ramakrishnan, Srinivasan (2019) In silico modeling of FDA-approved drugs for discovery of anticandida agents: a drug-repurposing approach. In: In silico drug design. Elsevier, pp 463–526

Cheng, Liu, Jiang, Lu, Li, Liu et al (2012) Prediction of drug-target interactions and drug repositioning via network-based inference. PLoS Comput Biol 8(5):e1002503

Dhorje, Lavhate, Srivastav (2020) In silico drug repurposing: an antifungal drug, itraconazole, repurposed as an anticancer agent using molecular docking. MGM J Med Sci 7(3):110–118

Djokovic, Ruzic, Djikic, Cvijic, Ignjatovic, Ibric et al (2021) An integrative in silico drug repurposing approach for identification of potential inhibitors of SARS-CoV-2 main protease. Mol Inform 40(5):2000187

Durant, Leland, Henry, Nourse (2002) Reoptimization of MDL keys for use in drug discovery. J Chem Inf Comput Sci 42(6):1273–1280

Ferraz, Gomes, Novaes S, Trossini G (2020) Ligand and structure-based virtual screening applied to the SARS-CoV-2 main protease: an in silico repurposing study. Future Med Chem 12(20):1815–1828

Gaulton, Bellis, Bento, Chambers, Davies, Hersey et al (2012) ChEMBL: a large-scale bioactivity database for drug discovery. Nucleic Acids Res 40(D1):D1100–D11D7. https://doi.org/10.1093/nar/gkr777

Ghavami, Adibzadeh, Amiri, Sardari (2023) Combined in silico strategy for repurposing DrugBank entries towards introducing potential anti-SARS-CoV-2 drugs. Can J Physiol Pharmacol 101(6):268–285

Haupt, Schroeder (2011) Old friends in new guise: repositioning of known drugs with structural bioinformatics. Brief Bioinform 12(4):312–326

Jamir, Sarma, Priyadarsinee, Kiewhuo, Nagamani, Sastry (2023) Polypharmacology guided drug repositioning approach for SARS-CoV2. PLoS One 18(8):e0289890

Kamboj, Rajput, Rastogi, Thakur, Kumar (2022) Targeting non-structural proteins of Hepatitis C virus for predicting repurposed drugs using QSAR and machine learning approaches. Comput Struct Biotechnol J 20:3422–3438

Kellenberger, Rodrigo, Muller, Rognan (2004) Comparative evaluation of eight docking tools for docking and virtual screening accuracy. Proteins 57(2):225–242

Konc, Janežič (2014) ProBiS-ligands: a web server for prediction of ligands by examination of protein binding sites. Nucleic Acids Res 42(W1):W215–WW20

Kumar, Kumar (2019) Molecular docking: a structure-based approach for drug repurposing. In: In Silico drug design Repurposing Techniques and Methodologies. Elsevier, pp 161–189

Kumar, Singh, Patel (2020a) In silico prediction of potential inhibitors for the main protease of SARS-CoV-2 using molecular docking and dynamics simulation based drug-repurposing. J Infect Public Health 13(9):1210–1223

Kumar, Chandel, Raj, Rathi (2020b) In silico identification of potent FDA approved drugs against Coronavirus COVID-19 main protease: a drug repurposing approach. Chemical Biology Letters 7(3):166–175

Lo, Torres (2019) In Silico repurposing of cell cycle modulators for cancer treatment. In: In silico drug design. Elsevier, United Kingdom, pp 255–279

Lo, Senese, Li, Hu, Huang, Damoiseaux et al (2015) Large-scale chemical similarity networks for target profiling of compounds identified in cell-based chemical screens. PLoS Comput Biol 11(3):e1004153

Lo, Senese, Damoiseaux, Torres (2016) 3D chemical similarity networks for structure-based target prediction and scaffold hopping. ACS Chem Bilo 11(8):2244–2253

Macip, Garcia-Segura, Mestres-Truyol, Saldivar-Espinoza, Ojeda-Montes, Gimeno et al (2022) Haste makes waste: a critical review of docking-based virtual screening in drug repurposing for SARS-CoV-2 main protease (M-pro) inhibition. Med Res Rev 42(2):744–769

MacMahon, Hwang, Yim, MacMahon, Abraham, Barton et al (2023) An in silico drug repurposing pipeline to identify drugs with the potential to inhibit SARS-CoV-2 replication. Inform Med Unlock 43:101387

Maffucci, Contini (2020) In silico drug repurposing for SARS-CoV-2 main proteinase and spike proteins. J Proteome Res 19(11):4637–4648

March-Vila, Pinzi, Sturm, Tinivella, Engkvist, Chen et al (2017) On the integration of in silico drug design methods for drug repurposing. Front Pharmacol 8:298

Matondo, Dendera, Isamura, Ngbolua, Mambo, Muzomwe et al (2022) In silico drug repurposing of anticancer drug 5-FU and analogues against SARS-CoV-2 main protease: molecular docking, molecular dynamics simulation, pharmacokinetics and chemical reactivity studies. Adv Appl Bioinforma Chem 15:59–77

Mohebbi, Ghorbanzadeh, Naderifar, Khalaj, Askari, Sammak (2023) A fragment-based drug discovery developed on ciclopirox for inhibition of Hepatitis B virus core protein: an in silico study. PLoS One 18(5):e0285941

Mullins (2022) Drug repurposing in silico screening platforms. Biochem Soc Trans 50(2):747–758

Onyango, Odhiambo, Angwenyi, Okoth (2022) In silico identification of new anti-SARS-CoV-2 main protease (M pro) molecules with pharmacokinetic properties from natural sources using molecular dynamics (MD) simulations and hierarchical virtual screening. J Trop Med 2022:3697498

Pan, Lin, Cao, Zeng, Yu, He et al (2022) Deep learning for drug repurposing: methods, databases, and applications. Wiley Interdiscipl Rev Comput Mol Sci 12(4):e1597

Parihar, Zafar, Khandia, Parihar, Dhote, Mishra (2022) In silico analysis for the repurposing of broad-spectrum antiviral drugs against multiple targets from SARS-CoV-2: a molecular docking and ADMET approach. Arch Proteom Bioinform 3(1):3–14
Park (2019) A review of computational drug repurposing. Transl Clin Pharmacol 27(2):59
Parveen, Alnoman, Aljohani, Aljohani, Alotaibi, Al-Juhani et al (2022) In Silico drug repurposing of 9H-Thioxanthene based FDA-approved drugs as potent chemotherapeutics targeting VEGFR-2 and COX-2. Biointerf Res Appl Chem 13:372
Parveen (2020) In silico drug repurposing of fda-approved artemisinins as potent chemotherapeutics targeting BCL-2, CDK-6 & VEGFR-2: density functional exploration and molecular docking study. Biointerf Res Appl Chem 11(2):9604–9618
Rácz, Bajusz, Héberger (2018) Life beyond the Tanimoto coefficient: similarity measures for interaction fingerprints. J Chem 10(1):48. https://doi.org/10.1186/s13321-018-0302-y
Repasky, Shelley, Friesner (2007) Flexible ligand docking with Glide. Curr Protoc Bioinformatics 18(1):8.12.1–8.. 36
Santos, Guedes, Karl, Dardenne (2020) Highly flexible ligand docking: benchmarking of the DockThor program on the LEADS-PEP protein–peptide data set. J Chem Inf Model 60(2):667–683
Schwede, Kopp, Guex, Peitsch (2003) SWISS-MODEL: an automated protein homology-modeling server. Nucleic Acids Res 31(13):3381–3385
Snell, Johnston, Srinivasan, Zhou, Gao, Skolnick (2016) Repurposing FDA-approved drugs for anti-aging therapies. Biogerontology 17:907–920
Studer, Rempfer, Waterhouse, Gumienny, Haas, Schwede (2020) QMEANDisCo—distance constraints applied on model quality estimation. Bioinformatics 36(6):1765–1771
Studer, Tauriello, Bienert, Biasini, Johner, Schwede (2021) ProMod3—a versatile homology modelling toolbox. PLoS Comput Biol 17(1):e1008667
Trott, Olson (2010) AutoDock Vina: improving the speed and accuracy of docking with a new scoring function, efficient optimization, and multithreading. J Comput Chem 31(2):455–461
Vogt, Mestres (2010) Drug-target networks. Mol Inform 29(1-2):10–14
Wallace, Laskowski, Thornton (1995a) LIGPLOT: a program to generate schematic diagrams of protein-ligand interactions. Protein Eng 8(2):127–134. https://doi.org/10.1093/protein/8.2.127
Wallace, Laskowski, Thornton (1995b) LIGPLOT: a program to generate schematic diagrams of protein-ligand interactions. Protein Eng Des Sel 8(2):127–134
Walters, Stahl, Murcko (1998) Virtual screening—an overview. Drug Discov Today 3(4):160–178. https://doi.org/10.1016/S1359-6446(97)01163-X
Wu, Cheng, Li, Li, Liu, Tang (2017) SDTNBI: an integrated network and chemoinformatics tool for systematic prediction of drug–target interactions and drug repositioning. Brief Bioinform 18(2):333–347
Yang, Cai, Zhao, Xie, Chen (2022) Concepts and applications of chemical fingerprint for hit and lead screening. Drug Discov Today 27(11):103356
Yuan, Chan, Hu (2017) Using PyMOL as a platform for computational drug design. Wiley Interdiscipl Rev Comput Mol Sci 7(2):e1298
Zapata-Cardona, Florez-Alvarez, Guerra-Sandoval, Chvatal-Medina, Guerra-Almonacid, Hincapie-Garcia et al (2023a) In vitro and in silico evaluation of antiretrovirals against SARS-CoV-2: a drug repurposing approach. AIMS Microbiol 9(1):20–40. https://doi.org/10.3934/microbiol.2023002
Zapata-Cardona, Florez-Alvarez, Guerra-Sandoval, Chvatal-Medina, Guerra-Almonacid, Hincapie-Garcia et al (2023b) In vitro and in silico evaluation of antiretrovirals against SARS-CoV-2: a drug repurposing approach. AIMS Microbiol 9(1):20

Repurposing of Therapeutics for Anticancer Activity

11

Samipta Singh, Nidhi Mishra, Alka, Abhishek Verma, Raquibun Nisha, Neelu Singh, and Shubhini A. Saraf

Abstract

This chapter delves into the innovative and strategic re-evaluation of existing pharmaceuticals to discover novel anticancer applications. This approach leverages the extensive clinical and pharmacological data available for approved drugs, significantly accelerating drug development and reducing associated costs and risks. The chapter outlines the drug repurposing methodologies, including computational approaches, high-throughput screening, and biological assays. It highlights case studies of successful repurposing endeavors that have led to groundbreaking advancements in cancer therapy. By exploring the potential of repurposing existing therapeutics, this chapter underscores the promising horizon of discovering practical and safer cancer treatments, ultimately contributing to the evolving landscape of oncology.

Repurposing therapeutics for anticancer activity has developed as a promising approach to fast-track the development of novel cancer treatments. This entails identifying existing drugs, initially designed for other therapeutic indications that exhibit potent anticancer effects. Drug repurposing offers advantages such as reduced development timelines and costs compared to de novo drug discovery as it leverages existing clinical data and safety profiles. This chapter explores various examples of repurposed therapeutics with anticancer potential, highlighting mechanisms of action, preclinical and clinical evidence, and challenges and

S. Singh · N. Mishra · Alka · A. Verma · R. Nisha · N. Singh
Department of Pharmaceutical Sciences, Babasaheb Bhimrao Ambedkar University Lucknow (A Central University), Uttar Pradesh, Lucknow, Uttar Pradesh, India

S. A. Saraf (✉)
Department of Pharmaceutical Sciences, Babasaheb Bhimrao Ambedkar University Lucknow (A Central University), Uttar Pradesh, Lucknow, Uttar Pradesh, India

National Institute of Pharmaceutical Education and Research, Raebareli (NIPER-R), Lucknow, Uttar Pradesh, India

R. K. Kaundal et al. (eds.), *Drug Repurposing: Exploring Newer Therapeutic Potentials*,
https://doi.org/10.1007/978-981-95-4522-3_11

opportunities in this rapidly evolving field. Critical considerations for successful repurposing strategies, including drug screening methodologies and regulatory pathways, are also discussed to provide insights into the future directions of anticancer drug development.

Keywords

Metastasis · Chemotherapy · Drug resistance · Therapeutic modalities · Drug screening

11.1 Introduction

Cancer is defined by the abnormal and unregulated proliferation of cells, disturbance in the regular control of the cell cycle, and cell death. The process by which cancer cells spread to different tissues and organs is termed metastasis; the unchecked proliferation results in an escalated dissemination, reducing treatment efficacy (Siddiqui et al. 2022). In 2018, around 18 million new cases of cancer were logged globally, and about 10 million lives were lost to cancer in the year 2020. World Health Organization reports these numbers to increase nearly fourfold by 2040 (Bray et al. 2018).

Cancer treatment is quite a challenging concern due to the acquired resistance by cancer cells and the appearance of other serious complications. Cancer cells develop resistance to drugs, which ultimately results in treatment failure, cancer advancement, low oxygen levels, high pressure of fluid within the tumor, inadequate blood supply, and resistance to drugs due to their toxic effects. Various therapeutic modalities have emerged in recent years for cancer management, encompassing surgical intervention, chemotherapy, targeted therapy, radiation therapy, and immunotherapy. Surgery is the prime treatment choice for all solid malignancies. Radiation therapy, which involves high-frequency radiation such as gamma radiation, X-rays, etc., is employed to destroy tumor cells (Wang et al. 2018). The use of drugs in cancer therapy can be traced back to the exploitation of mustard gas against skin cancers. Chemotherapy drugs, in particular, often have adverse side effects on patients in addition to drug resistance. With the increasing incidence of cancer, there is a pressing need to develop novel and safe drugs with fewer side effects. However, this development progress has decelerated due to the failure of several anticancer drugs to pass clinical trials successfully (Wang et al. 2018). The time required for conducting trials, evaluating safety, reviewing, and approving new drugs is substantial, contributing to poor global cancer management (Bray et al. 2018; Wang et al. 2018). State-of-the-art treatments, like targeted chemotherapy, eradicate cancer cells that have spread to different body parts (Barzaman et al. 2020). However, cancer treatment often results in treatment failure and relapse despite the wide range of drug availability (Mansoori et al. 2017). Therefore, emphasis on enhancing cancer management approaches, explicitly focusing on prioritizing cancer diagnosis and treatment, is required.

Subsequently, a new approach called drug repurposing/reprofiling/repositioning (DR) has gained momentum. Here, new uses of existing xenobiotics in the market are utilized by understanding their various mechanisms of action. DR helps to cut the long time required to take a drug from bench to bedside, saving millions of dollars spent on drug research and clinical trials.

11.1.1 Advantages and Disadvantages of DR

11.1.1.1 Advantages

- DR leads to cost-cutting in drug discovery research.
- Since the drug is not a novel chemical entity, its toxicities are already identified.
- Timelines are much shorter to develop required drug delivery systems.
- The associated big data helps access safety profiling.

11.1.1.2 Disadvantages

- DR requires much money to investigate newer uses of existing drugs.
- Protecting the intellectual property of drugs that are no longer under patent through repurposing can be done.
- Pharma companies are not interested in repurposing medications due to the challenges in acquiring rights.
- Repurposed medications show more anticancer activity when used alongside established cytotoxic drugs rather than on their own, more like a synergistic effect (Bertolini et al. 2015; Low et al. 2020).

Cancer is a disease that has an enormous burden in terms of morbidity as well as the emotional status of the patients and the caregivers. Patients in third-world countries cannot afford the cost of treatment, and the quest for the perfect treatment continues even in advanced countries (Xue et al. 2018). This chapter investigates the premise that repurposing can improve cancer treatment outcomes.

11.2 DR Approaches

Several principles, tools, and approaches are employed to identify potential candidates efficiently and effectively during DR. Figure 11.1 illustrates the various DR sources for identifying repurposed drugs. These principles guide selecting and validating existing drugs for new therapeutic indications.

Some fundamental principles and tools used in DR are described below.

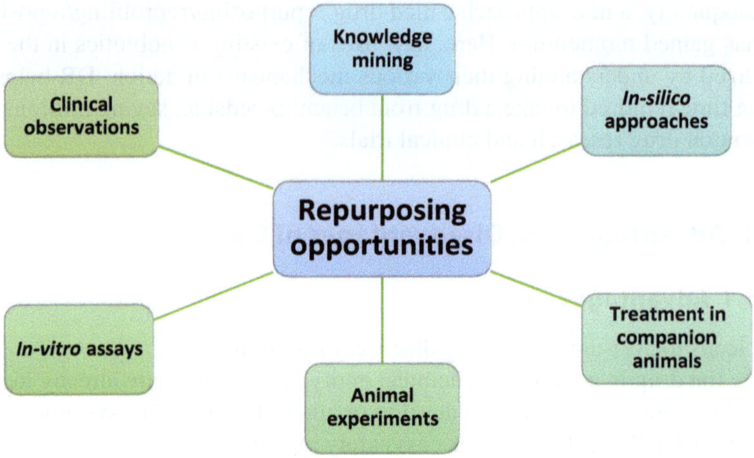

Fig. 11.1 Drug repurposing sources for identifying repurposed drugs

11.2.1 In Silico Repurposing Approaches

11.2.1.1 Computational Methods

Omics-Created Methods

"Omics" studies various biological science fields (Dai and Shen 2022; Hasin et al. 2017). The suffix "ome" in these terms denotes the specific objective of each field, like genome, transcriptome, proteome, and metabolome (Manzoni et al. 2018). Genomics centers on arranging, operating, developing, and charting genomes. Meanwhile, proteomics examines proteins and their biochemical properties. The field of transcriptomics focuses on gathering all the messenger RNA molecules in a cell, tissue, or organism, while metabolomics examines all the metabolites and related chemical processes (Vailati-Riboni et al. 2017). As research has progressed, it has become clear that addressing complex questions in biology requires a multi-omics approach. This involves integrating and analyzing datasets from different omics fields. Each omics data type offers unique insights into diseases like cancer, serving as markers of disease processes and providing information about pathways that differ between diseases and healthy controls (Momeni et al. 2020; Subramanian et al. 2020). With advancing technologies, researchers can quantify various biological elements. Multi-omics approaches, especially integrative analyses, are crucial for gaining deeper insights into oncological mechanisms. The complexity of cancer arises from the interactions among multiple biological entities, both cancer-related and non-cancer elements (Cruz-Acuña et al. 2021; Lambert et al. 2017; de Anda-Jauregui and Hernandez-Lemus 2020; Zhou et al. 2022; Mbemi et al. 2020; Shadkhoo and Mani 2019; Sompairac et al. 2019; Yang et al. 2014; Nunez et al. 2021). Integrative analysis aims to answer important biological and clinical questions and understand molecular behavior, relationships between molecular structures, disease pathology, and clinical features. Multi-omics is employed for tumor subtyping, prognosis, and diagnosis (Ahmed et al. 2023a; Vahabi and Michailidis 2022; Wang et al. 2022a; Zeng and Lumley 2018). A case study in hepatocellular carcinoma (HCC) demonstrates multi-omics approaches. The Bayesian network mixture model, bnClustOmics, was employed in HCC patients. The analysis revealed three HCC subtypes with molecular features linked to survival and clinical stage (Suter et al. 2022). Similar studies, such as the MOGONET framework and a convolutional neural network (CNN)-based cancer type prediction, emphasize the importance of fast computational frameworks for integrative multi-omics analysis (Wang et al. 2021a; Chuang et al. 2021). These approaches identify biomarkers, perform patient classification, and contribute better information for prudent decision-making. In conclusion, a comprehensive understanding of omics datasets and the integration of multi-omics approaches offer valuable insights into cancer research. This includes identifying risk factors, improving diagnostic capabilities, enhancing therapeutic effects, and refining prognosis in cancer studies.

Network Pharmacology

The rise of extensive biomedical datasets has paved the way for significant advancements in the exploration of repurposing. However, the abundance of diverse data sources poses a challenge in organizing intricate biological interactions within these datasets. While data-driven network-based models have demonstrated impressive performances in DR, they have limitations. These knowledge graph-based models, equipped with embedding techniques, have been established to recognize repurposed candidate drugs for specific disorders. Notably, these knowledge graph-based approaches are generic and can be readily applied to DR for other conditions, such as opioid use disorder (OUD) (Gao et al. 2022).

11.2.1.2 Text Mining and Literature Analysis

Literature mining serves as a valuable platform for researchers to access potential data, facilitating the establishment of hypotheses for DR. It involves extracting hidden, indirect relationships between drug–target diseases from a wealth of scientific literature. It often employs Medical Subject Heading (MeSH) terms to systematically retrieve pertinent and relevant information.

ABC Model (an Indirect DR Approach)

In 2016, Jung-Hsien Chiang and team investigated a new method for repurposing drugs by using data from the MEDLINE database. They created the ABC model, which examines the connection between a disease (A), a gene (B), and a drug (C) to establish an indirect relationship between the disease and the drug.

MeSHDD Designing (MeSH Terms-Based Drug–Drug Similarity Database)

A detailed examination of the MeSH terms that supplemented similarity between Food and Drug Administration (FDA)-approved drugs and diseases was done by Chirag and Adam. They created a database named MeSHDD via literature mining, involving the identification of MeSH and drug terms from the MEDLINE repository. Additionally, 2142 FDA-approved drugs were selected from the DrugBank database. The researchers mapped these elements, resulting in a refined dataset of MeSH–drug pairs. Utilizing a hypergeometric Bonferroni enrichment test, the analysis focused on the significance of drug–MeSH term pairs. The proximity between MeSH–drug pairs was determined in binary digits, where high similarity corresponded to values close to 0 and dissimilarity to values near 1 (Brown and Patel 2017).

11.2.1.3 Systematic Clinical Trials

Prospective Analysis

Prospective studies investigate events or trends from the present to the future. These studies are meticulously planned with specific methods for data collection, allowing for the targeted gathering of exposure data, which can result in more comprehensive information. However, the drawback of prospective cohort studies is the extended time required for follow-up and waiting for events or diseases to manifest. This makes the design inefficient for studying diseases with prolonged latency periods and increases the risk of high loss rates to follow-up. Despite the landmark success of the Framingham Heart Study, initiated in 1948 and ongoing, prospective cohort studies are generally considered impractical and inefficient in plastic surgery literature. Instead, retrospective cohort studies are more suitable due to their timeliness and cost-effectiveness (Song and Chung 2010).

Retrospective Analysis

Sildenafil, famously known as the "blue pill" or Viagra®, illustrates successful DR through retrospective clinical analyses. Initially designed for angina, the phosphodiesterase 5 (PDE5) inhibitor unexpectedly revealed efficacy in dilating penile blood vessels during a 1990s clinical trial. This led to its repositioning as a treatment for erectile dysfunction despite initial failure in its intended use. This strategy of repurposing, seen in examples like propranolol for osteoporosis and melanoma, played a vital role in comprehending SARS-CoV-2 pathogenesis, identifying potential repurposing candidates such as melatonin, carvedilol, and umifenovir. Studies comparing COVID-19 drugs retrospectively have hinted at the possible use of hydroxychloroquine, with combinations like oseltamivir and hydroxychloroquine showing promise. Repurposing existing drugs through retrospective analyses breathes new life into studies abandoned due to lack of competitiveness. The difficulties in obtaining and evaluating past medical data, such as maintaining patient privacy and navigating legal barriers, pose significant challenges. Data from sources like WHO's VigiBase offer rich information for identifying potential DR leads, underscoring the significance of this approach in medical research (Ng et al. 2021).

11.2.1.4 Pharmacological Insights

Mechanism-Based Repurposing

It combines treatment with omics data, available signal knowledge, and protein relations to reveal newer insights. These methods offer the advantage of uncovering diseases or drug-associated mechanisms and identifying drugs directly linked to treating explicit ailments. In essence, they provide a comprehensive understanding of how drugs exert their effects and contribute to advancing targeted therapeutic approaches (Mithun et al. 2020).

Target-Based Repurposing

Target-based methods include in vitro and in vivo drug molecule screenings for a specific protein target or biomarker identification using in silico screening techniques like QSAR. This method provides results quickly in a short period (Mithun et al. 2020).

11.2.1.5 Drug Combination Approaches

Combination Therapies

Combining two to three drugs with distinct action mechanisms effectively addresses DR challenges. This approach enhances the success rate by leveraging synergistic effects from targeting different biological pathways and reduces the required concentrations of individual drugs, making them more effective in combination. To implement this strategy, a two-step process was proposed: first, a DR screen using approved drug collections, and second, a drug combination test using identified hits to determine effective combinations for new clinical applications. Practically, two approaches for discovering synergistic combinations from drug repositioning screens are applied. The first addresses concerns about weak compound potency and toxicity, demonstrating a three-drug combination's synergistic effect that minimizes individual drug concentrations. The second involves correlating effective drug concentrations with attainable concentrations in human blood, ensuring identified combinations are both potent and within a clinically feasible range (Sun et al. 2016).

Polypharmacology

Recognizing the importance of multifactorial and polygenic factors in diseases like neurodegenerative conditions, cancer, diabetes, and hypertension has grown. High drug trial failure rates may be due to underestimating the complex nature of these diseases, which are often influenced by multiple factors and genetic determinants (Palop et al. 2006; Barabasi et al. 2011). Age, tissue type, and environmental stimuli also affect disease manifestation. Conventional single-drug approaches have proven ineffective. Polypharmacology addresses these complexities. One approach involves developing a single drug that can modulate multiple targets relevant to the disease (Bolognesi 2013). Another strategy uses various drugs, each interacting with different disease-associated targets. Drug combination therapy, involving two or three drugs, has gained acceptance. The debate over single-target drugs versus combinations continues.

11.2.1.6 Patient Data Analysis

Electronic Health Records (EHRs)

These serve as a goldmine, complementing cheminformatic, phenotypic, and transcriptomic datasets. Natural language processing (NLP) breakthroughs now empower researchers to extract insights from unstructured EHR data like clinic notes and web forum posts. Li et al. harnessed deep learning on EHR notes to enhance ADE detection, surpassing existing methods like the FDA Adverse Events Reporting System (Li et al. 2018). Mining off-label use data from EHR and social media unveils valuable DR prospects. Jung et al. employed a support vector machine (SVM) on clinical text, revealing 10,765 potential novel drug-indication pairs (Jung et al. 2013). Xu et al. utilized machine learning on EHRs, discovering the link between metformin (Xu et al. 2015). Integrating data, they identified metformin's association with decreased mortality post-cancer diagnosis, substantiated by preclinical and epidemiological studies (Tang et al. 2018). Advancements in NLP, deep learning, and vast EHR and online data hint at future DR breakthroughs. Giants like Google and Amazon have supercomputing capabilities that accelerate identifying new repurposing opportunities. The convergence of technology and healthcare data propels drug discovery into a dynamic era of efficiency and innovation (Xu et al. 2015).

11.2.1.7 Biological Pathway Analysis

Pathway-Based Repurposing

Dysregulated interactions are detected through three main approaches. First is the over-representation analysis (ORA) technique (Pomyen et al. 2015), confirming pathway over-representation or under-representation in a list of seed genes. The second approach involves scoring functional classes and utilizing gene expression values to compute gene-level statistics for each pathway. The third approach, pathway topology-based, considers pathway topology in computing gene-level statistics (Ahmed et al. 2023b). Understanding pathways and networks becomes crucial, given the tight connection between cancer and genes. Both pathways and networks involve focusing on processes on a smaller dimension. Recent studies have delved into capturing changes in genes, disease-associated pathways, and other essential interactions, shedding light on the dynamic nature of these biological systems (Zhong et al. 2009; Liu et al. 2012). Identifying altered pathways in cancer mutations is a challenge addressed by methods like gene set enrichment analysis (GSEA) (Lee et al. 2013). Other pathway-based studies, such as the one using a Pathifier algorithm, focus on deriving pathway deregulation scores specific to tumor samples. This algorithm, tested on datasets, demonstrates robustness and identifies pathways crucial to the survival of glioblastoma patients (Drier et al. 2013; Vishnubalaji et al. 2019). These approaches aid in discovering novel biomarkers and therapeutic targets, providing valuable insights into the mutated molecular networks. In summary, pathway-based approaches unravel the complexities of cancer biology.

11.2.1.8 Network Biology Methods

Cancer is a fallout of environmental and genetic influences, which poses challenges in comprehending the specific pathways involved (Hong et al. 2019; Hideyuki and Toshikazu 2019). Recent advances in network-based or systems biology methods have provided powerful tools for studying complex diseases such as cancer (Fiscon et al. 2018; Jin et al. 2019). These approaches utilize knowledge to denote connections between them (Ahmed et al. 2022; Song et al. 2018; Feng et al. 2017; Galan-Vasquez and Perez-Rueda 2021). To handle the complexity of more extensive networks, they are often made into smaller, less complex nodes that can be studied easily. Specific examples of systems biology approaches include node prediction (predicting individual molecule behavior) (Doshi and Chepuri 2022), link prediction (predicting interactions between pairs of molecules) (Abbas et al. 2021), and graph prediction (predicting overall network behavior) (Zhang et al. 2022a). AI/ML models help reveal disease subtypes and uncover critical molecular pathways involved in diseases like cancer (Chang et al. 2016; Liu and Chance 2013). Two notable models, GraphRepur (Cui et al. 2021) and DT2Vec (Amiri Souri et al. 2022), have shown success in DR against breast cancer (Amjad et al. 2020). For instance, a study applied systems biology to analyze gene expression profiles in HER2-negative breast cancer, identifying differentially expressed genes and constructing a network of interacting proteins (Karn et al. 2011; Raju et al. 2014). Similar approaches have been used in lung adenocarcinoma and lung cancer studies to discover critical genes, biomarkers, and mechanisms involved in disease initiation and progression (Shiroma et al. 2020; Anusewicz et al. 2020; Kaushik et al. 2020; Liu et al. 2021). In conclusion, these approaches help understand and predict complex diseases like cancer. These methods provide insights into the intricate interactions within biological systems, aiding in improved cancer therapies and critical biomarkers.

11.2.1.9 Machine Learning (ML) and Artificial Intelligence (AI)

Suppose a model is trained on small inhibitor sets of a specific chemical structure; it may not be able to predict accurately. It may excel at predicting compounds with similar structures but struggle with diverse ones, referred to as overfitting. As the network complexity increases, it becomes difficult to predict new chemical entities since these complex structures have not been fed into the system. Methods like reducing network size and accurately halting the experiment when the validation performances decline help mitigate this overfitting problem. Various studies have applied these strategies, such as the DEEP Screen model for drug-target interactions and a complex neural network predicting tumor response to drugs (Rifaioglu et al. 2020; Srivastava et al. 2014; Angermueller et al. 2016). Descriptors represent the physicochemical parameters of moieties and their interaction. Fingerprints, vectors of 0s and 1s, are commonly used to represent these features, including two- and three-dimensional aspects. (Sastry et al. 2010). Pharmacophore fingerprints, representing 3D chemical functional groups and distances between them, and circular fingerprints (e.g., ECFP) are effective input choices (Rogers and Hahn 2010). Different studies have utilized various descriptors, such as structural protein-ligand interaction fingerprints and even the 2D figures of xenobiotics (Rifaioglu et al. 2020). Dataset size is crucial, with larger datasets generally leading to better model performance. Overfitting is reduced with more data, especially for deep neural networks (Allen et al. 2019). However, data quality is equally important, like those in PubChem; data from high-throughput screens must adhere to FAIR standards (findability, accessibility, interoperability, and reusability) (Wilkinson et al. 2016; Sansone et al. 2019; Kim et al. 2019). Metadata, describing experimental conditions and providing context, is vital for model learning. Proper data management with community-accepted standards ensures accuracy and facilitates future integration into ML- and AI-driven drug discovery efforts (Stathias et al. 2018). Successful ML- and AI-driven models in drug discovery require careful consideration of data quality, type, and quantity, as well as the choice of input descriptors and effective validation methods. The adherence to community standards ensures that future efforts will benefit from high-quality, easily integrable data, leading to more accurate modeling and potential DR opportunities for cancer therapy.

11.2.1.10 Virtual Screening (VS)

The VS method is employed to sift through extensive chemical libraries, encompassing both commercial and freely available databases of chemical structures, intending to identify promising compounds (Gimeno et al. 2019; Patel et al. 2021). Traditional experimental screening of chemical libraries is constrained by limited reagent availability and assay throughput. In contrast, VS allows for exploring vast chemical libraries for probable bioactivities without costly experimental bioassays. This technique plays a crucial role in drug discovery by uncovering novel hits and prioritizing compounds for further testing, leading to significant cost and attrition rate reductions (Turon et al. 2023). There are two primary categories of VS approaches: structure-based (SB) and ligand-based (LB). The choice between them depends on the availability of three-dimensional (3D) information (Sabe et al. 2021). Instead of focusing on the structure of the target, LBVS derives and uses information from active compounds (Sabe et al. 2021; Lin 2022; Kwon et al. 2019). While SBVS and LBVS differ in computational methods' complexity and execution speed, both approaches advance research.

Ligand-Based Virtual Screening (LBVS)

LBVS is a frequently employed technique in computer-aided drug design (CADD), mainly when the three-dimensional structure of the target receptor is unknown. It relies on the principle of molecular structure resemblance, positing that compounds with analogous structures exhibit similar biological functions and target interactions (Macalino et al. 2015). Traditional methods within LBVS include the following.

Quantitative Structure-Activity Relationships (QSARs)

QSAR utilizes mathematical and statistical methods to establish correlations between the structures and functions of a group of compounds. This allows chemical databases to be examined to identify potential leads with desired biological activity. Cross-validation assesses model performance within the training dataset, preventing overfitting, while AD specifies the chemical space where predictions are reliable (Shayanfar and Shayanfar 2022). QSAR studies have been instrumental in discovering antimicrobial agents against drug-resistant pathogens, offering insights into DR and optimization of existing drugs (Chalkha et al. 2022; Bueso-Bordils et al. 2015; Suay-Garcia et al. 2020a, b; Zanni et al. 2017; Bueso-Bordils et al. 2017).

Pharmacophore Modeling

It elaborates the obligatory spatial arrangement of structural features for an active ligand to bind to its biological target. These features, including hydrogen-bond acceptors, donors, hydrophobic regions, and ionizable groups, are crucial for identifying molecule similarities and aiding drug design (Muhammed and Esin 2021; Giordano et al. 2022; Jade et al. 2022). Pharmacophore models, created through ligand-based or structure-based approaches, help understand how various compounds can bind to a specific target, assisting drug candidate discovery (Muhammed and Esin 2021; Jade et al. 2022). Ligand-based pharmacophore modeling is applied when the 3D structure of the target protein is unknown, but information on active ligands is available. In contrast, structure-based pharmacophore modeling uses the 3D structure of the protein to identify potential leads. Numerous programs and databases are utilized for pharmacophore modeling, including Catalyst, LigandScout, Pharmer, MOE, and PharmaGist (Muhammed and Esin 2021; Vuorinen and Schuster 2015; Hasman et al. 2015; Prachayasittikul et al. 2015; Wang et al. 2022b). Pharmacophore modeling has been employed in discovering and designing antimicrobial agents against antimicrobial resistance (AMR). Studies have reported successful identification of lead compounds using ligand-based pharmacophore approaches, contributing to the fight against AMR (Barakat et al. 2018; Zhang et al. 2018; Shaker et al. 2021). However, a drawback of pharmacophore-based LBVS is its reliance on pre-computed databases with limited conformations, potentially affecting active compounds' detection, especially those with small-molecular functionalities (Hassan Baig et al. 2016).

Structure-Based Virtual Screening (SBVS)

SBVS methods play a crucial role in drug discovery. This approach involves using computational tools to explore and identify potential hit compounds that interact with druggable targets (Li and Shah 2017). These are typically acquired from the Protein Data Bank (PDB). Advances in structural biology, including XRD, NMR, and cryo-EM, have facilitated the availability of these 3D structures. In cases where the structures are not available, researchers can use the amino acid sequence to construct 3D models via homology modeling (Sabe et al. 2021; Macalino et al. 2015; Hsieh et al. 2023; Oselusi et al. 2022). SBVS employs scoring functions to assess interactions among ligands and protein targets, helping identify top-hit compounds for experimental validation (Wu et al. 2019; Tomar et al. 2019a). In the context of antimicrobial resistance (AMR), SBVS can be helpful to identify possible drug targets among microbial genes associated with AMR (Sabe et al. 2021; Tomar et al. 2019b).

Molecular Docking

One essential technique within SBVS is molecular docking (MD), a computational method predicting how ligands interact with the binding site of a target protein, each addressing specific aspects of the ligand and target flexibility (Macalino et al. 2015; Tomar et al. 2019b; Baig et al. 2018).

Scoring Functions

Scoring functions like force field-based, knowledge-based, empirical-based, descriptor-based, etc., are essential in MD, as they estimate binding energies and help rank ligands (Zhang et al. 2022b). Despite their advancements, scoring functions still face challenges like limited ligand and receptor conformation sampling (Liu and Wang 2015; Chen et al. 2020; Varela-Rial et al. 2022).

Density Functional Theory (DFT)

DFT is useful for analyzing the electronic interactions between atoms, offering valuable information about the characteristics of ligands and their impact on protein target inhibitory activities. When used with other techniques, DFT expedites drug development by simulating the energy and geometry of protein-ligand complexes. (Tariq et al. 2022; Van Mourik et al. 2014; Ye et al. 2022; Tandon et al. 2019).

Molecular Dynamics Simulations (MDS)

MDS incorporate motion and a precise aqueous environment. These simulations validate MD results, offering insights into the rates and binding free energies over a specified time frame (Sabe et al. 2021; Sivakumar et al. 2020; Singh et al. 2022; Pinzi and Rastelli 2019).

While these techniques have significantly improved drug discovery, challenges persist, such as the need for enhanced scoring functions and addressing the computational intensity of simulations. Researchers often combine multiple methods to overcome these challenges and tailor their approaches to specific drug discovery projects.

11.2.1.11 Animal Models

Preclinical Testing

Animal models, such as rodents or other mammals, play a pivotal role in the search for therapeutic agents by contributing valuable insights into disease progression and treatment efficacy. These models provide crucial data on drug safety and efficacy before they undergo testing in human clinical trials. Animal models are utilized to assess the potential of repurposed drugs to treat different diseases and comprehend their mechanisms of action. These models allow for identifying promising candidates and optimizing dosage regimens, leading to more efficient and cost-effective drug development. Additionally, animal models help researchers uncover potential side effects and evaluate the overall impact of repurposed drugs on the physiological systems of living organisms, providing a bridge between in vitro studies and human trials in the pursuit of novel therapeutic applications. Despite their significance, animal models have limitations like genetics and species variations in rodents, hindering accurate disease progression representation. Non-human primate (NHP) models, like the rhesus macaque, while genetically closer to humans, present challenges due to high costs, limited accessibility, and potential variability in experimental outcomes. As researchers navigate these complexities, animal models are invaluable tools in the ongoing fight against emerging viral threats (Ng et al. 2021).

11.2.1.12 Experimental Screening Approaches

High-Throughput Screening (HTS)
HTS is a long-standing method in drug discovery and faces challenges such as low hit rates attributable to incompatible libraries and false-positive or false-negative hits. Successfully advancing hit compounds to Phase I clinical trials is crucial but challenging, given the high likelihood of drug candidate failures in the clinical development stage. Many pharmaceutical firms have shifted focus from drug development due to profitability concerns and the complexity of collaborative efforts. However, there is a potential resurgence in drug development with the adoption of pioneering computational approaches like molecular modeling, bioinformatics, AI, ML, and cheminformatics. These approaches revolutionize drug discovery, making it more data-driven, efficient, and cost-effective. They are crucial in addressing investment challenges, optimizing drug combinations, combating drug resistance, and contributing to global health (Christabel Chikodi et al. 2023).

11.2.2 Repurposing Non-oncology Drugs in Cancer Therapeutics

11.2.2.1 Antibacterial Drugs in Cancer Therapeutics

Autophagy involves carcinogenesis, cancer stem cell maintenance, tumor development, and chemotherapeutic resistance. Suppression of autophagy can suppress tumor growth. Macrolide antibiotics (such as clarithromycin and azithromycin) inhibit autophagy. (Takano et al. 2023). Ciprofloxacin, a second-generation fluoroquinolone antibiotic, treats various bacterial infections. Gupta et al. 2019 reported that ciprofloxacin inhibits the efflux function of ATP-binding cassette subfamily B member 1 (ABCB1), one of the reported pathways responsible for multiple drug resistance (MDR). With this, it can reverse MDR potentially (Gupta et al. 2019). It can inhibit liver cancer via upregulation of CD86+CD206− macrophages (Fan et al. 2020).

11.2.2.2 Antifungal Drugs Repurposed for Cancer

The efficacy of antifungal drugs against cancer can be credited to the overlapping molecular targets or pathways between cancer pathogenesis and antifungal action. Several "off-site" targets (such as G0/G1 arrest, Hedgehog (Hh) signaling pathway inhibition, decreasing angiogenesis, mitochondrial dysfunction, and apoptosis) from antifungal drugs are also discovered (Weng et al. 2023). Griseofulvin is used to treat dermatophytic infections. Griseofulvin (an antifungal medication for dermatophytic infections) has gained attention due to its mitotic inhibitory function. It is efficacious in myeloma and lymphoma cells (apoptotic cell death) (Schmeel et al. 2017). GF-15 (a derivative of griseofulvin) has been found to reduce spindle tension and spindle multipolarity and induce apoptosis. In vivo results have shown tumor growth inhibition and prolonged survival upon treatment of GF-15 in xenograft colon cancer and myeloma models (Raab et al. 2012). In 1957, 5-fluorocytosine (5-FC) was synthesized as a potential antitumor agent; however, it was not efficacious against tumors adequately. After 4 years, it was effective in *Candida* and *Cryptococcus* infections in mice. 5-FC has no antimycotic activity but results from converting 5-FC to 5-FU in the presence of cytosine deaminase (CD) in fungal cells. Fungi lacking this enzyme are not effective (Vermes et al. 2000). 5-FC is now repurposed for anticancer activity via a gene-directed enzyme prodrug therapy approach. This approach has three components: prodrug, suicide gene, and vector. Vectors are the number of delivery systems for gene transduction into the tumor cell. One successful example of this approach is the combination of Toca 511 (vocimagene amiretrorepvec) and Toca FC (extended-release 5-FC). Toca-511 is a replicating vector carrying a proactive gene. It can successfully deliver yeast cytosine deaminase genes to tumors after intravenous or intratumoral injection in glioma models (Yagiz et al. 2016). The approach (Toca-511 and 5-FU) also entered clinical trials (Cloughesy et al. 2016). Other antifungals that have shown some anticancer activity are terbinafine (hepatocellular carcinoma, colorectal cancer, promyelocytic leukemia, oral squamous cell carcinoma, and colon cancer), natamycin (hepatocellular carcinoma), itraconazole (medulloblastoma and basal cell carcinoma, endometrial cancer, melanoma, glioblastoma, endometrial cancer, hepatocellular carcinoma, colon cancer, pancreatic cancer, breast cancer, gastric cancer, and lung cancer), ketoconazole (colon cancer, breast cancer, hepatocellular carcinoma, and glioblastoma), miconazole (colon carcinoma and lung cancer), econazole (lung cancer, colon cancer, and pancreatic cancer), clotrimazole (endometrial cancer, lung cancer, colon cancer, and melanoma), and sertaconazole (lung cancer). Itraconazole and ketoconazole have passed Phase-I clinical trials (Weng et al. 2023).

11.2.2.3 Anthelmintic Drugs Repurposed for Cancer

Anthelmintic drugs primarily treat unicellular protozoa and parasitic worms of the intestine. Examples include flubendazole, albendazole, mebendazole, chloroquine, and ivermectin. Many studies have shown the beneficial effects of these drugs against cancer. They can work against critical oncogenic pathways, for instance, Wnt/β-catenin, STAT3, etc. The drugs evaluated under phase I/II/III trials are albendazole, mebendazole, ivermectin, levamisole, and niclosamide (Alavi and Ebrahimi

Shahmabadi 2021). Mebendazole was identified as well-tolerated and safe in advanced gastrointestinal cancer patients, but they developed rapid progressive disease (Mansoori et al. 2021).

11.2.2.4 Antimalarial Drugs Repurposed for Cancer

Aminoquinolones (antimalarials) may act as lysosomal alkalinizers, may overturn lysosomal-related barriers, avoid resistance, improve cytotoxic activity, and may work as chemosensitizers of chemotherapeutic drugs under specific microenvironmental conditions. Aminoquinolones induce ATM-ATR/p53/p21 signaling and exhibit overlapping apoptotic cascade capacity (Ferreira et al. 2021). Artemisinins have also shown anticancer properties via multiple modes of action (Augustin et al. 2020).

11.2.2.5 Antiviral Drugs Repurposed for Cancer

Among different non-nucleoside reverse transcriptase inhibitors, efavirenz was found to have the highest potential against cancer treatment (Hecht et al. 2015). Human papillomavirus vaccination is one of the effective ways to prevent cervical cancer (Kaarthigeyan 2012).

11.2.2.6 Antidiabetic Drugs Repurposed for Cancer

Different antidiabetic agents have different effects on cancer. Metformin is reported to decrease the risk of cancer (Evans et al. 2005), whereas sulfonylurea and insulin are linked with increased risk (Home 2013; Thakkar et al. 2013). Inhibition of overexpressed sodium-glucose cotransporter 2 (SGLT2) in cancers can prevent its progression. Though glucose uptake suppression is the chief mode of antidiabetic action of SGLT2 inhibitors, they have also revealed the potential to alleviate cancer by mechanisms like β-catenin, PI3K-Akt suppression, apoptosis, and oxidative phosphorylation downregulation (Basak et al. 2023).

11.2.2.7 Cardiovascular Drugs

Cardiac Glycosides

It was identified that cardiac glycosides can induce immunogenic cell death, which was associated with the inhibition of Na+/K+-ATPase. They intensify the antineoplastic effect of DNA-damaging agents in immunocompetent mice. Digoxin administration during chemotherapy exhibited a positive effect on survival in colorectal, breast, hepatocellular carcinoma, and head and neck cancer patients, mainly when treated with drugs other than oxaliplatin and anthracyclines (Menger et al. 2012). However, there are reports that digoxin can escalate the cancer incidences (Biggar et al. 2011). Another cardiac glycoside, ouabain (Na+/K+-ATPase inhibitor), exhibits its anticancer effects via apoptosis, autophagic cell death, and migration inhibition. It has been found to exhibit anticancer activity against various types of cancers (Wang et al. 2021b).

Antihyperlipidemic Drugs Repurposed for Cancer

Statins (rosuvastatin, simvastatin, pravastatin, fluvastatin, etc.) are cholesterol-lowering drugs that inhibit a rate-limiting enzyme (3-hydroxy-3-methylglutaryl-coenzyme A (HMG-CoA)) of the mevalonate pathway, thus lowering the biosynthesis of cholesterol. The mechanism of action pathway overlaps with some cancer pathways, which makes it a potential anticancer agent. Evidence shows that specific cancer cells depend on the mevalonate pathway for survival and growth, which makes them vulnerable to statin therapy (Longo et al. 2020). In addition to cholesterol, the pathway leads to the formation of several other byproducts. Statins, thus, also inhibit geranylgeranyl pyrophosphate and farnesyl pyrophosphate and ultimately cause apoptosis in the tumor cells (Martirosyan et al. 2010). Other drugs, such as fenofibrate, have also been effective in treating cancer (Lian et al. 2018).

Antihypertensive Drugs Repurposed for Cancer

Angiotensin II type 1 (AT1) receptors are identified in renal cell carcinoma, melanoma, ovarian cancer, pancreatic cancer, breast cancer, laryngeal carcinoma, and intrahepatic cholangiocarcinoma. Studies have also shown the colocalization of VEGF and AT1 receptors in pancreatic cancer. Renin-angiotensin system controlled VEGF and progression of the tumor. Several in vitro and in vivo cancer studies have demonstrated its blockage to be a promising cancer treatment method (Ishida et al. 2016).

11.2.2.8 Anti-inflammatory Drugs Repurposed for Anticancer

One of the hallmarks of cancer is cancer-related inflammation (Colotta et al. 2009). Several commercial drugs have exhibited anticancer activity in in vitro, in vivo, and epidemiological studies (Table 11.1). Several mechanisms have been proposed. Aspirin works through COX-2 pathways as well as modulating other cancer-associated pathways (Sankaranarayanan et al. 2020).

11.2.2.9 Antipsychotic Drugs in Cancer Therapeutics

Antipsychotic drugs such as chlorpromazine, perphenazine, prochlorperazine, and fluphenazine can inhibit the viability of various cancer cell lines. Possible targets of the phenothiazine derivatives are AKT and Wnt pathways and efflux pumps (ABCB1 and P-gp). The drugs have the potential to stimulate apoptosis, fragment DNA, inhibit migration and invasiveness of cells, and impair ROS production (Otręba and Kośmider 2021).

11.2.2.10 Antidepressant Drugs in Cancer Therapeutics

Cancer-related depression deteriorates the quality of life and increases physical distress. Antidepressants are often used in conjunction with anticancer therapy. Various antidepressants, including monoamine oxidase inhibitors, tricyclic antidepressants, selective serotonin reuptake inhibitors, tetracyclic antidepressants, selective 5-HT reuptake inhibitors, and norepinephrine reuptake inhibitors, have exhibited anticancer potential. Antidepressants have exhibited antitumor action through different mechanisms, which may include Ca^{2+}-mediated cell apoptosis, Fas death

Table 11.1 Different non-oncology repurposed against cancer

Original category of drug	Drug	Repurposed mechanism	Repurposed against	Study type	References
Antibacterial	Roxithromycin	Suppression of angiogenesis via VEGF production inhibition	Hepatoma	In vivo, in vitro	(Aoki et al. (2005)
	Azithromycin	Disrupts dynamics of intracellular keratin-18, suppression of lysosomal trafficking along microtubules	Lung cancer, multiple myeloma, adenocarcinoma	In vitro, in vivo	Takano et al. (2023)
	Clarithromycin	Blocks autophagic flux	Colorectal cancer	In vivo	Petroni et al. (2019)
	Cephalexin	Increases tumor cell-reactive oxygen species; Enhanced therapeutic ratio of radiotherapy	Breast cancer, head and neck cancer, melanoma	In vitro, in vivo	Labay et al. (2016)
	Clofoctol	Krüppel-like factor 13 upregulation	Glioma	In vivo	Hu et al. (2019)
Antifungal	5-Fluorocytosine	Gene-directed enzyme prodrug therapy; conversion to 5-FU	Glioblastoma	In vivo (mice)	Miller et al. (2002)
	Itraconazole	Hedgehog signaling pathway inhibition	Basal cell carcinoma	Phase-II trial	Weng et al. (2023)
Anthelmintic	Flubendazole	Caspases 3,7 activation G2/M phase arrest STAT3 dysregulation	Triple-negative breast cancer	In vitro	Mohi-Ud-Din et al. (2023)
	Albendazole	–	Advanced malignancies	Phase-I trial	Pourgholami et al. (2010)
	Niclosamide	Abrogated WNT7A/β-catenin signaling	Ovarian cancer	In vivo	King et al. (2015)
	Ivermectin	AKT-mTOR signaling pathway inhibition	Breast cancer	In vitro, in vivo	Wang et al. (2016)

(continued)

Table 11.1 (continued)

Original category of drug	Drug	Repurposed mechanism	Repurposed against	Study type	References
Antimalarial	Pyronaridine	Apoptosis, Interference with cell cycle progression, Intercalation with cellular DNA	Breast cancer Hematological cancer cells	In vitro	Villanueva et al. (2018)
	Pyronaridine	Topoisomerase II inhibition	Breast cancer	In vitro, in vivo	Villanueva et al. (2021)
	Amodiaquine (as inhalable polymeric NPs)	Apoptosis induction	Non-small-cell lung cancer	In vitro	Parvathaneni et al. (2020)
	Amodiaquine	Overexpression of LC3B-II, NOXA, GADD34, DDIT4, BNIP3, BNIP3L Downregulation of genes such as POLR1A, POLR1D, POLR1E, CDKN1A TAF1B, LAMP1	Breast cancer	In vitro	Parvathaneni et al. (2022)
	Ferroquine	Negative regulation of Akt kinase, HIF-1α	Prostate tumor	In vivo	Kondratskyi et al. (2017)
	Artesunate	STAT-3 inhibition	Hepatocellular carcinoma	In vitro	Ilamathi et al. (2016)
	Chloroquine	Induction of p53-dependent apoptosis Inhibition of ATP-binding cassette family protein	Many types of cancer	–	Zhang et al. (2015)
	Chloroquine and hydrochloroquine	Affect p53, Toll-like receptor 9, and CXCR4-CXCL12 pathway in cancer cells	Different types of cancer	–	Verbaanderd et al. (2017)

Original category of drug	Drug	Repurposed mechanism	Repurposed against	Study type	References
	Hydrochloroquine	Reversed drug sequestration in lysosomes Enhanced CD8+ T-cell immune response	Non-small-cell lung cancer	In vitro, in vivo	Villanueva et al. (2018)
	Mefloquine	Inhibition of NF-κB signaling pathway	Colorectal cancer	In vitro	Xu et al. (2018)
		Autophagy inhibition	Breast cancer	In vitro	Sharma et al. (2012)
		G1 cell cycle arrest Cyclin D1 accumulation via p21 upregulation	Prostate cancer	In vitro, in vivo	Yan et al. (2013)
Antiviral	Abacavir (antiretroviral)	Triggers senescence	Prostate cancer	In vitro	Carlini et al. (2010)
		Senescence Downregulation of telomerase activity	Medulloblastoma	In vitro	Rossi et al. (2009)
	Efavirenz	Mitochondrial toxicity to cancer cells	Pancreatic cancer	In vitro	Hecht et al. (2015)
		Linked to fatty acid metabolism pathway	Triple-negative breast cancer	In vitro	Chiou et al. (2021)
	Indinavir	Inhibition of heat shock protein 90 function	HER2-positive breast cancer	In vitro, in vivo	Shim et al. (2012)
	Maraviroc	Capable of inhibiting MDR1/P-gp	Ovarian cancer	In vitro, in vivo	Casagrande et al. (2024)
	Nelfinavir	Induction of unfolded protein response, inhibition of mTOR, reduction in achaete-scute homolog-1 expression	Small-cell lung cancer	In vitro, in vivo	Kawabata et al. (2021)

(continued)

Table 11.1 (continued)

Original category of drug	Drug	Repurposed mechanism	Repurposed against	Study type	References
Anti-inflammatory	Aspirin	Targeted heparanase	Breast cancer	In vitro, in vivo	Dai et al. (2017)
	Aspirin	–	Colorectal adenomas	Randomized trial	Baron et al. (2003)
	Aspirin	Bax activation	Gastric cancer	In vitro	Gu et al. (2005)
	Celecoxib	Cox-2 inhibition, Apoptosis induction	Various cancer types	–	Jendrossek (2013)
		Apoptosis	Breast cancer	In vitro	Wang et al. (2017)
		Inhibition of L1CAM, STAT3, and the NF-κB signaling pathway	Pancreatic cancer	In vitro, in vivo	Zuo et al. (2018)
	Dexamethasone	DR5 downregulation, c-FLIP(L) upregulation	Renal, lung, and breast cancer cells	In vitro	Jeon et al. (2020)
	Piroxicam	Regulating membrane properties	Colon cancer	In vivo	Saini and Sanyal (2014)
	Ibuprofen	Activation of nuclear β-catenin	Colon adenomas	Clinical (adenoma sections)	Greenspan et al. (2011)
	Sulindac	Induction of NF-κB, AP-1 signaling, and apoptosis	Colon cancer	In vitro	Mladenova et al. (2013)
Antidiabetic	Acarbose	Enhancement of CD8 T-cell antitumor immunity	Kidney tumor	In vivo	Orlandella et al. (2020)
	Canagliflozin (SGLT-2 inhibitor)	Blocks mitochondrial respiration, inhibits MAPK and mTOR-p70S6k/4EBP1 pathways, promotes AMPK activity, HIF-1α suppression	Prostate cancer	In vitro, in vivo	Ali et al. (2023)
	Canagliflozin (SGLT-2 inhibitor)	Reduction in mitochondrial respiration	Lung, prostate, breast, colon, liver, and ovarian cancer	In vitro	Villani et al. (2016)

Original category of drug	Drug	Repurposed mechanism	Repurposed against	Study type	References
	Dapagliflozin (SGLT-2 inhibitor)	Apoptosis via caspase-dependent pathway, Downregulation of $cFLIP_L$ Increases $cFLIP_s$ stability	Kidney cancer	In vitro	Jang et al. (2022)
	Epalrestat (aldose reductase inhibitor)	Inhibition of AKR1B	Cervical caner	In vitro	Jeon et al. (2020)
	Glibenclamide	G0/G1 arrest p27 levels upregulation Cyclin E diminution	Breast cancer	In vitro	Núñez et al. (2013)
	Glipizide	Angiogenesis inhibition by upregulating natriuretic peptide receptor A expression	Breast and renal cancer	In vivo (chick embryo, mice)	Qi et al. (2014)
	Linagliptin	Decreases Ki67 expression (a nuclear protein in proliferative cells) Promotion of cell apoptosis	Colorectal, breast, and liver cancer	In vitro	Li et al. (2020)
	Liraglutide	Activation of natural killer cells Suppression of IL-6/STAT3 signaling pathway	Hepatocellular carcinoma	In vitro, in vivo	Lu et al. (2021b)
	Metformin	Inhibition of mitochondrial complex I Energy depletion Activation of AMPK Inhibition of mTOR	Many types of cancer	–	Hua et al. (2023)
	Pioglitazone	Affects TGFβ pathway	Non-small-cell lung cancer	In vitro, ex vivo	Ciaramella et al. (2019)

(continued)

Table 11.1 (continued)

Original category of drug	Drug	Repurposed mechanism	Repurposed against	Study type	References
	Repaglinide	Via autophagic, apoptotic, and immune checkpoint signaling	Glioblastoma multiforme	In vitro, in vivo	Xiao et al. (2017)
	Rosiglitazone	Facilitated antitumor effect of 5-FU via PPARγ signaling pathway activation	Hepatocellular carcinoma	In vitro	Cao et al. (2009)
	Sitagliptin	Dipeptidyl peptidase 4 inhibition in glioma	Glioma	In vitro, in vivo	You et al. (2023)
Cardiac glycoside	Ouabain	Apoptosis induction G2/M arrest p53 and MAPK signaling pathway can play a role	Melanoma	In vitro	Wang et al. (2021b)
Antihypertensive	Candesartan	VEGF expression inhibition, angiotensin II type 1 receptor inhibition	Prostate cancer	In vivo, in vitro	Alhusban et al. (2014)
Antihyperlipidemic	Fenofibrate	Apoptosis	Breast cancer, liver cancer, glioma, prostate cancer,	–	Lian et al. (2018)
	Fluvastatin	Inhibition of HMG-CoA reductase	Breast cancer	In vitro, in vivo	Yu et al. (2020)
	Lovastatin	Apoptosis by blocking HMG-CoA reductase activity Increased sensitivity to chemotherapeutics in resistant cells	Ovarian cancer	In vitro	Martirosyan et al. (2010)
	Pravastatin	Upregulation of apolipoprotein A1	Gastric cancer, colorectal cancer	In vitro, in vivo	Yeh et al. (2016)

Original category of drug	Drug	Repurposed mechanism	Repurposed against	Study type	References
Antipsychotic	Chlorpromazine	Blocks autophagic flux Induced cytotoxic autophagy G2/M cell cycle arrest	Colorectal cancer, pulmonary metastasis	In vitro, in vivo	Xu et al. (2022)
	Fluphenazine	Targets ABCB1 and P-glycoprotein	Liver, lung, colon, brain, breast, skin, ovarian cancer, and leukemia	–	Duarte and Vale (2022)
	Fluspirilene	CDK2 inhibition	Hepatocellular carcinoma	In silico, in vitro, in vivo	Shi et al. (2015)
Anti-Parkinsonism	Benztropine	Reducing STAT3, NF-κB, and β-catenin Reducing cancer stem cell properties	Colorectal cancer	In vitro, In vivo	Ramachandran and Srivastava (2020)
	Pimavanserin	Inhibition of Akt/Gli1 signaling axis	Pancreatic cancer	In vivo	Ramachandran and Srivastava (2020)
Antidepressants	Citalopram	Apoptosis, cytochrome c release, activation of NFκB	Hepatocellular carcinoma	In vitro	Ahmadian et al. (2017)
	Fluvoxamine	Suppression of FAK and Akt	Glioblastoma	In vitro, in vivo	Hayashi et al. (2016)
	Imipramine	Blocking ER-α signaling Inhibition of RAD51 expression (TNBC)	ER+ breast cancer	In vitro, in vivo	Timilsina et al. (2022)
	Maprotiline	Cellular retinoic acid binding protein 1 targeting	Hepatocellular carcinoma	In vitro, in vivo	Zheng et al. (2021)
	Phenelzine	–	Castrate-sensitive prostate cancer	Phase II	Gross et al. (2021)
	Sertraline	Autophagy	Leukemia	In vitro	Xia et al. (2017)
		Apoptosis	Colorectal cancer	In vitro, in vivo	Gil-Ad et al. (2008)

(continued)

Table 11.1 (continued)

Original category of drug	Drug	Repurposed mechanism	Repurposed against	Study type	References
Phytonutrient	Curcumin	Tumor suppressor upregulation, oncogene downregulation, apoptosis induction, invasion, metastasis inhibition, drug resistance reversion, miRNA modulation	Many cancer types	–	Amiri and Sahebkar (2024)
	Quercetin	Autophagy-mediated ferroptosis	Gastric cancer	In vitro	Huang et al. (2024)
		Apoptotic, p53, NF-κB, MAPK, JAK/STAT, PI3K/AKT, and Wnt/β-catenin	Many cancer types	–	Asgharian et al. (2022)
	Piperine	Antimutagenic Inhibition of MRP-1	Various cancer types	–	Manayi et al. (2018)
	β-sitosterol	Affects several cancer pathways	Breast, lung, prostate, and colon cancer	–	Khan et al. (2022)
	Genistein	Affects multiple cellular signaling pathways such as Bax/Bcl-2, PTK, Akt, NF-κB, and MMP.	Ovarian cancer	–	Lee et al. (2012)
	Epigallocatechin-3-gallate	ROS production modulation NF-κB inhibition Epigenetic changes	Many cancer types	–	Min and Kwon (2014)
	Camptothecin	Targeting DNA topoisomerase I	Many types of cancer	–	Ghanbari-Movahed et al. (2021)

Original category of drug	Drug	Repurposed mechanism	Repurposed against	Study type	References
	Lycopene	Unproven, probably anti-oxidation	Lung, stomach, and prostate gland and is suggestive of cancers of the cervix, breast, oral cavity, pancreas, colorectum, and esophagus.	–	Giovannucci (1999)
		Inactivating ROS	Many cancer types	–	Trejo-Solís et al. (2013)
	Luteolin	Modulating various signaling pathways	Liver cancer, glioblastoma, gastric cancer, prostate cancer, colon cancer	–	Çetinkaya and Baran (2023)
	Resveratrol	Downregulation of β-catenin Suppression of TGF-β/Smad-induced epithelial-mesenchymal transition Lowers IKK-induced TNF-β expression Inhibition of p-PI3K/p-AKT-mediated FOXO3a nuclear accumulation Suppression of phosphorylation of Src-STAT3 Inhibition of AKT/MAPK-induced HIF-1α activation	Breast, prostate, colon, hepatic, skin, lung, and hematological cancer	–	Ren et al. (2021)

receptor-mediated cell apoptosis, nerve growth factor receptor-mediated cell apoptosis, NK-kappa B, ERK signaling pathway, by alternating immune response or cancer cell microenvironment (Zheng et al. 2023).

Monoamine oxidase (MAO) is a mitochondrial-bound protein that is best known for neurotransmitter regulation. It has two isoenzymes (MAO-A and MAO-B). MAO-A inhibitors are used in the treatment of atypical depression. MAO-A is correlated with CD44, found to be highly expressed in high-grade prostate cancer, and may contribute to the growth of high-grade cancer. Antidepressant drugs targeting MOA may help treat prostate cancer (Peehl et al. 2008). MAO-A inhibitors (particularly clorgyline and phenelzine) are found to be effective in decreasing MOA-A activity and growth of LNCaP, C4-2B, and 22Rv1 cells of prostate cancer (Gaur et al. 2019). The expression of MAO-A is also found to be increased in human glioma cells or tissue specimens. MAO inhibitors (clorgyline) reduced the progression of temozolomide-resistant glioma. Its efficacy is expected to result from increased cytotoxicity and decreased tumoral cell invasion. It can be used either alone or in combination with low-dose temozolomide for temozolomide-resistant gliomas (Kushal et al. 2016). MOA-A is strongly expressed in classical Hodgkin's lymphoma, weak expression in mediastinal large B-cell lymphoma, and mediastinal gray-zone lymphoma. No expression of MAO-A was found in nodular lymphocyte-predominant Hodgkin's lymphoma, non-neoplastic lymphoid tissues, or any other non-Hodgkin's lymphoma. Clorgyline (MAO-A inhibitor) reduced the growth of Hodgkin's lymphoma cell lines (L1236 cells and U-HO1 cells). The combined treatment of clorgyline with doxorubicin/dacarbazine/vinblastine/bleomycin was found to be more effective (Li et al. 2017). MOA-A inhibitors also inhibited breast cancer proliferation and suppressed MOA-A. MAO-A is reported to be involved in several genetic regulations in breast cancer. The inhibitors suppressed not only MAO-A but also the gene expressions of VEGF and Bcl-2 (Alkhawaldeh and Bardaweel 2023). However, in hepatocellular carcinoma (HCC), the expression of MAO-A was downregulated in the clinical samples of HCC. It was closely related to metastasis, vasoinvasion, and poor prognosis. Adrenergic receptor-mediated transactivation of EGFR was also involved in NE-induced invasion of HCC and inhibition of anoikis (Li et al. 2014). In vitro cell culture results have shown the potentiality of amitriptyline against HepG2 cells (mitochondria dysfunction and oxidative stress) (Villanueva-Paz et al. 2016). Selective serotonin reuptake inhibitors (citalopram, fluoxetine, and sertraline) have exhibited a reversal effect on multidrug resistance proteins (MRP-1 and MRP-7) in the presence of chemotherapies. This makes them a potential agent to target multidrug resistance in cancer cells (Bin Kanner et al. 2023).

11.2.2.11 Antineurodegenerative Drugs

Some of the drugs primarily used for treating neurodegenerative disorders (such as anti-Parkinson's disease) are also found to have antitumor activity. Examples include benztropine (Fu et al. 2022) and pimavanserin (Table 11.1).

11.2.2.12 Phytonutrients in Cancer Therapeutics

Phytonutrients can be divided into different chemical classes, such as alkaloids, polyphenols (phenolic acids, lignans, stilbenes, tannins, coumarins, and flavonoids), organosulfur compounds, nitrogen-containing compounds, phytosterols, and carotenoids. Several phytonutrient molecules have been shown to exhibit cancer-preventive and therapeutic activities (Heber and Seeram 2021). The anticancer activity of these phytonutrients depends on multitarget mechanisms. The anticancer phytonutrients/bioactive include curcumin, quercetin, piperin, vincristine, vinblastine, β-sitosterol, genistein, epigallocatechin-3-gallate, camptothecin, lycopene, luteolin, resveratrol, sulforane, 3,3′-diindolylmethane, and indole-3-carbinol (Adetunji et al. 2023).

11.2.2.13 Miscellaneous

PDE-5 Inhibitors

Sildenafil, vardenafil, and tadalafil are phosphodiesterase-5 inhibitors. Sildenafil was initially designed for treating angina and hypertension but is most commonly commercialized for the treatment of erectile dysfunction. They have an additional role in cancer. They can act as adjuvants to chemotherapy (Cruz-Burgos et al. 2021). Nitric oxide/phosphodiesterase-5-based pro-apoptotic effect may be one of the most common mechanisms. Autophagy and antitumor immunity modulation are the other pathways triggering the anticancer mechanism of sildenafil (Iratni and Ayoub 2021).

Disulfiram

Disulfiram is currently used in chronic alcoholism treatment. It has also been found to exhibit anticancer activity. It is found to be effective in non-small-cell lung cancer, glioblastoma, breast cancer, pancreatic cancer, liver cancer, prostate cancer, melanoma, and glioblastoma. Several proposed anticancer mechanisms include mitogen-activated protein kinase pathway activation, ubiquitin-proteosome system suppression, superoxide dismutase inhibition, and triggering oxidative stress by reactive oxygen species generation. It has also been shown to reverse chemotherapeutic resistance by inhibiting the P-glycoprotein efflux pump and suppressing NF-kB signaling activation. Disulfiram inhibits aldehyde dehydrogenase activity and proteasome activity. It also reportedly reduces angiogenesis due to its metal-chelating properties and inactivation of matrix metalloproteinases and Cu/Zn superoxide dismutase. Disulfiram not only kills cancer cells but can also target cancer stem cells. It also can bind to nuclear protein localization protein, inducing its immobilization and dysfunction, ultimately causing cell death (Ekinci et al. 2019; Lu et al. 2021a).

β-Adrenergic Agonists

Salbutamol (β2-agonist) has inhibited breast cancer cell lines and decreased pro-metastatic gene expression in MDA-MB-231 cells. It has also reduced lung metastases in vivo (Rivero et al. 2017). It has also been found to synergize with paclitaxel against breast cancer (Jabloñski et al. 2023). It has been found to improve the prognosis and reduce the risk of prostate cancer (Lehrer and Rheinstein 2024). However, salbutamol is also found to promote tumorigenesis of gastric cancer cells via β2-AR/ERK/EMT pathway (Lu et al. 2022).

11.2.3 Clinical Trials in Cancer DR

Despite advancements in cancer drug development, repurposing existing drugs presents a promising strategy. This method offers expedited, safer, and cost-effective drug options, which are crucial in addressing challenges such as side effects, drug resistance, and therapy expenses. However, not all drug categories are suitable for long-term or individualized use. Consequently, repurposing medications being used chronically appears more favorable. Notably, over 92 clinical trials are underway to assess the anticancer properties of 14 antidiabetic and antihypertensive drugs across 15 cancer types or more. Furthermore, some drugs have progressed to Phase IV trials, indicating potential approval as official anticancer treatments (Hijazi et al. 2023). However, it is noteworthy that an increasing number of incidents of repurposed pharmaceuticals, including some historic medications like sildenafil, aspirin, dimethyl fumarate, and others, have left an indelible mark on medical history (Jourdan et al. 2020). Metformin is the most prominent example of DR owing to its effectiveness in cancer prevention and therapy, either alone or in co-therapy. It has been confirmed by several systematic reviews of case-control, cohort investigations, observational, and clinical trials (Heckman-Stoddard et al. 2017). The doses employed in preclinical studies to achieve antitumor effects may surpass the quantities issued in the initial prescription, which is a limitation for repurposing non-oncology drugs as chemotherapy for cancer. Higher doses are sometimes necessary, but this does not always indicate that they harm people; instead, a dose-finding experiment in the intended population is needed for safety evaluation. Additionally, combination therapy with additional drugs may reduce the doses required to achieve antitumor effects, hence avoiding the issue of raising the initial dosages of each medication (De Lellis et al. 2021). Table 11.2 provides an overview of repurposed drugs that underwent clinical trials for cancer management.

11.3 Future Perspectives and Conclusion

The landscape of cancer therapy is poised for transformative changes as the practice of repurposing existing therapeutics continues to gain momentum. As evidenced by numerous successful case studies highlighted in this chapter, repurposing offers a pragmatic pathway to identify new therapeutic applications for existing drugs, bypassing many of the traditional hurdles associated with drug development. Future directions in this field are anticipated to be driven by AI and ML advancements, enabling more precise and rapid identification of repurposing candidates. Integrating big data analytics with omics technologies like proteomics, genomics, and metabolomics will improve comprehension of pharmacological mechanisms and cancer

Table 11.2 Overview of medications repurposed for cancer treatment

Repurposed drug name	Existing target/disease	Repurposed cancer target	Findings and status	References
Mebendazole	Antiparasitic drug	Metastatic colorectal cancer	After 12 weeks following therapy, there was an overall improvement in responsiveness, but not much after 12 months.	Hegazy et al. (2022)
Metformin	Diabetes	Advanced or metastatic non-small-cell lung cancer (NSCLC)	Metformin treatment was associated with a substantial improvement in progression-free survival ($p = 0.024$), but overall survival did not change appreciably.	Marrone et al. (2018)
Raltegravir	HIV-1 integrase; HIV-infection management	Metnase: adjuvant cancer therapy	There is an ongoing pilot clinical study for squamous cell carcinoma of the head and neck.	Oprea et al. (2011)
Astemizole	Histamine H1 receptors; seasonal allergy treatment	Autophagy-inducing agent used as a prostate cancer adjuvant treatment	There are ongoing in vivo investigations for prostate cancer.	
Ketorolac	Cyclooxygenases; NSAID (nonsteroidal anti-inflammatory drug) for short-term treatment of pain	RAC and CDC42 GTPases; as adjuvant therapy in cancer	Clinical trial to assess peritoneal distribution in ovarian cancer.	
Phenothiazines	Antipsychotics	VLA-4 (anti-adhesion inhibitors for inflammation and cancer management)	There are ongoing in vivo investigations for hematologic malignancies.	
Pravastatin	Hypercholesterolemia	Hepatocellular carcinoma (HCC)	Overall survival, 19 months (Pravastatin) v 9 months.	Kawata et al. (2001)

(continued)

Table 11.2 (continued)

Repurposed drug name	Existing target/disease	Repurposed cancer target	Findings and status	References
Simvastatin	Hypercholesterolemia	Multiple myeloma	Because of osteoclast activity, the study was stopped early.	Sondergaard et al. (2009)
Itraconazole	Fungal infections	Advanced or recurrent non-small-cell lung cancer (NSCLC)	Improved overall survival vs. pemetrexed monotherapy (median of 8 months vs. 32 months; trial discontinued owing to increased use of pemetrexed in the first-line setting.	Rudin et al. (2013)
Aspirin	Painkillers, strokes, and prevention of coronary problems	Early-stage colorectal carcinoma	A reduction in the probability of developing new adenomas after receiving CRC therapy. More cardiovascular randomized trial results indicate improved overall survival and decreased metastasis in people with CRC and other malignancies.	Sandler et al. (2003)
Thalidomide	Sedative, morning sickness during pregnancy	Multiple myeloma	Thalidomide, in combination with dexamethasone, was used in multiple myeloma patients.	Rehman et al. (2011)
Cimetidine	Duodenal and benign gastric ulceration	Resected colorectal cancer	Improved survival when used with colorectal cancer surgery.	Deva and Jameson (2012)

(continued)

Table 11.2 (continued)

Repurposed drug name	Existing target/disease	Repurposed cancer target	Findings and status	References
Raloxifene	Osteoporosis	Breast cancer	Over 3 years of therapy, the invasive risk was reduced by 76% in postmenopausal women with osteoporosis. Raloxifene was given FDA approval in 2007 to prevent invasive breast cancer.	Cummings et al. (1999)
Zoledronic acid	Osteoporosis	Breast cancer	Reduced recurrence and death from hormone receptor-positive breast cancer in postmenopausal women	Coleman et al. (2020)
Propranolol	High blood pressure, irregular heartbeat	Angiosarcoma	Propranolol by itself may be effective in treating angiosarcoma before regular therapy, according to pilot research. To gauge its effect, propranolol is administered immediately upon diagnosis until regular treatment starts.	Embaby et al. (2022)
Digoxin	Arrhythmia and heart failure	Melanoma	Digoxin and trametinib were administered parallelly to individuals with metastatic BRAF wild-type melanoma.	Frankel et al. (2017)

(continued)

Table 11.2 (continued)

Repurposed drug name	Existing target/disease	Repurposed cancer target	Findings and status	References
Doxycycline	Antibiotic	Melanoma	Patients with melanoma who received temozolomide, ipilimumab, and doxycycline concurrently did not see any significant clinical improvement (NCT01590082).	Cortés et al. (2020)
Itraconazole	Antimycotic	Skin cancer	The effects of itraconazole on skin cancer are being investigated in three clinical studies. Two (NCT02120677 and NCT02735356) concentrate on its molecular effects when administered locally to basal cell carcinoma (BCC), while another (NCT02354261) assesses its safety and effectiveness when taken orally by BCC patients.	

biology, hence facilitating the development of personalized medicine strategies. Moreover, collaborative efforts across academia, industry, and regulatory bodies will be crucial in streamlining the repurposing pipeline from discovery to clinical implementation. Expanding the repertoire of repurposed drugs through international consortia and public-private partnerships can further catalyze innovation and expedite the delivery of novel cancer therapies. Additionally, exploring combination therapies that pair repurposed drugs with existing cancer treatments holds promise for overcoming resistance mechanisms and improving patient outcomes.

Repurposing therapeutics for anticancer activity represents a paradigm shift in oncology, offering a promising path to fast-track the discovery of effective and affordable treatments. This approach harnesses existing pharmacological knowledge, mitigating the expensive and lengthy timelines typically accompanying the de novo drug development. This chapter has illuminated the potential and practicalities of DR in cancer therapy through detailed insights.

In conclusion, while challenges such as intellectual property rights, funding constraints, and the need for robust clinical validation remain, the paradigm of DR offers a transformative approach to addressing the complexities of cancer treatment. Building upon existing knowledge and resources, DR promises to deliver innovative therapies and embraces the potential to redefine the landscape of cancer care, offering renewed optimism to patients worldwide. As researchers continue to explore and refine these strategies, the future holds tremendous promise for leveraging existing therapies to combat cancer more effectively and efficiently than ever before. Ultimately, this innovative approach promises to deliver new hope to cancer patients worldwide, driving the evolution of cancer treatment toward more effective, personalized, and accessible therapies.

References

Abbas K et al (2021) Application of network link prediction in drug discovery. BMC Bioinform 22(1):187

Adetunji OA et al (2023) Role of phytonutrients in chemoprevention of cancer: a review. In: Curative and preventive properties of medicinal plants, pp 353–368

Ahmadian E et al (2017) Anti-cancer effects of citalopram on hepatocellular carcinoma cells occur via cytochrome C release and the activation of NF-kB. Anti Cancer Agents Med Chem 17(11):1570–1577

Ahmed F et al (2022) A comprehensive review of artificial intelligence and network based approaches to drug repurposing in Covid-19. Biomed Pharmacother 153:113350

Ahmed F et al (2023a) Drug repurposing for viral cancers: a paradigm of machine learning, deep learning, and virtual screening-based approaches. J Med Virol 95(4):e28693

Ahmed F et al (2023b) A systematic review of computational approaches to understand cancer biology for informed drug repurposing. J Biomed Inform 142:104373

Alavi SE, Ebrahimi Shahmabadi H (2021) Anthelmintics for drug repurposing: opportunities and challenges. Saudi Pharm J 29(5):434–445

Alhusban A et al (2014) Clinically relevant doses of candesartan inhibit growth of prostate tumor xenografts in vivo through modulation of tumor angiogenesis. J Pharmacol Exp Ther 350(3):635–645

Ali A et al (2023) The SGLT2 inhibitor canagliflozin suppresses growth and enhances prostate cancer response to radiotherapy. Commun Biol 6(1):919

Alkhawaldeh A, Bardaweel S (2023) Molecular investigation of the antitumor effects of monoamine oxidase inhibitors in breast cancer cells. Biomed Res Int 2023:2592691

Allen BK, Ayad NG, Schürer SC (2019) Kinome-wide activity classification of small molecules by deep learning. bioRxiv:512459

Amiri A, Sahebkar A (2024) 5 - an overview of cellular, molecular, and biological aspect(s) of curcumin in cancer. In: Kesharwani P, Sahebkar A (eds) Curcumin-based nanomedicines as cancer therapeutics. Academic Press, pp 93–111

Amiri Souri E et al (2022) Novel drug-target interactions via link prediction and network embedding. BMC Bioinform 23(1):121

Amjad E et al (2020) Systems biology comprehensive analysis on breast cancer for identification of key gene modules and genes associated with TNM-based clinical stages. Sci Rep 10(1):10816

Angermueller C et al (2016) Deep learning for computational biology. Mol Syst Biol 12(7):878

Anusewicz D, Orzechowska M, Bednarek AK (2020) Lung squamous cell carcinoma and lung adenocarcinoma differential gene expression regulation through pathways of Notch, Hedgehog, Wnt, and ErbB signalling. Sci Rep 10(1):21128

Aoki D et al (2005) Roxithromycin inhibits angiogenesis of human hepatoma cells <em>in vivo</em> by suppressing VEGF production. Anticancer Res 25(1A):133–138

Asgharian P et al (2022) Potential mechanisms of quercetin in cancer prevention: focus on cellular and molecular targets. Cancer Cell Int 22(1):257

Augustin Y, Staines HM, Krishna S (2020) Artemisinins as a novel anti-cancer therapy: targeting a global cancer pandemic through drug repurposing. Pharmacol Ther 216:107706

Baig MH et al (2018) Computer aided drug design and its application to the development of potential drugs for neurodegenerative disorders. Curr Neuropharmacol 16(6):740–748

Barabasi AL, Gulbahce N, Loscalzo J (2011) Network medicine: a network-based approach to human disease. Nat Rev Genet 12(1):56–68

Barakat A et al (2018) Synthesis, antimicrobial activity, pharmacophore modeling and molecular docking studies of new pyrazole-dimedone hybrid architectures. Chem Cent J 12(1):29

Baron JA et al (2003) A randomized trial of Aspirin to prevent colorectal adenomas. N Engl J Med 348(10):891–899

Barzaman K et al (2020) Breast cancer: biology, biomarkers, and treatments. Int Immunopharmacol 84:106535

Basak D, Gamez D, Deb S (2023) SGLT2 inhibitors as potential anticancer agents. Biomedicine 11(7):1867

Bertolini F, Sukhatme VP, Bouche G (2015) Drug repurposing in oncology—patient and health systems opportunities. Nat Rev Clin Oncol 12(12):732–742

Biggar RJ et al (2011) Digoxin use and the risk of breast cancer in women. J Clin Oncol 29(16):2165–2170

Bin Kanner Y et al (2023) Cytotoxicity and reversal effect of sertraline, fluoxetine, and citalopram on MRP1-and MRP7-mediated MDR. Front Pharmacol 14:1290255

Bolognesi LM (2013) Polypharmacology in a single drug: multitarget drugs. Curr Med Chem 20(13):1639–1645

Bray F et al (2018) Global cancer statistics 2018: GLOBOCAN estimates of incidence and mortality worldwide for 36 cancers in 185 countries. CA Cancer J Clin 68(6):394–424

Brown AS, Patel CJ (2017) MeSHDD: literature-based drug-drug similarity for drug repositioning. J Am Med Inform Assoc 24(3):614–618

Bueso-Bordils JI et al (2015) Topological model for the search of new antibacterial drugs. 158 theoretical candidates. Curr Comput Aided Drug Des 11(4):336–345

Bueso-Bordils JI et al (2017) Topological pattern for the search of new active drugs against methicillin resistant Staphylococcus aureus. Eur J Med Chem 138:807–815

Cao L-Q et al (2009) Rosiglitazone sensitizes hepatocellular carcinoma cell lines to 5-fluorouracil antitumor activity through activation of the PPARγ signaling pathway. Acta Pharmacol Sin 30(9):1316–1322

Carlini F et al (2010) The reverse transcription inhibitor abacavir shows anticancer activity in prostate cancer cell lines. PLoS One 5(12):e14221

Casagrande N et al (2024) In ovarian cancer maraviroc potentiates the antitumoral activity and further inhibits the formation of a tumor-promoting microenvironment by trabectedin. Biomed Pharmacother 172:116296

Çetinkaya M, Baran Y (2023) Therapeutic potential of Luteolin on cancer. Vaccine 11(3):554

Chalkha M et al (2022) Design, synthesis, characterization, in vitro screening, molecular docking, 3D-QSAR, and ADME-Tox investigations of novel pyrazole derivatives as antimicrobial agents. New J Chem 46(6):2747–2760

Chang Y et al (2016) COPD subtypes identified by network-based clustering of blood gene expression. Genomics 107(2–3):51–58

Chen G, Seukep AJ, Guo M (2020) Recent advances in molecular docking for the research and discovery of potential marine drugs. Mar Drugs 18(11):545

Chiou PT et al (2021) Efavirenz as a potential drug for the treatment of triple-negative breast cancers. Clin Transl Oncol 23(2):353–363
Christabel Chikodi E et al (2023) Computational approaches in drug repurposing. In: Mithun R (ed) Drug repurposing. Rijeka, IntechOpen, p Ch. 10
Chuang Y-H et al (2021) Convolutional neural network for human cancer types prediction by integrating protein interaction networks and omics data. Sci Rep 11(1):20691
Ciaramella V et al (2019) Activity and molecular targets of pioglitazone via blockade of proliferation, invasiveness and bioenergetics in human NSCLC. J Exp Clin Cancer Res 38(1):178
Cloughesy TF et al (2016) Phase 1 trial of vocimagene amiretrorepvec and 5-fluorocytosine for recurrent high-grade glioma. Sci Transl Med 8(341):341ra75-341ra75
Coleman R et al (2020) Bone health in cancer: ESMO clinical practice guidelines. Ann Oncol 31(12):1650–1663
Colotta F et al (2009) Cancer-related inflammation, the seventh hallmark of cancer: links to genetic instability. Carcinogenesis 30(7):1073–1081
Cortés H et al (2020) Repurposing of drug candidates for treatment of skin cancer. Front Oncol 10:605714
Cruz-Acuña R et al (2021) Emerging technologies provide insights on cancer extracellular matrix biology and therapeutics. iScience 24(5):102475
Cruz-Burgos M et al (2021) New approaches in oncology for repositioning drugs: the case of PDE5 inhibitor Sildenafil. Front Oncol 11:627229
Cui C et al (2021) Drug repurposing against breast cancer by integrating drug-exposure expression profiles and drug–drug links based on graph neural network. Bioinformatics 37(18):2930–2937
Cummings SR et al (1999) The effect of raloxifene on risk of breast cancer in postmenopausal women: results from the MORE randomized trial. Multiple Outcomes of Raloxifene Evaluation. JAMA 281(23):2189–2197
Dai X, Shen L (2022) Advances and trends in omics technology development. Front Med (Lausanne) 9:911861
Dai X et al (2017) Aspirin inhibits cancer metastasis and angiogenesis via targeting Heparanase. Clin Cancer Res 23(20):6267–6278
de Anda-Jauregui G, Hernandez-Lemus E (2020) Computational oncology in the multi-omics era: state of the art. Front Oncol 10:423
De Lellis L et al (2021) Drug repurposing, an attractive strategy in pancreatic cancer treatment: preclinical and clinical updates. Cancers (Basel) 13(16):3946
Deva S, Jameson M (2012) Histamine type 2 receptor antagonists as adjuvant treatment for resected colorectal cancer. Cochrane Database Syst Rev 8:Cd007814
Doshi S, Chepuri SP (2022) A computational approach to drug repurposing using graph neural networks. Comput Biol Med 150:105992
Drier Y, Sheffer M, Domany E (2013) Pathway-based personalized analysis of cancer. Proc Natl Acad Sci USA 110(16):6388–6393
Duarte D, Vale N (2022) Antipsychotic drug Fluphenazine against human cancer cells. Biomolecules 12(10):1360
Ekinci E et al (2019) Repurposing disulfiram as an anti-cancer agent: updated review on literature and patents. Recent Pat Anticancer Drug Discov 14(2):113–132
Embaby A et al (2022) Beta-adrenergic receptor blockade in angiosarcoma: which beta-blocker to choose? Front Oncol 12:940582
Evans JM et al (2005) Metformin and reduced risk of cancer in diabetic patients. BMJ 330(7503):1304–1305
Fan M et al (2020) Ciprofloxacin promotes polarization of CD86+CD206- macrophages to suppress liver cancer. Oncol Rep 44(1):91–102
Feng Y, Wang Q, Wang T (2017) Drug target protein-protein interaction networks: a systematic perspective. Biomed Res Int 2017:1289259
Ferreira PMP et al (2021) Aminoquinolines as translational models for drug repurposing: anticancer adjuvant properties and toxicokinetic-related features. J Oncol 2021:3569349

Fiscon G et al (2018) Network-based approaches to explore complex biological systems towards network medicine. Genes (Basel) 9(9):437
Frankel AE et al (2017) Digoxin Plus Trametinib therapy achieves disease control in BRAF wild-type metastatic melanoma patients. Neoplasia 19(4):255–260
Fu L et al (2022) Repurposing non-oncology small-molecule drugs to improve cancer therapy: current situation and future directions. Acta Pharm Sin B 12(2):532–557
Galan-Vasquez E, Perez-Rueda E (2021) A landscape for drug-target interactions based on network analysis. PLoS One 16(3):e0247018
Gao Z, Ding P, Xu R (2022) KG-Predict: a knowledge graph computational framework for drug repurposing. J Biomed Inform 132:104133
Gaur S et al (2019) Effect of Monoamine oxidase A (MAOA) inhibitors on androgen-sensitive and castration-resistant prostate cancer cells. Prostate 79(6):667–677
Ghanbari-Movahed M et al (2021) Recent advances in improved anticancer efficacies of Camptothecin Nano-formulations: a systematic review. Biomedicine 9(5):480
Gil-Ad I et al (2008) Evaluation of the potential anti-cancer activity of the antidepressant sertraline in human colon cancer cell lines and in colorectal cancer-xenografted mice. Int J Oncol 33(2):277–286
Gimeno A et al (2019) The light and dark sides of virtual screening: what is there to know? Int J Mol Sci 20(6):1375
Giordano D et al (2022) Drug design by pharmacophore and virtual screening approach. Pharmaceuticals (Basel) 15(5):646
Giovannucci E (1999) Tomatoes, Tomato-based products, lycopene, and cancer: review of the epidemiologic literature. J Natl Cancer Inst 91(4):317–331
Greenspan EJ et al (2011) Ibuprofen inhibits activation of nuclear β-catenin in human colon adenomas and induces the phosphorylation of GSK-3β. Cancer Prev Res 4(1):161–171
Gross ME et al (2021) Phase 2 trial of monoamine oxidase inhibitor phenelzine in biochemical recurrent prostate cancer. Prostate Cancer Prostatic Dis 24(1):61–68
Gu Q et al (2005) Activation of the caspase-8/Bid and Bax pathways in aspirin-induced apoptosis in gastric cancer. Carcinogenesis 26(3):541–546
Gupta P et al (2019) Ciprofloxacin enhances the chemosensitivity of cancer cells to ABCB1 substrates. Int J Mol Sci 20(2):268
Hasin Y, Seldin M, Lusis A (2017) Multi-omics approaches to disease. Genome Biol 18(1):83
Hasman H et al (2015) Detection of mcr-1 encoding plasmid-mediated colistin-resistant Escherichia coli isolates from human bloodstream infection and imported chicken meat, Denmark 2015. Euro Surveill 20(49):30085
Hassan Baig M et al (2016) Computer aided drug design: success and limitations. Curr Pharm Des 22(5):572–581
Hayashi K et al (2016) Fluvoxamine, an anti-depressant, inhibits human glioblastoma invasion by disrupting actin polymerization. Sci Rep 6(1):23372
Heber D, Seeram N (2021) Phytonutrients and cancer. In: Nutritional oncology, pp 45–68
Hecht M et al (2015) Efavirenz has the highest anti-proliferative effect of non-nucleoside reverse transcriptase inhibitors against pancreatic cancer cells. PLoS One 10(6):e0130277
Heckman-Stoddard BM et al (2017) Repurposing metformin for the prevention of cancer and cancer recurrence. Diabetologia 60(9):1639–1647
Hegazy SK et al (2022) Mebendazole; from an anti-parasitic drug to a promising candidate for drug repurposing in colorectal cancer. Life Sci 299:120536
Hideyuki T, Toshikazu U (2019) Accumulation of genetic and epigenetic alterations in normal cells and cancer risk. NPJ Precis Oncol 3(1):7
Hijazi MA, Gessner A, El-Najjar N (2023) Repurposing of chronically used drugs in cancer therapy: a chance to grasp. Cancers (Basel) 15(12):3199
Home P (2013) Insulin therapy and cancer. Diabetes Care 36 Suppl 2(Suppl 2):S240–S244
Hong SJ et al (2019) Factors affecting breast cancer patients' need for genetic risk information: from information insufficiency to information need. J Genet Couns 28(3):543–557

Hsieh CJ et al (2023) Computational chemistry for the identification of lead compounds for radiotracer development. Pharmaceuticals (Basel) 16(2):317
Hu Y et al (2019) The antibiotic clofoctol suppresses glioma stem cell proliferation by activating KLF13. J Clin Invest 129(8):3072–3085
Hua Y et al (2023) Metformin and cancer hallmarks: shedding new lights on therapeutic repurposing. J Transl Med 21(1):403
Huang J, Chen J, Li J (2024) Quercetin promotes ATG5-mediating autophagy-dependent ferroptosis in gastric cancer. J Mol Histol 55(2):211–225
Ilamathi M, Santhosh S, Sivaramakrishnan V (2016) Artesunate as an anti-cancer agent targets stat-3 and favorably suppresses hepatocellular carcinoma. Curr Top Med Chem 16(22):2453–2463
Iratni R, Ayoub MA (2021) Sildenafil in combination therapy against cancer: a literature review. Curr Med Chem 28(11):2248–2259
Ishida J et al (2016) Repurposing of approved cardiovascular drugs. J Transl Med 14(1):269
Jabłoński M et al (2023) The Beta2-adrenergic agonist salbutamol synergizes with paclitaxel on cell proliferation and tumor growth in triple negative breast cancer models. Cancer Chemother Pharmacol 92(6):485–499
Jade DD et al (2022) Ligand-based pharmacophore modeling of TNF-alpha to design novel inhibitors using virtual screening and molecular dynamics. J Biomol Struct Dyn 40(4):1702–1718
Jang JH et al (2022) Dapagliflozin induces apoptosis by downregulating cFILP L and increasing cFILP S instability in Caki-1 cells. Oncol Lett 24(5):1–11
Jendrossek V (2013) Targeting apoptosis pathways by Celecoxib in cancer. Cancer Lett 332(2):313–324
Jeon M-Y et al (2020) Dexamethasone inhibits TRAIL-induced apoptosis through c-FLIP(L) upregulation and DR5 downregulation by GSK3β activation in cancer cells. Cancer 12(10):2901
Jin S et al (2019) A network-based approach to uncover microRNA-mediated disease comorbidities and potential pathobiological implications. NPJ Syst Biol Appl 5(1):41
Jourdan JP et al (2020) Drug repositioning: a brief overview. J Pharm Pharmacol 72(9):1145–1151
Jung K, Lependu P, Shah N (2013) Automated detection of systematic off-label drug use in free text of electronic medical records. AMIA Jt Summits Transl Sci Proc 2013:94–98
Kaarthigeyan K (2012) Cervical cancer in India and HPV vaccination. Indian J Med Paediatr Oncol 33(1):7–12
Karn T et al (2011) Homogeneous datasets of triple negative breast cancers enable the identification of novel prognostic and predictive signatures. PLoS One 6(12):e28403
Kaushik AC et al (2020) Systems biology integration and screening of reliable prognostic markers to create synergies in the control of lung cancer patients. Front Mol Biosci 7:47
Kawabata S et al (2021) Nelfinavir inhibits the growth of small-cell lung cancer cells and patient-derived xenograft tumors. Anticancer Res 41(1):91–99
Kawata S et al (2001) Effect of pravastatin on survival in patients with advanced hepatocellular carcinoma. A randomized controlled trial. Br J Cancer 84(7):886–891
Khan Z et al (2022) Multifunctional roles and pharmacological potential of β-sitosterol: emerging evidence toward clinical applications. Chem Biol Interact 365:110117
Kim S et al (2019) PubChem 2019 update: improved access to chemical data. Nucleic Acids Res 47(D1):D1102–D1109
King ML et al (2015) WNT7A/β-catenin signaling induces FGF1 and influences sensitivity to niclosamide in ovarian cancer. Oncogene 34(26):3452–3462
Kondratskyi A et al (2017) Ferroquine, the next generation antimalarial drug, has antitumor activity. Sci Rep 7(1):15896
Kushal S et al (2016) Monoamine oxidase A (MAO A) inhibitors decrease glioma progression. Oncotarget 7(12):13842–13853
Kwon S et al (2019) Comprehensive ensemble in QSAR prediction for drug discovery. BMC Bioinform 20(1):521
Labay E et al (2016) Repurposing cephalosporin antibiotics as pro-senescent radiosensitizers. Oncotarget 7(23):33919–33933

Lambert AW, Pattabiraman DR, Weinberg RA (2017) Emerging biological principles of metastasis. Cell 168(4):670–691
Lee J-Y, Kim HS, Song Y-S (2012) Genistein as a potential anticancer agent against ovarian cancer. J Tradit Complement Med 2(2):96–104
Lee D et al (2013) Pathway-based analysis using genome-wide association data from a Korean non-small cell lung cancer study. PLoS One 8(6):e65396
Lehrer S, Rheinstein PH (2024) Salbutamol, a short acting beta-2 agonist, reduces risk and improves prognosis of prostate cancer. medRxiv. p. 2024.02. 16.24302956
Li Q, Shah S (2017) Structure-based virtual screening. Methods Mol Biol 1558:111–124
Li J et al (2014) Monoamine oxidase a suppresses hepatocellular carcinoma metastasis by inhibiting the adrenergic system and its transactivation of EGFR signaling. J Hepatol 60(6):1225–1234
Li PC et al (2017) Monoamine oxidase A is highly expressed in classical Hodgkin lymphoma. J Pathol 243(2):220–229
Li F, Liu W, Yu H (2018) Extraction of information related to adverse drug events from electronic health record notes: design of an end-to-end model based on deep learning. JMIR Med Inform 6(4):e12159
Li Y et al (2020) Repositioning of hypoglycemic drug linagliptin for cancer treatment. Front Pharmacol 11:187
Lian X et al (2018) Anticancer properties of fenofibrate: a repurposing use. J Cancer 9(9):1527–1537
Lin Y (2022) Review of modern computer-aided drug design methods. Int J Biol Life Sci 1(1):47–50
Liu Y, Chance MR (2013) Pathway analyses and understanding disease associations. Curr Genet Med Rep 1(4):230–238
Liu J, Wang R (2015) Classification of current scoring functions. J Chem Inf Model 55(3):475–482
Liu Y et al (2012) Gene interaction enrichment and network analysis to identify dysregulated pathways and their interactions in complex diseases. BMC Syst Biol 6:1–14
Liu B et al (2021) iCAV: an integrative database of cancer-associated viruses. Database (Oxford) 2021(2021):baab079
Longo J et al (2020) Statins as anticancer agents in the era of precision medicine. Clin Cancer Res 26(22):5791–5800
Low ZY, Farouk IA, Lal SK (2020) Drug repositioning: new approaches and future prospects for life-debilitating diseases and the COVID-19 pandemic outbreak. Viruses 12(9):1058
Lu C et al (2021a) Disulfiram: a novel repurposed drug for cancer therapy. Cancer Chemother Pharmacol 87(2):159–172
Lu X et al (2021b) Liraglutide activates nature killer cell-mediated antitumor responses by inhibiting IL-6/STAT3 signaling in hepatocellular carcinoma. Transl Oncol 14(1):100872
Lu Y et al (2022) Chronic stress model simulated by salbutamol promotes tumorigenesis of gastric cancer cells through β2-AR/ERK/EMT pathway. J Cancer 13(2):401–412
Macalino SJ et al (2015) Role of computer-aided drug design in modern drug discovery. Arch Pharm Res 38(9):1686–1701
Manayi A et al (2018) Piperine as a potential anti-cancer agent: a review on preclinical studies. Curr Med Chem 25(37):4918–4928
Mansoori B et al (2017) The different mechanisms of cancer drug resistance: a brief review. Adv Pharm Bull 7(3):339–348
Mansoori S et al (2021) A phase 2a clinical study on the safety and efficacy of individualized dosed mebendazole in patients with advanced gastrointestinal cancer. Sci Rep 11(1):8981
Manzoni C et al (2018) Genome, transcriptome and proteome: the rise of omics data and their integration in biomedical sciences. Brief Bioinform 19(2):286–302
Marrone KA et al (2018) A randomized phase II study of Metformin plus Paclitaxel/Carboplatin/Bevacizumab in patients with chemotherapy-Naïve advanced or metastatic nonsquamous non-small cell lung cancer. Oncologist 23(7):859–865
Martirosyan A et al (2010) Lovastatin induces apoptosis of ovarian cancer cells and synergizes with doxorubicin: potential therapeutic relevance. BMC Cancer 10(1):103

Mbemi A et al (2020) Impact of gene-environment interactions on cancer development. Int J Environ Res Public Health 17(21):8089

Menger L et al (2012) Cardiac glycosides exert anticancer effects by inducing immunogenic cell death. Sci Transl Med 4(143):143ra99

Miller CR et al (2002) Intratumoral 5-fluorouracil produced by Cytosine Deaminase/5-Fluorocytosine gene therapy is effective for experimental human Glioblastomas1. Cancer Res 62(3):773–780

Min K-J, Kwon TK (2014) Anticancer effects and molecular mechanisms of epigallocatechin-3-gallate. Integ Med Res 3(1):16–24

Mithun R, Shubham JK, Anil GJ (2020) Drug Repurposing (DR): an emerging approach in drug discovery. In: Farid AB (ed) Drug Repurposing. IntechOpen, Rijeka, p Ch. 1

Mladenova D et al (2013) Sulindac activates NF-κB signaling in colon cancer cells. Cell Commun Signal 11(1):73

Mohi-Ud-Din R et al (2023) Repurposing approved non-oncology drugs for cancer therapy: a comprehensive review of mechanisms, efficacy, and clinical prospects. Eur J Med Res 28(1):345

Momeni Z et al (2020) A survey on single and multi omics data mining methods in cancer data classification. J Biomed Inform 107:103466

Muhammed MT, Esin A-Y (2021) Pharmacophore modeling in drug discovery: methodology and current status. J Turkish Chem Soc Sect A Chem 8(3):749–762

Ng YL, Salim CK, Chu JJH (2021) Drug repurposing for COVID-19: approaches, challenges and promising candidates. Pharmacol Ther 228:107930

Núñez M et al (2013) Glibenclamide inhibits cell growth by inducing G0/G1 arrest in the human breast cancer cell line MDA-MB-231. BMC Pharmacol Toxicol 14(1):6

Nunez SK et al (2021) Identification of gene co-expression networks associated with consensus molecular subtype-1 of colorectal cancer. Cancers (Basel) 13(22):5824

Oprea TI et al (2011) Drug repurposing from an academic perspective. Drug Discov Today Ther Strateg 8(3–4):61–69

Orlandella RM et al (2020) The antidiabetic agent Acarbose improves anti-PD-1 and Rapamycin efficacy in preclinical renal cancer. Cancers (Basel) 12(10):2872

Oselusi SO et al (2022) Computational target-based screening of anti-MRSA natural products reveals potential multitarget mechanisms of action through peptidoglycan synthesis proteins. ACS Omega 7(42):37896–37906

Otręba M, Kośmider L (2021) In vitro anticancer activity of fluphenazine, perphenazine and prochlorperazine. A review. J Appl Toxicol 41(1):82–94

Palop JJ, Chin J, Mucke L (2006) A network dysfunction perspective on neurodegenerative diseases. Nature 443(7113):768–773

Parvathaneni V et al (2020) Development of pharmaceutically scalable inhaled anti-cancer nanotherapy—repurposing amodiaquine for non-small cell lung cancer (NSCLC). Mater Sci Eng C 115:111139

Parvathaneni V et al (2022) Exploring amodiaquine's repurposing potential in breast cancer treatment—assessment of in-vitro efficacy & mechanism of action. Int J Mol Sci:23. https://doi.org/10.3390/ijms231911455

Patel AR et al (2021) Virtual screening in drug discovery. J Veterin Pharmacol Toxicol 20(2):1–9

Peehl DM et al (2008) The significance of Monoamine Oxidase-A expression in high grade prostate cancer. J Urol 180(5):2206–2211

Petroni G et al (2019) Data describing the effects of the macrolide antibiotic Clarithromycin on preclinical mouse models of Colorectal cancer. Data Brief 26:104406

Pinzi L, Rastelli G (2019) Molecular docking: shifting paradigms in drug discovery. Int J Mol Sci 20(18):4331

Pomyen Y et al (2015) Over-representation of correlation analysis (ORCA): a method for identifying associations between variable sets. Bioinformatics 31(1):102–108

Pourgholami MH et al (2010) Phase I clinical trial to determine maximum tolerated dose of oral albendazole in patients with advanced cancer. Cancer Chemother Pharmacol 65(3):597–605

Prachayasittikul V et al (2015) Computer-aided drug design of bioactive natural products. Curr Top Med Chem 15(18):1780–1800
Qi C et al (2014) Glipizide, an antidiabetic drug, suppresses tumor growth and metastasis by inhibiting angiogenesis. Oncotarget 5(20):9966–9979
Raab MS et al (2012) GF-15, a novel inhibitor of Centrosomal clustering, suppresses tumor cell growth in vitro and in vivo. Cancer Res 72(20):5374–5385
Raju R et al (2014) The Triple-Negative Breast Cancer Database: an omics platform for reference, integration and analysis of triple-negative breast cancer data. Breast Cancer Res 16(6):490
Ramachandran S, Srivastava SK (2020) Repurposing pimavanserin, an anti-Parkinson drug for pancreatic cancer therapy. Molecul Ther Oncolyt 19:19–32
Rehman W, Arfons LM, Lazarus HM (2011) The rise, fall and subsequent triumph of thalidomide: lessons learned in drug development. Ther Adv Hematol 2(5):291–308
Ren B et al (2021) Resveratrol for cancer therapy: challenges and future perspectives. Cancer Lett 515:63–72
Rifaioglu AS et al (2020) DEEPScreen: high performance drug–target interaction prediction with convolutional neural networks using 2-D structural compound representations. Chem Sci 11(9):2531–2557
Rivero EM et al (2017) The β 2-Adrenergic agonist salbutamol inhibits migration, invasion and metastasis of the human breast cancer MDA-MB-231 cell line. Curr Cancer Drug Targets 17(8):756–766
Rogers D, Hahn M (2010) Extended-connectivity fingerprints. J Chem Inf Model 50(5):742–754
Rossi A et al (2009) The antiretroviral nucleoside analogue Abacavir reduces cell growth and promotes differentiation of human medulloblastoma cells. Int J Cancer 125(1):235–243
Rudin CM et al (2013) Phase 2 study of pemetrexed and itraconazole as second-line therapy for metastatic nonsquamous non-small-cell lung cancer. J Thorac Oncol 8(5):619–623
Sabe VT et al (2021) Current trends in computer aided drug design and a highlight of drugs discovered via computational techniques: a review. Eur J Med Chem 224:113705
Saini MK, Sanyal SN (2014) Piroxicam and c-phycocyanin prevent colon carcinogenesis by inhibition of membrane fluidity and canonical Wnt/β-catenin signaling while up-regulating ligand dependent transcription factor PPARγ. Biomed Pharmacother 68(5):537–550
Sandler RS et al (2003) A randomized trial of aspirin to prevent colorectal adenomas in patients with previous colorectal cancer. N Engl J Med 348(10):883–890
Sankaranarayanan R et al (2020) Mechanisms of colorectal cancer prevention by Aspirin—a literature review and perspective on the role of COX-dependent and -independent pathways. Int J Mol Sci 21(23):9018
Sansone S-A et al (2019) FAIRsharing as a community approach to standards, repositories and policies. Nat Biotechnol 37(4):358–367
Sastry M et al (2010) Large-scale systematic analysis of 2D fingerprint methods and parameters to improve virtual screening enrichments. J Chem Inf Model 50(5):771–784
Schmeel LC et al (2017) Griseofulvin efficiently induces apoptosis in <em>in vitro</em> treatment of lymphoma and multiple myeloma. Anticancer Res 37(5):2289–2295
Shadkhoo S, Mani M (2019) The role of intracellular interactions in the collective polarization of tissues and its interplay with cellular geometry. PLoS Comput Biol 15(11):e1007454
Shaker B et al (2021) In silico methods and tools for drug discovery. Comput Biol Med 137:104851
Sharma N et al (2012) Inhibition of autophagy and induction of breast cancer cell death by mefloquine, an antimalarial agent. Cancer Lett 326(2):143–154
Shayanfar S, Shayanfar A (2022) Comparison of various methods for validity evaluation of QSAR models. BMC Chem 16(1):1–9
Shi XN et al (2015) In silico identification and in vitro and in vivo validation of anti-psychotic drug Fluspirilene as a potential CDK2 inhibitor and a candidate anti-cancer drug. PLoS One 10(7):e0132072
Shim JS et al (2012) Selective inhibition of HER2-positive breast cancer cells by the HIV protease inhibitor nelfinavir. J Natl Cancer Inst 104(20):1576–1590

Shiroma Y et al (2020) Targeting DNA binding proteins for cancer therapy. Cancer Sci 111(4):1058–1064

Siddiqui S et al (2022) Drug repurposing: re-inventing therapies for cancer without re-entering the development pipeline-a review. J Egypt Natl Canc Inst 34(1):33

Singh S, Baker QB, Singh DB (2022) Molecular docking and molecular dynamics simulation. In: Bioinformatics. Elsevier, pp 291–304

Sivakumar KC et al (2020) Prospects of multitarget drug designing strategies by linking molecular docking and molecular dynamics to explore the protein–ligand recognition process. Drug Dev Res 81(6):685–699

Sompairac N et al (2019) Metabolic and signalling network maps integration: application to cross-talk studies and omics data analysis in cancer. BMC Bioinform 20(Suppl 4):140

Sondergaard TE et al (2009) A phase II clinical trial does not show that high dose simvastatin has beneficial effect on markers of bone turnover in multiple myeloma. Hematol Oncol 27(1):17–22

Song JW, Chung KC (2010) Observational studies: cohort and case-control studies. Plast Reconstr Surg 126(6):2234–2242

Song W et al (2018) Learning node and edge embeddings for signed networks. Neurocomputing 319:42–54

Srivastava N et al (2014) Dropout: a simple way to prevent neural networks from overfitting. J Machine Learn Res 15(1):1929–1958

Stathias V et al (2018) Sustainable data and metadata management at the BD2K-LINCS Data Coordination and Integration Center. Sci Data 5(1):1–14

Suay-Garcia B et al (2020a) Quantitative structure–activity relationship methods in the discovery and development of antibacterials. Wiley Interdiscipl Rev Comput Mol Sci 10(6):e1472

Suay-Garcia B et al (2020b) Tree-based QSAR model for drug repurposing in the discovery of new antibacterial compounds against Escherichia coli. Pharmaceuticals (Basel) 13(12):431

Subramanian I et al (2020) Multi-omics data integration, interpretation, and its application. Bioinform Biol Insights 14:1177932219899051

Sun W, Sanderson PE, Zheng W (2016) Drug combination therapy increases successful drug repositioning. Drug Discov Today 21(7):1189–1195

Suter P et al (2022) Multi-omics subtyping of hepatocellular carcinoma patients using a Bayesian network mixture model. PLoS Comput Biol 18(9):e1009767

Takano N et al (2023) Azithromycin, a potent autophagy inhibitor for cancer therapy, perturbs cytoskeletal protein dynamics. Br J Cancer 128(10):1838–1849

Tandon H, Chakraborty T, Suhag V (2019) A brief review on importance of DFT in drug design. Res Med Eng Stud 39:46

Tang GH et al (2018) Association of Metformin with breast cancer incidence and mortality in patients with type II diabetes: a GRADE-assessed systematic review and meta-analysis. Cancer Epidemiol Biomarkers Prev 27(6):627–635

Tariq K et al (2022) Role of DFT in drug design: a mini review. Drug Des 11:216

Thakkar B et al (2013) Metformin and sulfonylureas in relation to cancer risk in type II diabetes patients: a meta-analysis using primary data of published studies. Metabolism 62(7):922–934

Timilsina S et al (2022) The antidepressant imipramine inhibits breast cancer growth by targeting estrogen receptor signaling and DNA repair events. Cancer Lett 540:215717

Tomar V et al (2019a) Small molecule drug design. In: Ranganathan S et al (eds) Encyclopedia of bioinformatics and computational biology. Academic Press, Oxford, pp 741–760

Tomar V, et al (2019b) Small molecule drug design

Trejo-Solís C et al (2013) Multiple molecular and cellular mechanisms of action of lycopene in cancer inhibition. Evid Based Complement Alternat Med 2013:705121

Turon G et al (2023) First fully-automated AI/ML virtual screening cascade implemented at a drug discovery centre in Africa. Nat Commun 14(1):5736

Vahabi N, Michailidis G (2022) Unsupervised multi-omics data integration methods: a comprehensive review. Front Genet 13:854752

Vailati-Riboni M, Palombo V, Loor JJ (2017) What are omics sciences? In: Periparturient diseases of dairy cows, pp 1–7

Van Mourik T, Bühl M, Gaigeot M-P (2014) Density functional theory across chemistry, physics and biology. In: Philosophical transactions of the Royal Society a: mathematical, physical engineering sciences. The Royal Society Publishing, p 20120488
Varela-Rial A, Majewski M, De Fabritiis G (2022) Structure based virtual screening: fast and slow. Wiley Interdiscipl Rev Comput Mol Sci 12(2):e1544
Verbaanderd C et al (2017) Repurposing Drugs in Oncology (ReDO)-chloroquine and hydroxychloroquine as anti-cancer agents. Ecancermedicalscience 11:781
Vermes A, Guchelaar H-J, Dankert J (2000) Flucytosine: a review of its pharmacology, clinical indications, pharmacokinetics, toxicity and drug interactions. J Antimicrob Chemother 46(2):171–179
Villani LA et al (2016) The diabetes medication Canagliflozin reduces cancer cell proliferation by inhibiting mitochondrial complex-I supported respiration. Mol Metab 5(10):1048–1056
Villanueva PJ et al (2018) Pyronaridine exerts potent cytotoxicity on human breast and hematological cancer cells through induction of apoptosis. PLoS One 13(11):e0206467
Villanueva PJ et al (2021) The antimalarial drug Pyronaridine inhibits Topoisomerase II in breast cancer cells and hinders tumor progression in vivo. Clin Cancer Drugs 8(1):50–56
Villanueva-Paz M et al (2016) Amitriptyline induces mitophagy that precedes apoptosis in human HepG2 cells. Genes Cancer 7(7–8):260–277
Vishnubalaji R et al (2019) Integrated transcriptome and pathway analyses revealed multiple activated pathways in breast cancer. Front Oncol 9:910
Vuorinen A, Schuster D (2015) Methods for generating and applying pharmacophore models as virtual screening filters and for bioactivity profiling. Methods 71:113–134
Wang K et al (2016) Ivermectin induces PAK1-mediated cytostatic autophagy in breast cancer. Autophagy 12(12):2498–2499
Wang G et al (2017) Celecoxib induced apoptosis against different breast cancer cell lines by down-regulated NF-κB pathway. Biochem Biophys Res Commun 490(3):969–976
Wang Y et al (2018) Combining immunotherapy and radiotherapy for cancer treatment: current challenges and future directions. Front Pharmacol 9:185
Wang T et al (2021a) MOGONET integrates multi-omics data using graph convolutional networks allowing patient classification and biomarker identification. Nat Commun 12(1):3445
Wang L et al (2021b) Ouabain exhibited strong anticancer effects in melanoma cells via induction of apoptosis, g2/m phase arrest, and migration inhibition. Onco Targets Ther 14:1261–1273
Wang C et al (2022a) Network-based integration of multi-omics data for clinical outcome prediction in neuroblastoma. Sci Rep 12(1):15425
Wang H et al (2022b) ELIXIR-A: an interactive visualization tool for multi-target pharmacophore refinement. ACS Omega 7(15):12707–12715
Weng N et al (2023) Repurposing antifungal drugs for cancer therapy. J Adv Res 48:259–273
Wilkinson MD et al (2016) The FAIR guiding principles for scientific data management and stewardship. Sci Data 3(1):160018
Wu K-J et al (2019) Mimicking strategy for protein–protein interaction inhibitor discovery by virtual screening. Molecules 24(24):4428
Xia D et al (2017) Sertraline exerts its antitumor functions through both apoptosis and autophagy pathways in acute myeloid leukemia cells. Leuk Lymphoma 58(9):2208–2217
Xiao ZX et al (2017) Identification of repaglinide as a therapeutic drug for glioblastoma multiforme. Biochem Biophys Res Commun 488(1):33–39
Xu H et al (2015) Validating drug repurposing signals using electronic health records: a case study of metformin associated with reduced cancer mortality. J Am Med Inform Assoc 22(1):179–191
Xu X et al (2018) Antimalarial drug mefloquine inhibits nuclear factor kappa B signaling and induces apoptosis in colorectal cancer cells. Cancer Sci 109(4):1220–1229
Xu F et al (2022) Repurposed antipsychotic chlorpromazine inhibits colorectal cancer and pulmonary metastasis by inducing G2/M cell cycle arrest, apoptosis, and autophagy. Cancer Chemother Pharmacol 89(3):331–346
Xue H et al (2018) Review of drug repositioning approaches and resources. Int J Biol Sci 14(10):1232–1244

Yagiz K et al (2016) Toca 511 plus 5-fluorocytosine in combination with lomustine shows chemotoxic and immunotherapeutic activity with no additive toxicity in rodent glioblastoma models. Neuro-Oncology 18(10):1390–1401
Yan KH et al (2013) Mefloquine induces cell death in prostate cancer cells and provides a potential novel treatment strategy in vivo. Oncol Lett 5(5):1567–1571
Yang Y et al (2014) Gene co-expression network analysis reveals common system-level properties of prognostic genes across cancer types. Nat Commun 5:3231
Ye N, Yang Z, Liu Y (2022) Applications of density functional theory in COVID-19 drug modeling. Drug Discov Today 27(5):1411–1419
Yeh C et al (2016) Pravastatin inhibits tumor growth through elevating the levels of apolipoprotein A1. Adv Digest Med 3(1):3–10
You F et al (2023) Sitagliptin inhibits the survival, stemness and autophagy of glioma cells, and enhances temozolomide cytotoxicity. Biomed Pharmacother 162:114555
Yu JS, Shin DH, Kim JS (2020) Repurposing of Fluvastatin as an anticancer agent against breast cancer stem cells via encapsulation in a Hyaluronan-conjugated liposome. Pharmaceutics 12(12):1133
Zanni R et al (2017) Molecular topology: a new strategy for antimicrobial resistance control. Eur J Med Chem 137:233–246
Zeng ISL, Lumley T (2018) Review of statistical learning methods in integrated omics studies (an integrated information science). Bioinform Biol Insights 12:1177932218759292
Zhang Y et al (2015) The utility of chloroquine in cancer therapy. Curr Med Res Opin 31(5):1009–1013
Zhang M et al (2018) Pharmacophore modeling, synthesis, and antibacterial evaluation of Chalcones and derivatives. ACS Omega 3(12):18343–18360
Zhang F, Hu W, Liu Y (2022a) GCMM: graph convolution network based on multimodal attention mechanism for drug repurposing. BMC Bioinform 23(1):372
Zhang B et al (2022b) Molecular docking-based computational platform for high-throughput virtual screening. CCF Trans High Perform Comput 4(1):63–74
Zheng C et al (2021) Maprotiline suppresses cholesterol biosynthesis and hepatocellular carcinoma progression through direct targeting of CRABP1. Front Pharmacol 12:689767
Zheng Y et al (2023) The application of antidepressant drugs in cancer treatment. Biomed Pharmacother 157:113985
Zhong Q et al (2009) Edgetic perturbation models of human inherited disorders. Mol Syst Biol 5(1):321
Zhou K et al (2022) Integration of multimodal data from disparate sources for identifying disease subtypes. Biology 11. https://doi.org/10.3390/biology11030360
Zuo C et al (2018) Celecoxib suppresses proliferation and metastasis of pancreatic cancer cells by down-regulating STAT3 / NF-kB and L1CAM activities. Pancreatology 18(3):328–333

Drug Repurposing for Neurological Disorders

12

Saptapadi Saha, Arun Baidya, Shuvam Sar, Tanmoy Banerjee, Ankit Kumar, Axata Lama, Rituparna Chaki, Amit Halder, and Nilanjan Ghosh

Abstract

Numerous neurological illnesses lack adequate therapy, making them a major healthcare burden. Repurposing older drugs and re-evaluating them for novel therapeutic uses emerge as immediate alternative strategy for meeting the unresolved research gaps in the area of neuroscience since conventional drug research and development may not suffice to meet the demand for being expensive and tedious. Using the well-established safety profiles and pharmacological characterization of licensed drugs, the repurposing strategy can expedite the advancement of potentially beneficial compounds into clinical trials. An overview of current developments in medication repurposing for a range of neurological conditions, including multiple sclerosis (MS), epilepsy, Parkinson's disease (PD), Alzheimer's disease (AD), etc. has been covered in this chapter. The strategies employed to find the most suitable and potent repurposable drug candidates and their most probable mechanisms of action to alleviate the respective neurological diseases have also been reviewed and documented in the following text. We further emphasized noteworthy clinical studies to show how drug repurposing of currently available drugs shows hope in the treatment strategies for neurological illnesses intending to lessen the burden of unsurpassable medical threats while simultaneously giving the required therapy for better disease alleviation.

S. Saha · A. Baidya · S. Sar · T. Banerjee · A. Kumar · A. Lama · N. Ghosh (✉)
Department of Pharmaceutical Technology, Jadavpur University, Kolkata, West Bengal, India
e-mail: nilanjanghosh.phamacy@jadavpuruniversity.in

R. Chaki · A. Halder
Dr. B.C. Roy College of Pharmacy & Allied Health Sciences, Durgapur, India

R. K. Kaundal et al. (eds.), *Drug Repurposing: Exploring Newer Therapeutic Potentials*,
https://doi.org/10.1007/978-981-95-4522-3_12

Keywords

Drug repurposing · Neurological illnesses · Multiple sclerosis · Epilepsy · Parkinson's disease · Alzheimer's disease.

Abbreviation

5-HT	5-Hydroxytryptamine
AD	Alzheimer's Disease
ALS	Amyotrophic Lateral Sclerosis
AMPK	AMP-activated protein kinase
APP	Amyloid precursor protein
BBB	Blood-brain barrier
CNS	Central nervous system
EGFR	Epidermal growth factor receptor
GABA	Gamma-aminobutyric acid
GLP	Glucagon-like peptide
GSK	Glycogen synthase kinase
HDAC	Histone Deacetylase
HD	Huntington's disease
HMG-CoA	3-hydroxy-3-methylglutaryl-coenzyme A
Ig	Immunoglobulin
IL-1β	Interleukin 1β
MAO	Monoamine oxidase
MOA	Mechanism of action
MS	Multiple sclerosis
NAC	N-acetyl cysteine
NMDA	N-methyl-D-aspartate
PD	Parkinson's Disease
PDE	Phosphodiesterase
PKC	Protein kinase C
TBI	Traumatic brain injury
TNF	Tumor necrosis factor

12.1 Introduction

Drug repurposing, often referred to as drug repositioning or drug reprofiling, is a technique used in pharmaceutical research and development that entails finding new therapeutic applications for medications or chemicals that were formerly created for unrelated medicinal reasons. Repositioning aims to fill the deficiencies in the treatment options and has been very beneficial in treating uncommon disorders by investigating alternate therapies for a range of disorders by working on the established toxicity profiles, pharmacokinetics, and mechanisms of action of current

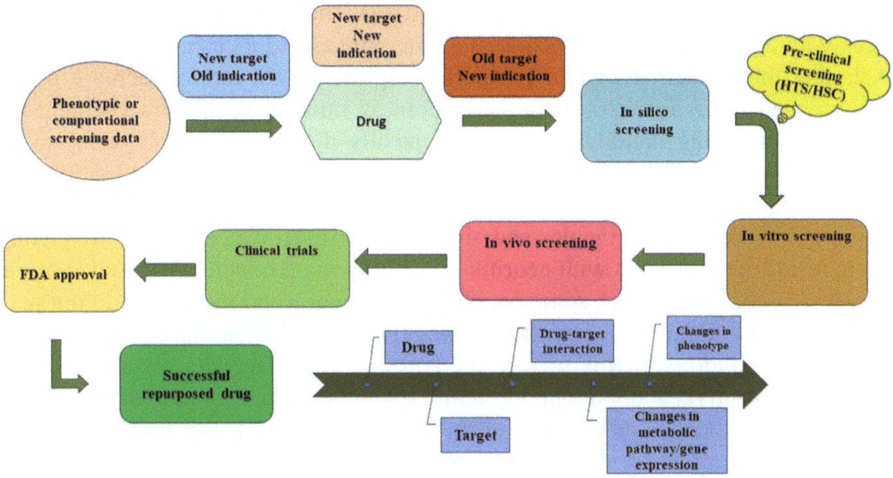

Fig. 12.1 Drug repositioning process

medications rather than designing and developing new ones from scratch. Repurposed medicines have benefits like briefer development times and cheaper research costs. This strategy makes use of the wealth of information and chemicals that are prevalent in the pharmaceutical industry to possibly provide more effective and easily accessible treatments for a range of medical ailments. The overall process of drug repurposing is depicted in Fig. 12.1.

Neurological disorders related to CNS include a broad variety of ailments, such as MS, epilepsy, AD, PD, and more. Scientists and healthcare providers have been searching for the possibility of repurposing medications authorized for other ailments to meet unfulfilled needs in the area of neurological disorders. For instance, studies have been carrying on the possibility of potency of antipsychotic drugs, which were first created to treat schizophrenia, to help individuals with AD, lessening their symptoms. In a similar vein, several antihypertensive medications have shown potential in treating migraine headache symptoms. It makes use of the richness of information and currently accessible pharmacological substances to treat the complicated traits of various neurological diseases.

12.2 Neurological Disorders and Their Complexity

12.2.1 Alzheimer's Disease

AD is a slowly developing neurodegenerative condition that begins in the entorhinal cortex of the hippocampal region and is characterized by a progressive loss of cognitive and behavioral abilities, such as attention, comprehension, judgment, language, memory, and reasoning (Kumar et al. 2022). Dementia is the broad term for this reduction in cognitive abilities, particularly when it becomes severe enough to affect day-to-day functioning (Duong et al. 2017). The likelihood of developing AD rises with age. Usually, older people in China, India, and their neighbors in South Asia and the Western Pacific are experiencing the fastest rate of increase in the occurrence of AD. In 2020, there were about 55 million AD sufferers around the globe. In emerging nations, this figure typically doubles every 20 years and is expected to reach 78 million in 2030 and 139 million in 2050, but by that time, it will account for 71% of the total population (Mayeux and Stern 2012). In AD, abnormal neurofibrillary tangles and amyloid multifocal deposition are frequently seen, resulting in plaques with neuritis in the gray matter, meningeal, and cerebral arteries. On the other hand, some people with dementia and those who do not may have amyloid plaques. A transmembrane protein APP produces extracellular β-amyloid peptide, which is seen in the cores of neurodegenerative plaques. The lipid metabolism regulator apolipoprotein E is another genetic marker that raises the risk of AD. It shows a preference for the protein β-amyloid. Increased concentrations of the 42 amino acid peptide β-amyloid 42 cause amyloid to aggregate and subsequently damage neurons. Accumulated fibrillary amyloid protein is preferred over normal APP breakdown by β-amyloid 42 (Kumar et al. 2022). On the other hand, intracytoplasmic neurofibrillary structures found in the tangled form are twisted paired helical filaments formed by tau protein. When extracellular β-amyloid aggregates in AD, tau is hyperphosphorylated, leading to the formation of tau aggregates, also referred to as neurofibrillary tangles. These tangles originate in the hippocampus and may spread to other areas of the cerebral cortex (Muralidar et al. 2020). Plaques and tangles, deterioration of pyramidal cells granulovacuolar in the hippocampus by amyloid angiopathy, lowered density, and diminished neurons in Meynert's nucleus basalis, which result from low levels of acetylcholine, are some of the factors that greatly influence the pathogenesis of AD (Lopez and DeKosky 2003). Classification of drugs and their mechanism are discussed in Table 12.1.

12.2.2 Parkinson's Disease

The second most prevalent and rapidly expanding neurological syndrome worldwide is PD, characterized by gradual selective death of nigrostriatal dopaminergic neurons in substantia nigra, which causes individuals to have motor deficits, which is the primary clinical characteristic (Agostini et al. 2021). Over 65-year-olds worldwide represent about 1% to 2% of the population that have PD. Motor symptoms including rigidity, akinesia, bradykinesia, and tremors are what characterize PD (Skibinski and Finkbeiner 2011). Parkinsonism results from an imbalance between the cholinergic (a facilitative system) and the dopaminergic (an inhibitory system) (Crans and Ciruela 2021). Although the exact etiology of this syndrome is unknown, it has been connected to genetic predisposition, aging, environmental irritants such as pesticides, and other toxins like N-methyl-4-phenyl tetrahydrogenase or iatrogenic factors. Other symptoms include cognitive impairment, anxiety,

Table 12.1 Currently used drugs for the treatment of AD

Category	Drugs	Mechanism of action (MOA)	References
Cholinesterase inhibitors	Donepezil, rivastigmine, and galantamine	Increase the level of acetylcholine by inhibiting the enzyme cholinesterase	Singh and Sadiq (2022)
NMDA antagonist	Memantine	Binds to NMDA receptors and prevents glutamate binding when present in excess, thus prevents the release of calcium into the nerve cells	Hewitt (2000)
Hydroxyquinolines (metal chelators)	Clioquinol and 2-(dimethylamino) methyl-5,7-dichloro-8-hydroxyquinoline	Form chelates with copper, zinc, and iron metals and attenuate metal-Aβ complex derived neuronal toxicity, inflammation, and damages	Matlack et al. (2014)
Immunotherapy	Bapineuzumab, solanezumab, and aducanumab (Aduhelm)	Igγ1(IgG1) the monoclonal antibody that targets the Aβ plaques Aducanumab preferentially binds to and lowers both soluble and insoluble types of Aβ aggregates in the AD brain after crossing the BBB	Bittar et al. (2018)

depression, hallucinations, autonomic dysfunction, pain, rapid eye movement sleep disorder, constipation, and fatigue (Parsons 2019). L-DOPA as a "Pro-drug," dopaminergic agonists (Bromocriptine, Ropinirole), peripheral decarboxylase inhibitors (Carbidopa, Benserazide), MAO-B inhibitors (Selegiline, Rasagiline), catechol O-methyl transferase inhibitors (Tolcapone, Entacapone), anticholinergics (Benztropine, Procyclidine), dopamine facilitator (Amantadine), surgery, brain stimulation, antihistaminics, antidepressants, and stimulants for motor symptoms are only a few of the therapies available for PD (Parsons 2019). Levodopa is frequently used to treat motor impairments in PD patients who have dementia or hallucinations. Enduring levodopa therapy for first-line dopamine agonists or MAO-B inhibitors may lead to less symptomatic improvements in patients, even with adjunctive levodopa introduced later (Salat and Tolosa 2013). Long-term effects from dopaminergic treatments for PD result in motor fluctuations between periods of excellent and poor regulation of symptoms. Peak-dose dyskinesia is a condition characterized by involuntary movements, often occurring when the concentration of levodopa is at its maximum, or less common at the starting and end, doses of levodopa. Other psychotic characteristics are illusions, frequently accompanied by paranoia, as well as hallucinations are constituted on the auditory, cognitive, tangible, or olfactory levels (Kalia and Lang 2015).

12.2.3 Epilepsy

Epilepsy is a neurological disorder that affects many aspects of daily life and can be challenging to identify and treat. It has a life-changing impact and can disrupt everyday activities. Symptomatic epilepsy is connected to brain injuries, loss of neurons, and gliosis. Epileptogenesis is the change of a normal neural network into a persistently hyperexcitable one. It may be impacted by developmental events. Idiopathic epilepsy may be induced by neurotransmitter receptor dysfunction (Sumadewi et al. 2023). Treatment options include anti-epileptic medicines, vagal nerve stimulation, diets, and surgery. However, over 30% of patients continue to have untreated epilepsy, and many are taking drugs with serious adverse effects (Mirza et al. 2017). The most popular method of treating epilepsy is the use of anti-seizure drugs. There are now over 40 distinct anti-seizure drugs in the market, each with unique advantages and disadvantages. One medication can usually manage most seizures. As combining drugs might worsen adverse effects including weariness and dizziness, physicians almost always prescribe a single medicine. Still, in some cases, combinations of medications are required for epilepsy types that are not responsive to a single medication (Goldenberg 2010). Barbiturates, namely, phenobarbital and primidone, valproic acid, lamotrigine, topiramate, and levetiracetam, are the first-choice drugs for epilepsy. The mechanism of lamotrigine action involves the inhibition of voltage-gated sodium channels, which lowers neuronal activity. As lamotrigine inhibits the release of glutamate and aspartate via its action on sodium channels, it may specifically affect neurons that produce these excitatory neurotransmitters (Loscher et al. 2021).

12.2.4 Mania and Bipolar Disorder

Mania is a symptom of bipolar illness, often known as manic depression. Bipolar illnesses feature recurring phases of both sadness and an elevated mood that are accompanied by adjustments in activity or energy. While hypomania is transient and can lead to severe and psychotic manic episodes. Manic and depressed symptoms frequently co-occur, leading to mixed states, which are the outcome of the two types of bipolar disorders: bipolar I and II (Anderson et al. 2012). An individual with bipolar illness often has episodes of mania, a high level of energy, and depression accompanied by extreme physical and mental excitation, impatience, aggressive conduct, decreased sleep, excessive optimism, rapid thinking, lack of judgment, and hazardous actions. Serotonin, dopamine, and norepinephrine are neurotransmitters that have been investigated as potential contributors to mania. In a manic episode, patients with bipolar illness have noticeably increased levels of the hormones norepinephrine and epinephrine (Santos et al. 2011). Mania and bipolar disorder are highly correlated with genetic, psychological, and social variables, while the exact cause is unknown. The start and recurrence of manic periods are also influenced by environmental factors and stressful living circumstances (Koyuncu et al. 2019). The pathophysiology of mania and bipolar disorder finds involvement of particular brain areas, with modifications in the basal ganglia, amygdala, prefrontal cortex, hippocampus, and anterior cingulate. Enhanced emotions and reduced executive function may result from amygdala hyperactivity and decreased activity in cortical areas. Lithium, valproic acid, and carbamazepine were the first drugs used to treat mania associated with bipolar illness. These drugs worked as anticonvulsants, which were good in regulating the state of mind. Second-generation neuroleptics are currently categorized as mood stabilizers. Atypical antipsychotics were shown to be more successful than mood stabilizers for treating acute mania, but not usually for maintaining bipolar disease, according to a meta-analysis. Antipsychotics produce adverse effects, some of which are intolerable for certain groups of patients. Risperidone, olanzapine, and haloperidol are a few successful drug examples. Comparatively successful medications include lithium, quetiapine, and aripiprazole. Although lithium is beneficial, drug monitoring is necessary. The effectiveness of valproic acid, carbamazepine, and ziprasidone was higher but not as high as placebo. Despite being less popular, clozapine and electroconvulsive therapy have demonstrated advantages for treating mania that has resisted treatment (Dailey and Saadabadi 2022). Acute therapy and sustained remission are two treatments for bipolar disorder and mania (Joshi et al. 2019).

12.2.5 Dementia

A person with dementia has cognitive deterioration in many areas, such as personality, praxis, abstract thought, language, executive functioning, complex attention, social and visuospatial abilities, and memory. Decreased control over motivation, behavior, and emotional and social reactions often precedes and coexists with cognitive impairment. Over 55 million individuals worldwide suffer from dementia, and approximately 10 million new cases are diagnosed each year. Furthermore, this figure is predicted to increase to 78 million by 2030 and to 139 million by 2050 as the percentage of older people in the global population keeps rising (Zagórska et al. 2023). Dementia patients should have supervision and help to complete their daily activities of living. Patients are often looked after by their families, who endure the psychological, physical, and financial burdens of unpaid caregiving. The magnitude of the issue is difficult to overestimate (Tisher and Salardini 2019). Vascular dementia is the second most common form of dementia after AD. Many different types of dementia such as PD dementia, dementia with Lewy bodies, frontotemporal dementia, HD dementia, and alcoholism dementia are some other prevalent types of dementia (Chang et al. 2016). There is a significant variation in the prevalence of dementia disorders: the frequency of dementia with Lewy bodies (with no or only mild AD lesions) ranged from 1.1 to 24.7%, while the prevalence of AD and vascular dementia ranged from 12 to 67% and 4.5 to 46.8%, respectively. The frequency of mixed dementia (definite AD with severe vascular encephalopathy) is between 2 and 86% (Jellinger and Attems 2011), and frontotemporal dementia, which contributes from 5% to 10% of all dementia patients (Zagórska et al. 2023). Dementia is linked to a variety of lifestyle behaviors, including drinking and smoking, as well as metabolic conditions including hypertension and hyperglycemia (Ogura and Yamaguchi 2022). The pathogenic mechanisms shared by dementia subtypes involve irregular deposition of misfolded proteins, such as alpha-synuclein in PD and Lewy Body disease, and tau and amyloid beta in AD. The aging process is characterized by the loss of proteostasis, which implies that the exponential increase in dementia prevalence with age may be partially explained by the age-related reduction in the capacity to refold or break down damaged proteins. Several drugs are approved including mammalian target of rapamycin inhibitors (sirolimus, rapamycin, and everolimus) and tyrosine kinase inhibitors (nilotinib) (Heard et al. 2018).

12.2.6 Depression

Depression is a chronic disease that is characterized by the presence of specific symptoms including the inability to experience positive emotions, significant changes in weight, hypersomnia or insomnia, excessive feelings of negative emotions such as guilt and worthlessness, having a hard time concentrating and reoccurring suicidal thoughts. In spite of specific treatments for chronic depressive disorder, one-third of the population of patients don't respond well to first-line therapy. Treatment may prove to be unsuccessful for about 10–30% of the patients even after they are given two therapies of different pharmacological classes these cases are called Treatment Resistant Depression. The European Medicines Agency has defined treatment-resistant depression as major depressive disorder where it does not respond to two lines of oral antidepressants which are used for a sufficient length of time at an adequate dose within a single episode. Depression can be life-threatening for its chronic nature and the symptoms associated with mental disorder which can reduce the quality of life (Denee et al. 2021). The monoaminergic theory proposes that depression is caused by a chemical imbalance in the brain related to insufficiency in neurotransmitters like norepinephrine, dopamine, and serotonin; these chemicals affect factors like mood, sleep, and appetite of an individual (Mrozek et al. 2023). Abnormal processes in the prefrontal cortex and the limbic system also lead to neuroplasticity disorders which is the ability of the neurons to restructure according to the environment; recent studies have shown that this can be directly correlated to depression in individuals. Tricyclic anti-depressants include secondary amines like nortriptyline and tertiary amines like Clomipramine. MAO inhibitors include first-generation drugs like phenelzine, second-generation drugs like selegiline, and third-generation drugs like moclobemide. Serotonin and norepinephrine reuptake inhibitors like venlafaxine and milnacipran. Atypical tetracyclic anti-depressants like mirtazapine and NMDA receptor antagonists like ketamine. Noradrenergic and specific serotoninergic antidepressants like aptazapine and selective serotonin reuptake inhibitors like Escitalopram (Mrozek et al. 2023).

12.2.7 Anxiety

General anxiety is a mood disorder characterized by the patient constantly feeling overwhelmed, scared, or worrying about things, and the patient having deep illogical fear of situations. This condition is mainly diagnosed by the Diagnostic and Statistical Manual of Mental Disorders and includes symptoms such as feeling fatigued very easily, inability to focus on tasks, getting irritated constantly, disturbed sleep and restlessness, and muscle tension (Munir and Takov 2022). Anxiety is generally accompanied by the individual not being able to function well in social settings and constantly feeling agitated. Although similar in some symptoms, general anxiety differs from the other forms of panic disorders. It may be explained as the pathological form of fear which is the result of excessive thinking and worrying about situations and deep-rooted fears. Other than psychological causes, many other factors may lead to anxiety in individuals such as a history of depression and anxiety in the family, diabetes, environmental factors, child abuse, drug abuse, etc. (Adwas et al. 2019). Noradrenergic and serotonergic are the most commonly responsible neurotransmitter system among various other systems that are responsible for the different steps that later cause the onset of anxiety in an individual. When the noradrenergic system is activated in a much greater amount and the serotonergic system is activated in a much lesser amount, it leads to disturbance in one's emotional state and may alter many physiological functions. Thus, low serotonin activity and a higher noradrenergic activity are mainly the cause of anxiety in individuals (Adwas et al. 2019). Corticotropin-releasing factor receptor 1 like antalarmin, serotonin receptor modulators, and reuptake inhibitors like vortioxetine. Neurokinin receptor antagonists like aprepitant, γ-Aminobutyric acid receptor modulators like benzodiazepine, and cholecystokinin B antagonists like RP-69758 (Adwas et al. 2019).

12.2.8 Traumatic Brain Injury

A disruption in the brain's normal function known as a TBI can result from a head injury or another type of head trauma that penetrates the brain (Hiskens 2022). The WHO estimates that low- and middle-income countries have been the scene of 90% of these injuries. Between one-third and half of casualties and disabilities globally are caused by TBI (Mahmoudi et al. 2022). The projected global incidence of TBI is 939 cases per 100,000 people, and 5.48 million people are predicted to suffer from severe TBI every year (73 cases per 100,000 people). Mild TBI is responsible for 81% and moderate TBI for 11%, and severe TBI accounts for 8% of all recorded TBIs globally (Iaccarino et al. 2018). TBI happens when the brain's regular function is interfered with by a blow, bump, or jolt to the head. As a result of the primary injury phase and subsequent secondary injury processes, neuronal tissues involved with TBI are damaged (Shakkour et al. 2021). Physical head trauma leads to primary TBI, which damages neurons and triggers neuroinflammatory processes (Ghiam et al. 2021). After a primary injury, neurons can no longer regulate ion homeostasis within minutes, which causes intracellular calcium to accumulate, cells to get depolarized, glutamate to be released in an excitotoxic manner, and subsequently ionic gradient disruptions (Mohamadpour et al. 2019). Multiple physiological processes, including free radical injury, inflammation, glutamatergic excitotoxicity, mitochondrial dysfunction, oxidative stress, lipid peroxidation, axon degeneration, and apoptotic cell death, have been hypothesized to contribute to secondary, or delayed TBI. Drug repurposing can offer a sped-up path to establishing TBI-effective medicines. NAC, minocycline, progesterone, propranolol, and valproic acid belong to the substances being examined for treatment (Ghiam et al. 2021).

12.2.9 Huntington's Disease

HD is a progressive neurological disorder marked by jerky involuntary movements, cognitive impairment, abnormal muscle tone, and disruptive behaviors. It is caused by an enlarged cytosine, adenine, and guanine (CAG) triplet repeat in the Huntington gene, leading to an increased polyglutamine stretch. Tetrabenazine for the management of chorea is associated with HD. The likelihood of depression worsening is one of the negative effects. Amantadine is a dopamine facilitator; Pridipidine, a medication from a unique family of dopaminergic stabilizers, has also been investigated for use as a symptomatic treatment for HD. HD cognitive impairment treatment options remain limited, with randomized studies involving cholinesterase inhibitors yielding unsatisfactory results and addressing behavioral issues can be challenging (Ha and Fung 2012).

12.2.10 Multiple Sclerosis

MS is a central nervous system (CNS) autoimmune illness that is persistent, characterized by fatigue, vision problems, numbness, muscle spasms, stiffness, impaired mobility, despaired thinking, learning, speech, and deglutition, and problems with sexual, bladder, and bowel functions, is usually caused by the destruction of myelin sheaths, commonly known as demyelination, causing axonal and neuronal degeneration (Tafti et al. n.d.). MS is one of the lousiest neurologic disorders in the world, affecting over 3 million individuals globally and being the primary cause of non-traumatic neuronal disability in several countries, particularly among adolescents, which was surprisingly undiagnosed and unidentified as a separate disease for a long time (Reich et al. 2018). While the condition often arises between the ages of twenty and forty, it can appear at any age (Romero-Pinel et al. 2022). Studies reveal that smoking encourages MS occurrence by about 50%. While most research indicates that among those of African and East Asian ancestry, the incidence is declining, the estimated prevalence of European ancestry in populations is 0.1% (Tafti et al. n.d.). Inflammation, neuronal demyelination, and axonal damage mostly in gray and white matter disrupt ventricles, retinal nerves, and tracts as a result of plaques that MS creates in the brain and spinal cord (Huang et al. 2017). The

Table 12.2 Currently used drugs for different phases of MS

Phases	Drugs/treatment	Effect	References
Attacks	Methylprednisolone, plasmapheresis, or plasma exchange	Immunosuppression and reduction of inflammation	Fymat (2023)
RRMS	Interferons, glatiramer acetate, alemtuzumab, natalizumab, and fingolimod	Decrease the frequency of attacks	Fymat (2023)
Secondary Progressive MS	Mitoxantrone, siponimod and cladribine	Prevents disability progression	Fymat (2023)
Primary progressive MS	Ocrelizumab	Treats relapsing forms of disease	Fymat (2023)

autoimmune destruction of nerve tissues crosses the BBB and reaches the CNS (Palmer 2013). The deletion of autoreactive immune cells may fail for either poor performance of the supervisory T cells or depress autoreactive T-cell tolerance. T1 hyperintense lesions, lessened brain volume, and elevated stimulation of B cell indicators are associated with neurological impairment in relapsing-remitting MS (RRMS) patients (Kaskow and Baecher-Allan 2018). Certain cytokines, such as Interleukin (IL)-17, interferon-gamma, and granulocyte-macrophage colony-stimulating factors, are produced by autoreactive T-cells and may play a role in the pathogenesis of MS (Kuwabara et al. 2017). Moreover, B cells in MS can cause damage to oligodendrocytes by the release of chemicals and the presentation of antigens to T cells. The main T cell subsets associated with MS include Th17 cells, CD8+ T cells, and CD4+ Th1 cells. An MS diagnostic feature is the intrathecal synthesis of oligoclonal immunoglobulins. IL-1β and TNF-α cytokines, released by macrophages and microglia, can cause neurodegeneration by inducing dysfunctional ribonucleic acid-binding proteins, inhibiting astrocytic glutamate reuptake, leading to cell death. Glutamate is released by macrophages and microglia, which results in reactive oxygen/nitrogen species. These species can trigger mitochondrial damage and oxidative stress, which can lead to excitotoxicity of glutamate and dementia (Dighriri et al. 2023). In Table 12.2, the drugs used in different phases of MS are highlighted.

12.3 Mechanisms of Drug Repurposing for Neurological Diseases

Network-based drug repurposing investigation, in silico studies, and computational methods are rapidly being used for more than 10 years to select suitable drugs for testing in vivo. Even artificial intelligence and machine learning techniques have also been included in this run. Genome-wide association studies along with "multi-omics" are extensively used to find the candidate drugs to run clinical trials with minimum failure.

12.3.1 Drug Repurposing in AD

To date, many of the older drugs have shown promising effects in several preclinical and clinical studies. Using the new-generation tools, CCND1, EGFR, GSK3β, and KDR have been found to be the most common and frequently expressed genes in AD. The best target for treating AD is EGFR since research indicates that it protects Aβ and ApoE-encouraged mental retardation. Erlotinib, Genefitinib, and Vandetanib are on our suggested list of repurposed drugs; their therapeutic potential is supported by physicochemical, BBB, and structural similarities. Recently, tetradecanoylphorbol acetate, cyclandelate, idebenone, amoxapine, apomorphine, citicoline, quetiapine, risperidone, propranolol, mefenamic Acid, polaprezinc, buflomedil, clozapine, prazosin, paroxetine, selenium, and cerebrolysin have been identified as potent treatment candidates for AD, which can be used in clinical trials (Zeng et al. 2019). Drug repurposing in AD is depicted in Fig. 12.2 and other drugs with action on AD discussed in Table 12.3.

The proposed machine learning setup can be used for discovering new drugs for PD based on the predicted increase of expression of PINK1. Our study identified nitazoxanide as well as representatives of imidazolidines, trifluoromethyl benzenes, anilides, nitriles, stilbenes, and steroid esters as the best potential drug candidates for PD with PINK1 expression-enhancing activity on or inside the cell's mitochondria.

12.3.2 Drug Repurposing in PD

The growth of clinical drug libraries and advancements in disease genomics, biology, and bioinformatics has accelerated the development of candidate drugs for repurposing to be used in PD. This strategy has become a robust alternative to

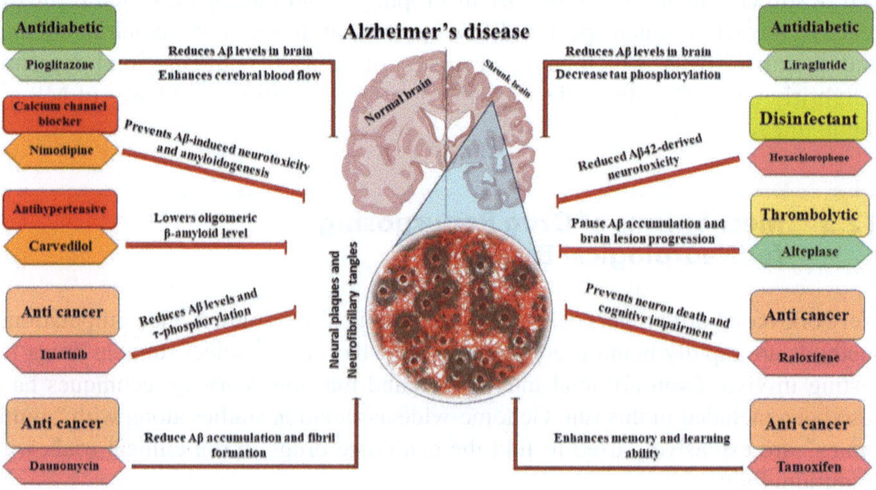

Fig. 12.2 Drug repurposing in AD

Table 12.3 List of drugs being tested for AD and being repurposed

Name of drug	Primary use	Effect on AD	Study type	References
Carvedilol	Antihypertensive	Attenuated brain oligomeric β-amyloid level and cognitive deterioration in AD mice models	Clinical and preclinical study	Wang et al. (2011)
Fluticasone	Antiasthmatic, non-steroidal anti-inflammatory drug (NSAID)	Prolonged use effectively reduced the likelihood of developing AD in older subjects	Clinical study	Lehrer and Rheinstein (2018)
Febuxostat	Antihyperuricemic (xanthine oxidase inhibitor)	Reduced the risk of AD-related dementia in adults	Clinical and epidemiological study	Singh and Cleveland (2018)
Pioglitazone	Anti-type 2 diabetic	Restored energy metabolism and reduced Aβ levels in the brain of APP/PS1 mice. Also improved cognition and regional cerebral blood flow in mild AD patients	In vitro, preclinical and clinical study	Savva et al. (2022)
Tamoxifen	Anticancer	Prolonged life of AD patients by enhancing memory, reducing impairment, and boosting learning ability	Clinical study	Soleimani Zakeri et al. (2021)
Alteplase/t-PA	Thrombolytic agent	Improved cognitive function, paused the progression of ischemic brain lesions, omits Aβ accumulation	Clinical study	Soleimani Zakeri et al. (2021)
Raloxifene	Anticancer	Prevented cell death in neurons and ameliorated cognitive impairment	Clinical study	Veenman (2020)

(continued)

Table 12.3 (continued)

Name of drug	Primary use	Effect on AD	Study type	References
Hexachlorophene	Topical antibacterial	Reduced Aβ42-derived neurotoxicity and prevents neural damage. Regulates tau protein levels in AD patients	Clinical and preclinical study	Terry et al. (2007), Vieira and Gamarra (2016)
Haloperidol	Antipsychotic	Improved psychosis and destructive behavior in AD patients	Clinical study	Dysken et al. (1994)
Daunomycin	Anticancer drug	Slows down the progression of AD by reducing Aβ accumulation and preventing Aβ fibril formation	Preclinical and in vitro study	Wu et al. (2019)
Clozapine	Atypical antipsychotic drug	Improves short-term memory loss in AD patients	Clinical study	Niculescu et al. (2020)
Estradiol	Estrogenic steroid	Enhanced cognition	Clinical study	Huang et al. (2009)
Progesterone	Progestins	Enhances memory and recall capacity by enhancing neuronal glucose metabolism as well as increases cognitive abilities and prevents Aβ inflammation	Preclinical and in vitro study	Wu et al. (2019)
Melatonin	Antioxidant	Prevented the formation and deposition of amyloid fibrils in the brain; decreased the secretion of soluble Aβ and enhanced memory function by decreasing tau hyperphosphorylation	In vitro, preclinical and clinical study	Norins (2021)
Rifamycin	Antituberculosis	Reduced Aβ production while increasing Aβ clearance	In vitro	Abuznait et al. (2011)

Minocycline	Antibiotic, anti-inflammatory	Inhibited c-Jun N-terminal kinase (JNK) activation associated with plaques and neurofibrillary tangles and thus decreased Aβ-induced cytotoxicity and cleavage of AβPP. Decreased soluble Aβ oligomers	In vitro, preclinical	Norins (2021)
Acyclovir	Anti-herpes virus	Lowers the risk of developing AD-related dementia	Cohort and clinical study	Tzeng et al. (2018)
Lithium	Antimanic	Promptly reduced tau phosphorylation and Aβ production by modulating AβPP processing in AD mouse models	Preclinical study	Forlenza et al. (2014)
Liraglutide	GLP-1 analog (antidiabetic)	Reduced cell death, AβPP, and Aβ levels through GSK3β-related mechanism and decreased tau phosphorylation	In vitro and in vivo	Perry et al. (2003), Li et al. (2012)
Sildenafil	PDE-5 inhibitor	Inhibits Cyclic guanosine 3′, 5′- monophosphate e(cGMP) degradation and increases the concentration in I hippocampal region which upregulates the expression of proliferator-activated receptor-γ coactivator 1α which indirectly suppresses Aβ generation	In vitro and in vivo	Sanders (2020)

(continued)

Table 12.3 (continued)

Name of drug	Primary use	Effect on AD	Study type	References
Dantrolene	Calcium channel blocker, muscle spasticity relaxer	Antagonizes ryanodine receptors (RyRs) which are increased in AD in the hippocampus, and release excess Ca^{2+} from the endoplasmic reticulum, leading to mitochondrial free radical release resulting in the loss of neurons and oxidative stress	In vitro and preclinical study	Kelliher et al. (1999)
Imatinib methane sulfonate	Anticancer	Lowered lipopolysaccharide inflammation, Aβ levels, τ-phosphorylation, and cognitive deficits in transgenic mice, while also inhibiting the pathogenic function of c-Abl/Cdk5 in AD	Preclinical study	Weintraub et al. (2013)
Metformin	Antidiabetic	Protects human neuronal stem cells by activating the AMPK pathway and enhances memory by upregulating PKC levels	Preclinical and clinical study	Padhi and Govindaraju (2022)
Valsartan	Angiotensin receptor blocker	Greater activity of the protein-degrading enzyme insulin-bound in membranes, leading to a better proteolytic breakdown of Aβ42 peptides	In vitro, in vivo and clinical study	Wang et al. (2007)

Candesartan	Angiotensin receptor blocker	Minimizes the risk factors for AD by causing Aβ42 to be phagocytosed by microglia while leaving the degrading enzymes alone	In vivo study	Torika et al. (2018)
Nimodipine	Calcium channel blocker, antihypertensive	Amyloidogenesis may be inhibited and Aβ-induced neurodegeneration can be attenuated by diminishing the quantity of open Ca^{2+} channels in cell membranes and restoring regulated Ca^{2+} inflow	In vivo study	Padhi and Govindaraju (2022)

developing new PD medications, with successful clinical trials highlighting the latest developments in this area (Chen et al. 2018). According to clinical and preclinical data of various drugs, drugs used for repurposing in PD are highlighted in Table 12.4 with their probable MOA (Fig. 12.3).

12.3.3 Drug Repurposing in Epilepsy

Drugs like anakinra, atorvastatin, brivaracetam, ceftriaxone, fingolimod, gabapentinoids, levetiracetam, losartan, n-acetylcysteine, perampanel, rapamycin, sulforaphane, topiramate, modulate neurogenesis, excitatory synaptogenesis, postsynaptic excitatory transmission, presynaptic transmitter release, and also loosen BBB disruption, minimize glutamatergic toxicity, and scavenge free radicals to lessen epileptic shocks. Axonal and dendritic regeneration, neurogenesis, and presynaptic transmitter release including anti-inflammatory properties are among the other impacts they put (Klein et al. 2020). Other drugs with action on epilepsy are discussed in Table 12.5.

12.3.4 Drug Repurposing in Mania and Bipolar Disorder

Tamoxifen and allopurinol can be safe and effective adjuvant treatment alternatives for manic episodes, according to meta-analytic evidence. When administered with valproate or lithium, the anti-cancer drug tamoxifen demonstrated improved response to treatment and a decrease in manic symptoms. Armodafinil was found to be more effective than a placebo for having less negative effects. Without raising any safety issues, pramipexole adjunctive therapy was more effective in symptom reduction for bipolar depression. As an adjuvant, celecoxib outperformed a placebo in terms of remission, while NAC was more successful in lowering depression scale scores without posing any safety risks. Aside from cytidine, levothyroxine, minocycline, medroxyprogesterone, and rivastigmine, these medications have also been used as supplementary therapy for bipolar disorder (Bartoli et al. 2021). Repurposed drugs are highlighted in Table 12.6.

12.3.5 Drug Repurposing in Dementia

Details of several drugs as repurposable candidates to be used in dementia are summarized in Table 12.7.

12.3.6 Drug Repurposing in Depression

According to transcriptomic data, ketamine shows mood improvement by working on miR-598-5p and miR-451. Vorinostat (HDAC inhibitor) and pioglitazone

Table 12.4 List of drugs being tested for PD and being repurposed

Name of drug	Primary use	Action on PD	Study type	References
Metformin	Activates the AMPK pathway to raise peripheral utilization of glucose, also decreases insulin resistance	Increased the phosphorylation of tyrosine hydroxylase, enhanced in brain-derived neurotrophic factor, glial cell line-derived neurotrophic factor, promoted dopaminergic neuron Enhanced motor performance, altered cognitive and locomotor function	Animal: 10-week-old Mice, 200 mg/kg/day; Animal: 4-month-old mice, 180 mg/kg/day	Katila et al. (2020), Li et al. (2019)
Tetracycline (Minocycline, Doxycycline)	Prevent translation by interacting with 30S ribosomal subunit	Lowered matrix metalloproteinase, protected dopaminergic cells, blocking α-synuclein's activation, microglia, and production of proinflammatory molecules	Clinical trial Phase II	Bortolanza et al. (2018)
Exenatide	GLP-1 receptor agonist, stimulates insulin release and β cell proliferation	Neuroprotective properties	Clinical Trial Phase II	Dusek et al. (2022)
Liraglutide	GLP-1 receptor agonist, reduces insulin resistance	Improved motor and cognitive symptoms	Clinical Trial Phase II 57 participants 1.2 or 1.8 mg	Zhang et al. (2020)
Lixisenatide	GLP-1 receptor agonist, lowers resistance of insulin	Enhanced motor and non-motor function, enhanced learning and memory by boosting hippocampus neurogenesis	Clinical Trial Phase II 156 participants 20 μg/day	https://clinicaltrials.gov/study/NCT03439943
MSDC-0160	Targeted mitochondrial pyruvate carrier, and modulates cellular function	Reduced nigral dopaminergic neurons' deterioration, α-synuclein protein, and glial cell activation, enhanced locomotor behavior, protected from 1-methyl-4-phenyl pyridinium, modulated mitochondrial function, and mTOR signaling		Volpicelli-Daley and Brundin (2018)

(continued)

Table 12.4 (continued)

Name of drug	Primary use	Action on PD	Study type	References
Semaglutide	GLP-1 receptor agonist, reduces insulin resistance	Improved motor and non-motor symptoms, cognitive function, nigrostriatal degeneration, neuroprotective and anti-inflammatory	Clinical trial Phase II	https://clinicaltrials.gov/ct2/show/NCT03659682
Ambroxol	Mucolytic	Enhanced glucocerebrosidase, lowered α -synuclein protein	Animal: Cynomolgus monkeys	Migdalska-Richards et al. (2017)
Ursodeoxych--olic acid	Reduces cholesterol absorption, dissolves gallstones,	Improved mitochondrial function in fibroblasts, decreased sporadic AD, dynamin-related protein 1, and manifesting LRRK2(G2019S) and non-manifesting-LRRK2 (G2019S), attenuated mitochondrial dysfunction		Bell et al. (2018), Mortiboys et al. (2015)
Simvastatin	HMG-CoA reductase enzyme inhibitor, promotes conversion of HMG-CoA to mevalonate	Anti-oxidative action, boosted superoxide dismutase, heme oxygenase-1, peroxisome proliferator-activated receptor (PPAR)-γ coactivator-1α, and glutamate-cysteine ligase, nicotinamide adenine dinucleotide phosphate oxidase/p38 mitogen-activated protein kinase pathways	Animal: C57BL/6 mice 8- to 10-week-old male	Tong et al. (2018)

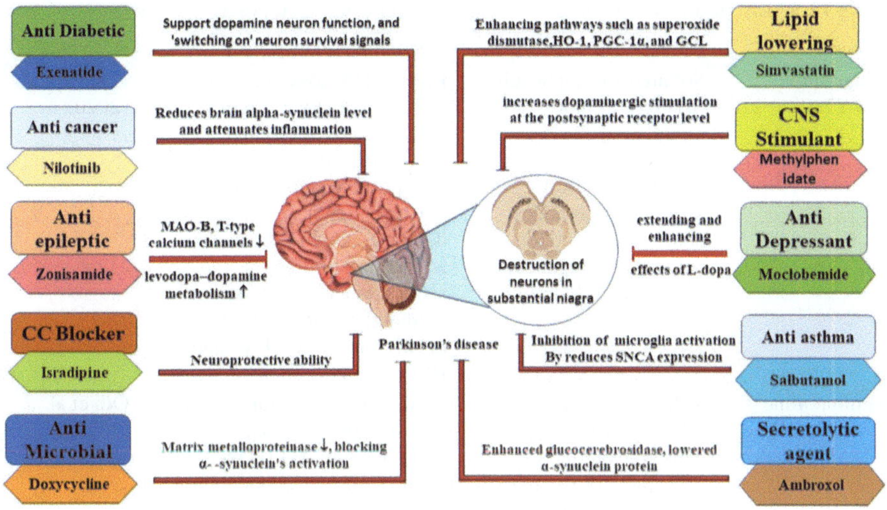

Fig. 12.3 Drug repurposing in PD

(PPAR-γ agonist) have also shown promising action against depression by modulating gene expressions. HMG-CoA reductase inhibitors (statins) in combination with SSRIs have shown potent antidepressant action by working as PPAR-α ligands, enhancing neuronal growth factor expression, whereas cabergoline and pergolide have been demonstrated to lessen hallucination and have antidepressant-like qualities in Parkinson's disease. When applied in combination with fluoxetine, the narcolepsy drug modafinil can help treat depression. Another anticonvulsant medication that works well for depression is phenytoin.

Computational data of molecular docking suggests that dextromethorphan, mecamylamine, and cyproheptadine can hypothetically work to lower depressive symptoms. In addition, genome-wide association studies suggest that alizapride, arcainesulfate, cycloserine, diethylstilbestrol, gabapentin, ifenprodil, ketoconazole, levonorgestrel, mesoridazine, nitrendipine, pregabalin, papaverine, risperidone, scopolamine, and sulpiride can also be repurposed for depressive disorders. Other studies imply that nimodipine and cannabidiol are also potent agents to be repositioned. The traditional Chinese medicine, Sho-saiko-to, has preclinically demonstrated antidepressant effects through its action on the serotonergic system in the CNS. Besides these, in general p38 mitogen-activated protein (p38-MAPK) signaling pathways inhibitory drugs like neflamapimod can have desirable effects on depression as they can work to reduce neuro-inflammation. Spermine has been proved to be useful in treatment-resistant depression (Mohammad Sadeghi et al. 2021). Several other drugs are mentioned in Table 12.8.

Table 12.5 List of drugs being tested for Epilepsy and being repurposed

Name of drug	Primary use	Action on epilepsy	References
Fingolimod (FTY720)	Sphingosine-1-phosphate (S1PR) receptor modulation (S1PR1/3/4/5) prevents lymphocytes from leaving lymph nodes, hence lowers the total amount of lymphocytes in peripheral blood	Lowered the loss of neural cells, as seen by a rise in NeuN-positive cells and a decreasing number in FJB-positive cells. Additionally, during epileptic seizures, it was demonstrated to decrease the activation of microglia and astrocytes and to activate/ modulate S1PR1 and SP1R3 (as evidenced by GFAP and IBA1 positive cells)	Paudel et al. (2020)
Fenfluramine	Releases serotonin or 5-HT by preventing the neurotransmitter's vesicular storage and preventing its reuptake	A combination of antagonistic effects at the σ1 receptor and 5-HT2C and 5-HT1D receptors. Positive allosteric modulation may be one of the most complicated effects of fenfluramine at the σ1-receptor	Odi et al. (2021)
Losartan	Angiotensin-type 1 receptor (AT1) antagonist	Reduced BBB permeability and stops astrocyte activation to prevent the development of spontaneous recurring seizures	Nozaki et al. (2018)
Isoflurane	Most likely binds to glutamate, GABA, and glycine receptors, having varying effects on each receptor	Same as Losartan	Ji et al. (2014), Larsen et al. (1998)
N-acetylcysteine	Binding to the harmful metabolites and scavenging free radicals, inactivates toxicants or reacts with electrophiles, and emerges as metabolite or lipid peroxidation products	Boosted the quantity of intact neurons in the hilar and hippocampal CA3 areas significantly after epilepsy was triggered	Mohammadi et al. (2023)
Atorvastatin	Lowers low-density lipoprotein cholesterol (LPL–C) and Triglyceride (TG) and raises High-density lipoprotein (HDL-C) by suppressing HMG-CoA reductase through competitive binding.	Atorvastatin therapy lowers behavioral comorbidities and neuroinflammation during epileptogenesis. Delays the deterioration of neurocognitive function and lowers proinflammatory indicators while raising levels of anti-inflammatory cytokines.	de Oliveira et al. (2018)

Table 12.6 List of drugs being tested for mania and bipolar disorder and being repurposed

Name of drug	Primary use	Action on mania and bipolar disorder	References
Doxycycline	Prevented translation by interacting with 30S ribosomal subunit	Lowered Iba1, neuroinflammation, Increased TNFα, MPO, and lipid peroxidation increased the phosphorylated (inactivated) form of GSK3β (Ser9), reversed cognitive impairment, neuroinflammation, induced oxidative imbalance, nitric oxide synthase, and nitrite, anti-inflammatory, potential procognitive	Chaves Filho et al. (2021)
Allopurinol	Inhibited xanthine oxidase, lowered uric acid level, prevented purine degradation	Reduced superoxide, and hydrogen peroxide mitochondrial dysfunction and oxidative stress	Machado-Vieira et al. (2007)
Tamoxifen	Non-steroidal antioestrogen	Inhibited the intracellular action of PKC, dopamine antagonists	Palacios et al. (2019)
Pramipexole	Non-ergot dopamine agonist	Pro-dopaminergic agent like action	Mah et al. (2011)
Memantadine	NMDA receptor blocker	Blocked the hyperactivity of dopamine receptor, prevented neurodegeneration and excitotoxic effect of stimulated glutamate NMDA receptor, enhanced neurotrophic factors and neurogenesis	Serra et al. (2014)

12.3.7 Drug Repurposing in Anxiety

Ditolylguanidine, an agonist of the Sigma receptor, and LY-2140023, an agonist of the glutamate receptor, has shown hope as novel treatments for anxiety (Woodward et al. 2023). Details of other candidate drugs are summarized in Table 12.9.

12.3.8 Drug Repurposing in TBI

NAC, minocycline, and (−)-phenserine can be useful repositioned candidate drugs to treat TBI according to clinical studies. NAC, used for acetaminophen overdose hepatotoxicity, has been approved to treat traumatic brain injury recently for its anti-inflammatory and antioxidant qualities. In a separate study, potent synergistic neuroprotective action has been observed with minocycline and (−)-phenserine drug having qualities (Ghiam et al. 2021). Other repurposed drugs are highlighted in Table 12.10.

Table 12.7 List of medicines being tested for dementia and being repurposed

Drugs	Class	MOA	Repurposed effects	References
Metformin	Biguanide class Antidiabetic drug	Anti-hyperglycemic property, mainly for T2DM	Neuroprotective effects prevent dopamine depletion, inhibit oxidative stress, improve mitochondrial viability, reduce inflammation, and improve synaptic function in several cell and animal models of PD and AD	O'Brien et al. (2022)
Valsartan	Angiotensin receptor blockers (ARBs)	Inhibition of inflammation, vasoconstriction, and mitochondrial dysfunction, and promotion of acetylcholine release	Reduced Aβ burden and improved cognitive function, reduction of incident dementia	Corbett et al. (2012)
Liraglutide	GLP1 analogues	Effects on oxidative stress and apoptotic pathways	Improved synaptic plasticity and cognitive function, and reduced AD pathology	McClean et al. (2011)
Sildenafil	PDE5 inhibitor	Augments nitric oxide-dependent vasodilation, by impeding PDE5-mediated cyclic guanosine monophosphate (cGMP) breakdown in vascular myocyte	Decreased spontaneous neural activity in the right hippocampus, increased cerebral metabolic rate of oxygen and CBF, decreased cerebral vascular reactivity, enhanced memory	Hainsworth et al. (2023)
Cilostazol	PDE inhibitors	Inhibition of PDE3 enzyme	Neuroprotective in cerebral ischemia models, prevents memory impairment after chronic cerebral hypoperfusion in middle-aged rats and rats with diabetes-induced vascular dementia	Godinho et al. (2015)
Baclofen	GABA-B receptor agonist	Inhibition of the transmission of both mono- and polysynaptic reflexes at the spinal cord, relaxing spasticity	Ameliorated cognitive impairment in a rat model of cerebral hypoperfusion by modulating expression of nonselective cyclic-nucleotide gated cation channels in the hippocampus	Li et al. (2014)

Table 12.8 List of medicines being tested for depression and being repurposed

Name of drug	Primary use	Proposed MOA	Study type	References
Omeprazole	Gastritis	Inhibits MAO-A	Animal 3 months old mice,100 mg/kg/day	Betari et al. (2020)
β-Sitosterol	Hypocholesterolemic agent	Causes specific knockdown effects in the hippocampus	Male mice 100 mg/kg/hr	Panayotis et al. (2021)
Ketamine	Anesthesia	NMDA receptor antagonist.	Female rodent 10 mg/kg/day	Polis et al. (2019)
Minocycline	Anti-biotic	Reduces brain inflammation	Male rat 80 mg/kg/day	Reis et al. (2019)
Riluzole	ALS	Prevents glutamate toxicity	Male mice 13.2 mg/kg/day	Gourley et al. (2012)
Metformin	Anti-diabetic drug	Activates 5′ adenosine monophosphate-activated protein kinase	Male mice 200 mg/kg/day	Chen et al. (2020)
Statins	Hypocholesterolemic agent	Anti-inflammatory properties	Male rat 10 mg/kg/day	Lin et al. (2014)
Buprenorphine	Analgesic	Kappa opioid receptor antagonism	WKY rats 2.25 mg/kg/day	Browne et al. (2015)
Celecoxib	Analgesic	Inhibits COX-2	CUMS rats 20 mg/kg/day	Song et al. (2018)
Modafinil	Used to treat sleep disorders	Inhibits dopamine reuptake	Female Sprague-Dawley rat 4 mg/kg/day	Yates et al. (2009)

12.3.9 Drug Repurposing in HD

Several new drugs have been tested for repositioning to be used in HD. Fenofibrate is primarily used as an antihyperlipidemic drug, but through a clinical trial, the gene analysis data pointed out that it also stops neurodegeneration by activating mitochondrial biogenesis of ATP via upregulating PGC-1α and thus helps control ROS generation in HD. Ursodiol is primarily used to treat gallstone, but its neuroprotective action was demonstrated in terms of neuronal apoptosis along with improved locomotor and sensorimotor activity in preclinical studies. Cysteamine has recently been developed to treat the effects of cystinosis, but recently it was found in clinical studies to increase brain-derived neurotrophic factors, which get usually reduced in HD patients for lower concentration of heat shock DnaJ-containing protein 1b. Riluzole is also a new drug for ALS, and its neuroprotective effect has been surfaced by clinical study outcomes, though further studies should be performed to confirm the protective role in HD. Sodium phenylbutyrate is primarily used was for treating

Table 12.9 List of medicines being tested for anxiety and being repurposed

Name of drug	Primary function	Proposed MOA	Subject used	References
Telmisartan	Inhibits angiotensin II	Activates the peroxisome proliferator-activated receptors (PPAR)-δ expression	Male rat 1 mg/kg/day	Li et al. (2017)
Propranolol	Inhibits the endogenous catecholamines	Endogenous catecholamines antagonist	Male Sprague-Dawley rat 10 mg/kg/day	Rodriguez-Romaguera et al. (2009)
Gabapentin	Inhibits neurotransmitters	Affected the GABAergic system	Male Wistar Rat30 mg/kg/day	Wallace et al. (2008)
Pregabalin	It binds to the α-2 delta subunit	Decreased release of neurotransmitters	Male Hooded Lister rats3.30 mg/kg/day	Field et al. (2001)
D-Cycloserine	Inhibits the cell wall synthesis in bacteria.	Partial NMDA glutamatergic receptor agonist	Male Wistar Rats30 mg/kg/day	Ho et al. (2005)
Ondansetron	Inhibition of 5-HT3 receptor.	Serotonin 5-HT3 receptor antagonist.	Male Swiss albino mice1 mg/kg/day	Jain et al. (2022)
Losartan	Inhibits angiotensin II receptors	Angiotensin II receptor antagonist	Long Evans rats10 mg/kg/day	Campos et al. (2020)
Baclofen	Influx of potassium into the neuron	GABA-B receptor agonist.	Sprague–Dawley rats5 mg/kg/day	Knapp et al. (2007)
mirtazapine	Adrenergic α-2 auto receptor antagonist	Enhanced noradrenergic and serotonergic activity.	Male Sprague–Dawley rats3 μg/kg/day	An et al. (2016)
Valproic acid	Inhibits voltage-gated sodium channel	Increased GABA levels in the brain	Male hooded rats320 mg/kg/day	Lal et al. (1980)

urea cycle disorders has been found to inhibit HDAC related chromatin remodeling leading to gene expressions seen in HD. Ethyl eicosapentaenoic acid was approved to treat severe hypertriglyceridemia, but it is found to improve motor function in HD patients through mitochondrial function stabilization (Devadiga and Bharate 2022). Further details of other drugs repurposed in HD with their probable MOAs are summarized in Table 12.11.

Table 12.10 List of medicines being tested for TBI and being repurposed

Name of drug	Drug class	Primary use	Effect on TBI	References
Ketamine	NMDA antagonist	Anesthetic	Decreases neurological deficits, inhibits neuronal apoptosis, decreases brain edema	Liang et al. (2018)
Fluoxetine	Selective serotonin reuptake inhibitor	Antidepressant, anxiolytic	Reactivates neuronal plasticity	Castrén and Antila (2017)
Dexmedetomidine	α_2 agonist	Sedative and anti-hypertensive drug	Decreases axonal injury and synaptic degeneration, brain edema reduction, suppression of apoptosis and oxidative stress	Wu et al. (2018), Humble et al. (2016)
PDE4	PDE4 inhibitor	Chronic obstructive broncho-pneumopathy	Enhances memory, improves cognitive dysfunction	Titus et al. (2018)
Curcumin	Derived from Curcuma longa plants	Anti-apoptotic, antioxidative antitumor, anti-inflammatory, anti-convulsive	Ameliorate water content in the brain, lowered oxidative stress and apoptosis, boost the neurological function and severity score (effects by maintaining glutathione levels)	Dai et al. (2018)
Icariin	Natural agent from *Epimediiherba*	Neuroprotective, neuroplastic	Upregulate the brain-derived neurotrophic factor, synaptophysin, and postsynaptic density protein 95 expressions	Cichon et al. (2020)
Gallic acid	Phenolic compound	Antioxidant, anti-inflammatory, anti-tyrosine	Reduce brain IL-1β, IL-6, and TNFα 48 hr after TBI, enhanced memory	Kratz et al. (2008), Mirshekar et al. (2018)

(continued)

Table 12.10 (continued)

Name of drug	Drug class	Primary use	Effect on TBI	References
Bazedoxifene	Selective estrogen receptor modulator	Postmenopausal osteoporosis	Neuroprotection reduces BBB destruction	Lan et al. (2019)
Metformin	Biguanide	Type 2 diabetes mellitus, polycystic ovary syndrome	Anti-inflammatory, antioxidative, increased vascular and endothelial functions	Taheri et al. (2019), Tao et al. (2018)
Maraviroc	Antiretroviral drug	C-C chemokine receptor (CCR5) receptor antagonist, human immunodeficiency virus (HIV), graft-versus-host disease	Enhanced motor function, modulates learning and memory, increased cognitive function	Zhou et al. (2016), Liao et al. (2022)

12.3.10 Drug Repurposing in MS

Various in vitro and in vivo models of MS have been used to study the effects of various drugs with other indications to evaluate their action against MS. In silico and network pharmacology studies have also been involved in selecting candidate drugs for clinical trials. Recent studies have identified several histamine drugs, including amodiaquine, rupatadine, and diphenhydramine, as potential therapeutic agents for MS, expanding the landscape beyond existing FDA-approved treatments. (Amadio et al. 2022) (Table 12.12).

12.4 Drug Repurposing Opportunities and Challenges

The worth of repurposing drugs in the pharmaceutical industry has been rising over the last several years. Medication repurposing is a viable way to accelerate the development of suitable drug for appropriate disease since it saves time and money, while new drug development typically takes 13–15 years. Despite the enormous expense of fundamental research and clinical testing, many new treatments are discontinued owing to ineffectiveness or adverse effects, making it a viable option. Synergistic medication combinations can boost drug repositioning success by reducing therapeutic doses compared to monotherapies. Pharmaceutical companies contain rich chemical libraries, including unsuccessful drug candidates, which may be saved and used to solve unmet medical needs. Drug rescue provides low-cost options to commercialize abandoned drug ideas, boosting the value of pharmaceutical corporations' portfolios (Napolitano et al. 2013). Network pharmacology emphasizes the potential advantages of poly-pharmacy in treating complicated

Table 12.11 List of medicines being tested for HD and being repurposed

Name of drug	Primary use	Proposed MOA	Study type	References
Valbenazine	Tardive dyskinesia	Reduced chorea	Clinical trial phase III participants -167 40 mg/day once daily for 4 weeks and 80 mg/day 48 weeks	https://clinicaltrials.gov/study/NCT04400331#publications
Riluzole	ALS	Increased of the motor score, reduced in the dyskinesia	Clinical trial phase III Participants – 537 50 mg 8 mg/kg/day	Schramm et al. (2016)
Glipizide	Type 2 diabetes mellitus	Prevented neurodegeneration pathway by alleviated poly(Q), raised histone acetylation		Tandon and Sarkar (2023)
Cysteamine	Nephropathic cystinosis	Raised glutathione level, inhibited thiol transglutaminase enzyme and caspase 3	Clinical trial phase III participants -96 cysteamine vs placebo	Lesort et al. (2003)
Minocycline	Antibiotics	Decreased free radical formation and nitric oxide synthetase production, caspase inhibitor	Clinical trial phase III participants – 114 Minocycline vs placebo	Bonelli et al. (2004)

Table 12.12 List of medicines being tested for MS and being repurposed

Name of drug	Primary use	Effect on MS	Study type	References
Miconazole	Antifungal	Remyelination through mitogen-activated protein kinase and glucocorticoid receptor signaling	In vitro and in vivo	Najm et al. (2015)
Clobetasol	Corticosteroid (anti-inflammatory)	Remyelination through mitogen-activated protein kinase and glucocorticoid receptor signaling. It also works as a potent immunosuppressant	In vitro and in vivo	Najm et al. (2015)
Ibudilast	Antiasthmatic	Non-selective PDE inhibitor and the migration of macrophages inhibitor factor inhibitor with several activities related to secondary progressive MS; these activities include reducing nitric oxide and reactive oxygen species, which attenuate the pro-inflammatory activity of microglia and astrocytes; encouraging the synthesis of neurotrophins, such as growth factor for nerves and glial cell line-derived neurotrophic factor; and encouraging the production of neurotrophins.	Clinical study	Rolan et al. (2009)
Riluzole		Reducing glutamate release and antagonism of voltage-dependent sodium channels	Clinical study	Gibson and Mark (2012)
Amiloride	Diuretic	Worked as acid-sensing ion channel blocker has been identified as myeloid- and cytoprotective effects	In vivo and clinical study	Chu and Xiong (2013)

(continued)

Table 12.12 (continued)

Name of drug	Primary use	Effect on MS	Study type	References
Pirfenidone	Antifibrotic/ anti-inflammatory	Improved neurological function	Clinical study	Walker et al. (2005)
Fluoxetine	Antidepressant	Enhancing glycogenolysis and increased the production of brain-derived neurotrophic factor in rodent astrocyte cultures; improved cerebral white matter N-acetyl aspartate/ creatine ratio was found on magnetic resonance imaging in patients, enhanced in axonal mitochondrial energy metabolism; lowered the presenting capacity of antigen in glial cells		Kong et al. (2002), Allaman et al. (2011), Mostert et al. (2006, 2008)
Oxcarbazepine	Anticonvulsant	Reduced paroxysmal pain in MS	Clinical study	Solaro et al. (2007)
R-α-lipoic acid	Dietary supplements, anti-diabetic neuropathy	Antioxidant, anti-inflammatory and neuroprotective	Clinical study	Spain et al. (2017)
Metformin	Antihyperglycemic	Anti-inflammatory and promotes remyelination and neuroprotection	Clinical study	Nath et al. (2009)
Niacin	Antihypercholes terolaemic	Promotes oligodendrocyte proliferation, remyelination, and neuroprotection	Clinical study	Kaneko et al. (2006)
Clemastine	Antihistamine	Promotes oligodendrocyte progenitor differentiation and remyelination	Clinical study	Green et al. (2017)
Lamotrigine	Anticonvulsant	Neuroprotective effects	In vivo study	Bechtold et al. (2006)

(continued)

Table 12.12 (continued)

Name of drug	Primary use	Effect on MS	Study type	References
Oxcarbazepine	Anticonvulsant	Neuroprotective effects	In vivo study	Al-Izki et al. (2014)
Nimodipine	Calcium channel blocker	Promotes remyelination, neuroprotection, and restores CNS perfusion and oxygenation	In vivo study	Cunniffe et al. (2021)
Flunarizine	Migraine prophylaxis	Neuroprotection	Clinical study	Bostanci et al. (2006)

illnesses, since network analysis employing genetic, protein, or disease data may assist in discovering repurposing targets. Systematic examination of health records, clinical trial data, and post-marketing monitoring data may assist in guiding drug repurposing (Greene and Voight 2016). New prospects for medication repurposing are presented by enormous-scale in vitro drug screens that make use of genetic data, biobanks connected to electronic health records, and self-reported patient data (Hsieh et al. 2016). Modern genomic and proteomic methods have identified novel pharmacological targets, offering prospects for therapeutic repurposing.

Drug repurposing has been successful in certain circumstances, although it often fails in phase III studies. These failures are more likely to occur in late-stage development owing to impediments such as patent concerns, regulatory considerations, and organizational difficulties, rather than toxicity. Incentives for medication repurposing might be reduced by certain regulatory factors that could hinder the enforcement of patent rights and/or the patentability of new medicinal uses (Hsieh et al. 2016). Novel indications for off-patent medications may be patentable, but if the new indication makes use of already-approved dosage forms and strengths, enforcement issues could arise. Having untapped strengths or special formulas is the best scenario. Given their substantial influence on the potential profit from the repurposed product, the obstacles associated with patenting and enforcing patent rights are crucial in providing incentives for medicinal repurposing (Greene and Voight 2016). The US and European regulatory frameworks for repurposed medications emphasize the significance of any potential exclusivity advantages. Patent rights are critical in motivating medication repurposing and influencing future profit. The European Union provides 8 years of data protection and 2 years of market exclusivity, but the United States provides 5 years, which might be extended by 3 years. Additional economic incentives are required for cost-effective medication repurposing (Nygaard et al. 2015). Regulatory exclusivities, medication withdrawals, and problems with label extension in approved medications provide difficulties for the creation of clinical trials, clinical adoption, and drug licensing. These problems may be linked to medication withdrawals, regulatory limitations, and difficulties extending the label on medications that have been approved by national protocols (Rebelo et al. 2021).

12.5 Conclusion

Repurposing medications is an exciting new treatment strategy for many illnesses, particularly uncommon ones. Pharmaceutical combinations and dosages that work best for a certain condition may be predicted by new technology. Using already available medications with well-established safety characteristics, drug repurposing for neurological illnesses provides a quicker, more affordable option. Evidence shows that this technique might enhance the quality of life for those afflicted, but further investigation and clinical studies are required. This study investigates medication repurposing for neurological illnesses, emphasizing its benefits such as decreased development time and cost, as well as the possibility for speedier delivery of innovative medicines. The "intended reframing approach" seeks repurposed pharmaceuticals that can modify the appearance of dysregulated genes. This technique requires powerful hybrid medication development, technical advancements, drug libraries' connectivity, evidence demonstrating clinical trials, and a knowledge of the MOA. Overall, medication reusing and hybrid drug emergence are potential techniques for discovering ailment-modifying pharmacological therapies in neurological illness therapy.

Conflict of Interest The authors declare no conflicts of interest.

Funding None.

References

Abuznait AH, Cain C, Ingram D, Burk D, Kaddoumi A (2011) Up-regulation of P-glycoprotein reduces intracellular accumulation of beta amyloid: investigation of P-glycoprotein as a novel therapeutic target for Alzheimer's disease. J Pharm Pharmacol 63(8):1111–1118. https://doi.org/10.1111/j.2042-7158.2011.01309.x

Adwas A, Jbireal J, Azab A (2019) Anxiety: insights into signs, symptoms, etiology, pathophysiology, and treatment. S Afr J Med Sci 2:80–91

Agostini F, Masato A, Bubacco L, Bisaglia M (2021) Metformin repurposing for Parkinson disease therapy: opportunities and challenges. Int J Mol Sci 23(1):398. PubMed. https://doi.org/10.3390/ijms23010398

Al-Izki S, Pryce G, Hankey DJR, Lidster K, von Kutzleben SM, Browne L, Clutterbuck L, Posada C, Edith Chan AW, Amor S, Perkins V, Gerritsen WH, Ummenthum K, Peferoen-Baert R, van der Valk P, Montoya A, Joel SP, Garthwaite J, Giovannoni G et al (2014) Lesional-targeting of neuroprotection to the inflammatory penumbra in experimental multiple sclerosis. Brain 137(1):92–108. https://doi.org/10.1093/brain/awt324

Allaman I, Fiumelli H, Magistretti PJ, Martin J-L (2011) Fluoxetine regulates the expression of neurotrophic/growth factors and glucose metabolism in astrocytes. Psychopharmacology 216(1):75–84. https://doi.org/10.1007/s00213-011-2190-y

An Y, Chen C, Inoue T, Nakagawa S, Kitaichi Y, Wang C, Izumi T, Kusumi I (2016) Mirtazapine exerts an anxiolytic-like effect through activation of the median raphe nucleus-dorsal hippocampal 5-HT pathway in contextual fear conditioning in rats. Prog Neuro-Psychopharmacol Biol Psychiatry 70:17–23. https://doi.org/10.1016/j.pnpbp.2016.04.014

Anderson IM, Haddad PM, Scott J (2012) Bipolar disorder. Br Med J 345:e8508. https://doi.org/10.1136/bmj.e8508

Bartoli F, Cavaleri D, Bachi B, Moretti F, Riboldi I, Crocamo C, Carrà G (2021) Repurposed drugs as adjunctive treatments for mania and bipolar depression: a meta-review and critical appraisal of meta-analyses of randomized placebo-controlled trials. J Psychiatr Res 143:230–238. https://doi.org/10.1016/j.jpsychires.2021.09.018

Bechtold DA, Miller SJ, Dawson AC, Sun Y, Kapoor R, Berry D, Smith KJ (2006) Axonal protection achieved in a model of multiple sclerosis using lamotrigine. J Neurol 253(12):1542–1551. https://doi.org/10.1007/s00415-006-0204-1

Bell SM, Barnes K, Clemmens H, Al-Rafiah AR, Al-Ofi EA, Leech V, Bandmann O, Shaw PJ, Blackburn DJ, Ferraiuolo L, Mortiboys H (2018) Ursodeoxycholic acid improves mitochondrial function and redistributes Drp1 in fibroblasts from patients with either sporadic or familial Alzheimer's disease. J Mol Biol 430(21):3942–3953. PubMed. https://doi.org/10.1016/j.jmb.2018.08.019

Betari N, Sahlholm K, Morato Arús X, Godoy-Marín H, Jáuregui O, Teigen K, Ciruela F, Haavik J (2020) Inhibition of tryptophan hydroxylases and monoamine oxidase-A by the proton pump inhibitor, omeprazole—in vitro and in vivo investigations. Front Pharmacol 11. https://doi.org/10.3389/fphar.2020.593416

Bittar A, Sengupta U, Kayed R (2018) Prospects for strain-specific immunotherapy in Alzheimer's disease and tauopathies. Npj Vaccines 3(1):9. https://doi.org/10.1038/s41541-018-0046-8

Bonelli RM, Hödl AK, Hofmann P, Kapfhammer H-P (2004) Neuroprotection in Huntington's disease: a 2-year study on minocycline. Int Clin Psychopharmacol 19(6) https://journals.lww.com/intclinpsychopharm/fulltext/2004/11000/neuroprotection_in_huntington_s_disease__a_2_year.4.aspx

Bortolanza M, Nascimento GC, Socias SB, Ploper D, Chehín RN, Raisman-Vozari R, Del-Bel E (2018) Tetracycline repurposing in neurodegeneration: focus on Parkinson's disease. J Neural Transm 125(10):1403–1415. https://doi.org/10.1007/s00702-018-1913-1

Bostanci MÖ, Bağirici F, Canan S (2006) A calcium channel blocker flunarizine attenuates the neurotoxic effects of iron. Cell Biol Toxicol 22(2):119–125. https://doi.org/10.1007/s10565-006-0037-9

Browne CA, van Nest DS, Lucki I (2015) Antidepressant-like effects of buprenorphine in rats are strain dependent. Behav Brain Res 278:385–392. https://doi.org/10.1016/j.bbr.2014.10.014

Campos GV, de Souza AMA, Ji H, West CA, Wu X, Lee DL, Aguilar BL, Forcelli PA, de Menezes RC, Sandberg K (2020) The Angiotensin Type 1 Receptor antagonist Losartan prevents Ovariectomy-induced cognitive dysfunction and anxiety-like behavior in long Evans rats. Cell Mol Neurobiol 40(3):407–420. https://doi.org/10.1007/s10571-019-00744-x

Castrén E, Antila H (2017) Neuronal plasticity and neurotrophic factors in drug responses. Mol Psychiatry 22(8):1085–1095. PubMed. https://doi.org/10.1038/mp.2017.61

Chang D, Liu J, Bilinski K, Xu L, Steiner GZ, Seto SW, Bensoussan A (2016) Herbal medicine for the treatment of vascular dementia: an overview of scientific evidence. Evid Based Complement Alternat Med 2016:7293626. https://doi.org/10.1155/2016/7293626

Chaves Filho AJM, Cunha NL, de Rodrigues PA, de Souza AG, Soares MVR, Jucá PM, de Queiroz T, da Clemente DCS, Mottin M, Andrade CH, Peixoto CA, Macedo DS (2021) Doxycycline reverses cognitive impairment, neuroinflammation and oxidative imbalance induced by D-amphetamine mania model in mice: a promising drug repurposing for bipolar disorder treatment? Eur Neuropsychopharmacol 42:57–74. https://doi.org/10.1016/j.euroneuro.2020.11.007

Chen X, Gumina G, Virga K (2018) Recent advances in drug repurposing for Parkinson's disease. Curr Med Chem 25. https://doi.org/10.2174/0929867325666180719144850

Chen J, Zhou T, Guo A-M, Chen W-B, Lin D, Liu Z-Y, Fei E-K (2020) Metformin ameliorates lipopolysaccharide-induced depressive-like behaviors and abnormal glutamatergic transmission. Biology 9(11). https://doi.org/10.3390/biology9110359

Chu X-P, Xiong Z-G (2013) Acid-sensing ion channels in pathological conditions. In: Annunziato L (ed) Sodium calcium exchange: a growing spectrum of pathophysiological implications: proceedings of the 6th international conference on sodium calcium exchange. Springer US, pp 419–431. https://doi.org/10.1007/978-1-4614-4756-6_36

Cichon N, Saluk-Bijak J, Gorniak L, Przyslo L, Bijak M (2020) Flavonoids as a natural enhancer of neuroplasticity-an overview of the mechanism of neurorestorative action. Antioxidants (Basel, Switzerland) 9(11):1035. PubMed. https://doi.org/10.3390/antiox9111035

Corbett A, Pickett J, Burns A, Corcoran J, Dunnett SB, Edison P, Hagan JJ, Holmes C, Jones E, Katona C, Kearns I, Kehoe P, Mudher A, Passmore A, Shepherd N, Walsh F, Ballard C (2012) Drug repositioning for Alzheimer's disease. Nat Rev Drug Discov 11(11):833–846. https://doi.org/10.1038/nrd3869

Crans RAJ, Ciruela F (2021) Dopaminergic-cholinergic imbalance in movement disorders: a role for the novel striatal dopamine D(2)- muscarinic acetylcholine M(1) receptor heteromer. Neural Regen Res 16(7):1406–1408. PubMed. https://doi.org/10.4103/1673-5374.300988

Cunniffe N, Vuong KA, Ainslie D, Baker D, Beveridge J, Bickley S, Camilleri P, Craner M, Fitzgerald D, de la Fuente AG, Giovannoni G, Gray E, Hazlehurst L, Kapoor R, Kaur R, Kozlowski D, Lumicisi B, Mahad D, Neumann B et al (2021) Systematic approach to selecting licensed drugs for repurposing in the treatment of progressive multiple sclerosis. J Neurol Neurosurg Psychiatry 92(3):295. https://doi.org/10.1136/jnnp-2020-324286

Dai W, Wang H, Fang J, Zhu Y, Zhou J, Wang X, Zhou Y, Zhou M (2018) Curcumin provides neuroprotection in model of traumatic brain injury via the Nrf2-ARE signaling pathway. Brain Res Bull 140:65–71. https://doi.org/10.1016/j.brainresbull.2018.03.020

Dailey MW, Saadabadi A (2022) Mania (29630220). StatPearls Publishing, Treasure Island (FL). PubMed. http://europepmc.org/abstract/MED/29630220

de Oliveira CV, Grigoletto J, Canzian JM, Duarte MMMF, Duarte T, Furian AF, Oliveira MS (2018) Effect of atorvastatin on behavioral alterations and neuroinflammation during epileptogenesis. Epilepsy Behav 78:109–117. https://doi.org/10.1016/j.yebeh.2017.10.021

Denee T, Kerr C, Ming T, Wood R, Tritton T, Middleton-Dalby C, Massey O, Desai M (2021) Current treatments used in clinical practice for major depressive disorder and treatment resistant depression in England: a retrospective database study. J Psychiatr Res 139:172–178. https://doi.org/10.1016/j.jpsychires.2021.05.026

Devadiga SJ, Bharate SS (2022) Recent developments in the management of Huntington's disease. Bioorg Chem 120:105642. https://doi.org/10.1016/j.bioorg.2022.105642

Dighriri I, Aldalbahi A, Albeladi F, Tahiri A, Kinani E, Almohsen R, Alamoudi N, Alanazi A, Alkhamshi S, Althomali N, Alrubaiei S, Altowairqi F (2023) An overview of the history, pathophysiology, and pharmacological interventions of multiple sclerosis. Cureus 15. https://doi.org/10.7759/cureus.33242

Duong S, Patel T, Chang F (2017) Dementia: what pharmacists need to know. Canad Pharm J/Revue Des Pharmaciens Du Canada 150:171516351769074. https://doi.org/10.1177/1715163517690745

Dusek P, Hofer T, Alexander J, Roos PM, Aaseth JO (2022) Cerebral iron deposition in neurodegeneration. Biomolecules 12(5):714. PubMed. https://doi.org/10.3390/biom12050714

Dysken MW, Johnson SB, Holden L, Vatassery G, Nygren J, Jelinski M, Kuskowski M, Schut L, McCarten JR, Knopman D, Maletta GJ, Skare S (1994) Haloperidol concentrations in patients with Alzheimer's dementia. Am J Geriatr Psychiatry 2(2):124–133. https://doi.org/10.1097/00019442-199405000-00006

Field M, Oles R, Singh L (2001) Pregabalin may represent a novel class of anxiolytic agents with a broad spectrum of activity. Br J Pharmacol 132:1–4. https://doi.org/10.1038/sj.bjp.0703794

Forlenza OV, De-Paula VJR, Diniz BSO (2014) Neuroprotective effects of lithium: implications for the treatment of Alzheimer's disease and related neurodegenerative disorders. ACS Chem Neurosci 5(6):443–450. https://doi.org/10.1021/cn5000309

Fymat AL (2023) Multiple sclerosis: III. Treatment and prognosis. J Neurol Psychol Res 4(2):1–46

Ghiam MK, Patel SD, Hoffer A, Selman WR, Hoffer BJ, Hoffer ME (2021) Drug repurposing in the treatment of traumatic brain injury. Front Neurosci 15:635483–635483. PubMed. https://doi.org/10.3389/fnins.2021.635483

Gibson SBB, Mark B (2012) Amyotrophic lateral sclerosis: drug therapy from the bench to the bedside. Semin Neurol 32(03):173–178. https://doi.org/10.1055/s-0032-1329193

Godinho J, de Oliveira JN, Ferreira EDF, Zaghi GGD, Bacarin CC, de Oliveira RMW, Milani H (2015) Cilostazol but not sildenafil prevents memory impairment after chronic cerebral hypoperfusion in middle-aged rats. Behav Brain Res 283:61–68. https://doi.org/10.1016/j.bbr.2015.01.026

Goldenberg MM (2010) Overview of drugs used for epilepsy and seizures: Etiology, diagnosis, and treatment. P T 35(7):392–415. PubMed

Gourley SL, Espitia JW, Sanacora G, Taylor JR (2012) Antidepressant-like properties of oral riluzole and utility of incentive disengagement models of depression in mice. Psychopharmacology 219(3):805–814. https://doi.org/10.1007/s00213-011-2403-4

Green AJ, Gelfand JM, Cree BA, Bevan C, Boscardin WJ, Mei F, Inman J, Arnow S, Devereux M, Abounasr A (2017) Clemastine fumarate as a remyelinating therapy for multiple sclerosis (ReBUILD): a randomised, controlled, double-blind, crossover trial. Lancet 390(10111):2481–2489

Greene CS, Voight BF (2016) Pathway and network-based strategies to translate genetic discoveries into effective therapies. Hum Mol Genet 25(R2):R94–R98. PubMed. https://doi.org/10.1093/hmg/ddw160

Ha AD, Fung VSC (2012) Huntington's disease. Curr Opin Neurol 25(4) https://journals.lww.com/co-neurology/fulltext/2012/08000/huntington_s_disease.16.aspx

Hainsworth AH, Arancio O, Elahi FM, Isaacs JD, Cheng F (2023) PDE5 inhibitor drugs for use in dementia. Alzheimer's Dementia (New York, N Y) 9(3):e12412. PubMed. https://doi.org/10.1002/trc2.12412

Heard DS, Tuttle CSL, Lautenschlager NT, Maier AB (2018) Repurposing Proteostasis-modifying drugs to prevent or treat age-related dementia: a systematic review. Front Physiol 9:1520–1520. PubMed. https://doi.org/10.3389/fphys.2018.01520

Hewitt D (2000) The use of NMDA-receptor antagonists in the treatment of chronic pain. Clin J Pain 16:S73–S79. https://doi.org/10.1097/00002508-200006001-00013

Hiskens MI (2022) Targets of neuroprotection and review of pharmacological interventions in traumatic brain injury. J Pharmacol Exp Ther 382(2):149. https://doi.org/10.1124/jpet.121.001023

Ho Y-J, Hsu L-S, Wang C-F, Hsu W-Y, Lai T-J, Hsu C-C, Tsai Y-F (2005) Behavioral effects of d-cycloserine in rats: the role of anxiety level. Brain Res 1043(1):179–185. https://doi.org/10.1016/j.brainres.2005.02.057

Hsieh Y-Y, Chou C-J, Lo H-L, Yang P-M (2016) Repositioning of a cyclin-dependent kinase inhibitor GW8510 as a ribonucleotide reductase M2 inhibitor to treat human colorectal cancer. Cell Death Discov 2:16027. PubMed–16027. https://doi.org/10.1038/cddiscovery.2016.27

Huang DW, Sherman BT, Lempicki RA (2009) Bioinformatics enrichment tools: paths toward the comprehensive functional analysis of large gene lists. Nucleic Acids Res 37(1):1–13. https://doi.org/10.1093/nar/gkn923

Huang W, Chen W, Zhang X (2017) Multiple sclerosis: pathology, diagnosis and treatments (review). Exp Ther Med 13(6):3163–3166. https://doi.org/10.3892/etm.2017.4410

Humble SS, Wilson LD, Leath TC, Marshall MD, Sun DZ, Pandharipande PP, Patel MB (2016) ICU sedation with dexmedetomidine after severe traumatic brain injury. Brain Inj 30(10):1266–1270. PubMed. https://doi.org/10.1080/02699052.2016.1187289

Iaccarino C, Carretta A, Nicolosi F, Morselli C (2018) Epidemiology of severe traumatic brain injury. J Neurosurg Sci 62(5):535–541. PubMed. https://doi.org/10.23736/s0390-5616.18.04532-0

Jain VK, Shilpa KN, Rajnish S, Megha J, Ritesh U, Dileep D (2022) Evaluation of anxiolytic activity of Ondansetron in mice models—an animal experimental research. J Med Sci 8(2):106

Jellinger KA, Attems J (2011) Prevalence and pathology of dementia with Lewy bodies in the oldest old: a comparison with other dementing disorders. Dement Geriatr Cogn Disord 31(4):309–316. https://doi.org/10.1159/000327360

Ji M, Dong L, Jia M, Liu W, Zhang M, Ju L, Yang J, Xie Z, Yang J (2014) Epigenetic enhancement of brain-derived neurotrophic factor signaling pathway improves cognitive impairments induced by isoflurane exposure in aged rats. Mol Neurobiol 50(3):937–944. PubMed. https://doi.org/10.1007/s12035-014-8659-z

Joshi A, Bow A, Agius M (2019) Pharmacological therapies in bipolar disorder: a review of current treatment options. Psychiatr Danub 31:595–603

Kalia LV, Lang AE (2015) Parkinson's disease. Lancet 386(9996):896–912. https://doi.org/10.1016/S0140-6736(14)61393-3

Kaneko S, Wang J, Kaneko M, Yiu G, Hurrell JM, Chitnis T, Khoury SJ, He Z (2006) Protecting axonal degeneration by increasing Nicotinamide Adenine Dinucleotide levels in experimental autoimmune encephalomyelitis models. J Neurosci 26(38):9794. https://doi.org/10.1523/JNEUROSCI.2116-06.2006

Kaskow BJ, Baecher-Allan C (2018) Effector T cells in multiple sclerosis. Cold Spring Harb Perspect Med 8(4). https://doi.org/10.1101/cshperspect.a029025

Katila N, Bhurtel S, Park P-H, Hong JT, Choi D-Y (2020) Activation of AMPK/aPKCζ/CREB pathway by metformin is associated with upregulation of GDNF and dopamine. Biochem Pharmacol 180:114193. https://doi.org/10.1016/j.bcp.2020.114193

Kelliher M, Fastbom J, Cowburn RF, Bonkale W, Ohm TG, Ravid R, Sorrentino V, O'Neill C (1999) Alterations in the ryanodine receptor calcium release channel correlate with Alzheimer's disease neurofibrillary and β-amyloid pathologies. Neuroscience 92(2):499–513. https://doi.org/10.1016/S0306-4522(99)00042-1

Klein P, Friedman A, Hameed MQ, Kaminski RM, Bar-Klein G, Klitgaard H, Koepp M, Jozwiak S, Prince DA, Rotenberg A, Twyman R, Vezzani A, Wong M, Löscher W (2020) Repurposed molecules for antiepileptogenesis: missing an opportunity to prevent epilepsy? Epilepsia 61(3):359–386. https://doi.org/10.1111/epi.16450

Knapp DJ, Overstreet DH, Breese GR (2007) Baclofen blocks expression and sensitization of anxiety-like behavior in an animal model of repeated stress and ethanol withdrawal. Alcohol Clin Exp Res 31(4):582–595. https://doi.org/10.1111/j.1530-0277.2007.00342.x

Komang Trisna, Sumadewi Saktivi, Harkitasari David Christopher, Tjandra (2023) Biomolecular mechanisms of epileptic seizures and epilepsy: a review Abstract Acta Epileptologica 5(1). https://doi.org/10.1186/s42494-023-00137-0

Kong E, Peng L, Chen Y, Yu A, Hertz L (2002) Up-regulation of 5-HT2B receptor density and receptor-mediated Glycogenolysis in mouse astrocytes by long-term fluoxetine administration. Neurochem Res 27:113–120. https://doi.org/10.1023/A:1014862808126

Koyuncu A, İnce E, Ertekin E, Tükel R (2019) Comorbidity in social anxiety disorder: diagnostic and therapeutic challenges. Drugs Cont 8:212573–212573. PubMed. https://doi.org/10.7573/dic.212573

Kratz JM, Andrighetti-Fröhner CR, Leal PC, Nunes RJ, Yunes RA, Trybala E, Bergström T, Barardi CRM, Simões CMO (2008) Evaluation of anti-HSV-2 activity of Gallic acid and Pentyl Gallate. Biol Pharm Bull 31(5):903–907. https://doi.org/10.1248/bpb.31.903

Kumar A, Sidhu J, Goyal A, Tsao JW (2022) Alzheimer Disease (29763097). StatPearls Publishing, Treasure Island (FL). PubMed. http://europepmc.org/abstract/MED/29763097

Kuwabara T, Ishikawa F, Kondo M, Kakiuchi T (2017) The role of IL-17 and related cytokines in inflammatory autoimmune diseases. Mediat Inflamm 2017:3908061. https://doi.org/10.1155/2017/3908061

Lal H, Shearman GT, Fielding S, Dunn R, Kruse H, Theurer K (1980) Effect of valproic acid on anxiety-related behaviors in the rat. Brain Res Bull 5:575–577. https://doi.org/10.1016/0361-9230(80)90093-3

Lan Y-L, Wang X, Zou Y-J, Xing J-S, Lou J-C, Zou S, Ma B-B, Ding Y, Zhang B (2019) Bazedoxifene protects cerebral autoregulation after traumatic brain injury and attenuates impairments in blood–brain barrier damage: involvement of anti-inflammatory pathways by blocking MAPK signaling. Inflamm Res 68(4):311–323. https://doi.org/10.1007/s00011-019-01217-z

Larsen M, Valø ET, Berg-Johnsen J, Langmoen IA (1998) Isoflurane reduces synaptic glutamate release without changing cytosolic free calcium in isolated nerve terminals. Eur J Anaesthesiol 15(2):224–229. Cambridge Core. https://doi.org/10.1111/j.0265-0215.1998.00275.x

Lehrer S, Rheinstein P (2018) Alzheimer's disease and intranasal fluticasone propionate in the FDA MedWatch adverse events database. J Alzheimer's Dis Rep 2:1–5. https://doi.org/10.3233/ADR-170033

Lesort M, Lee M, Tucholski J, Johnson GVW (2003) Cystamine inhibits caspase activity: IMPLICATIONS FOR THE TREATMENT OF POLYGLUTAMINE DISORDERS*. J Biol Chem 278(6):3825–3830. https://doi.org/10.1074/jbc.M205812200

Li L, Zhang Z-F, Holscher C, Gao C, Jiang Y-H, Liu Y-Z (2012) (Val8) glucagon-like peptide-1 prevents tau hyperphosphorylation, impairment of spatial learning and ultra-structural cellular damage induced by streptozotocin in rat brains. Eur J Pharmacol 674(2):280–286. https://doi.org/10.1016/j.ejphar.2011.11.005

Li C, Lu Y, Zhou M, Zong X, Li C, Xu X, Guo L, Lu Q (2014) Activation of GABAB Receptors ameliorates cognitive impairment via restoring the balance of HCN1/HCN2 surface expression in the Hippocampal CA1 area in rats with chronic cerebral Hypoperfusion. Mol Neurobiol 50(2):704–720. https://doi.org/10.1007/s12035-014-8736-3

Li Y, Cheng K-C, Liu K-F, Peng W-H, Cheng J-T, Niu H-S (2017) Telmisartan activates PPARδ to improve symptoms of unpredictable chronic mild stress-induced depression in mice. Sci Rep 7. https://doi.org/10.1038/s41598-017-14265-4

Li W, Chaudhari K, Shetty R, Winters A, Gao X, Hu Z, Ge W-P, Sumien N, Forster M, Liu R, Yang S-H (2019) Metformin alters locomotor and cognitive function and brain metabolism in Normoglycemic mice. Aging Dis 10(5):949–963. PubMed. https://doi.org/10.14336/AD.2019.0120

Liang J, Wu S, Xie W, He H (2018) Ketamine ameliorates oxidative stress-induced apoptosis in experimental traumatic brain injury via the Nrf2 pathway. Drug Des Devel Ther 12:845–853. PubMed. https://doi.org/10.2147/DDDT.S160046

Liao H, Zou Z, Liu W, Guo X, Xie J, Li L, Li X, Gan X, Huang X, Liu J, Li W, Zeng H, Chen Z, Jiang Q, Yao H (2022) Osteopontin-integrin signaling positively regulates neuroplasticity through enhancing neural autophagy in the peri-infarct area after ischemic stroke. Am J Transl Res 14(11):7726–7743. PubMed

Lin P-Y, Chang AYW, Lin T-K (2014) Simvastatin treatment exerts antidepressant-like effect in rats exposed to chronic mild stress. Pharmacol Biochem Behav 124:174–179. https://doi.org/10.1016/j.pbb.2014.06.006

Lopez O, DeKosky S (2003) Neuropathology of Alzheimer's disease and mild cognitive impairment. Rev Neurol 37(2):155–163

Machado-Vieira R, Andreazza AC, Viale CI, Zanatto V, Cereser V, da Vargas RS, Kapczinski F, Portela LV, Souza DO, Salvador M, Gentil V (2007) Oxidative stress parameters in unmedicated and treated bipolar subjects during initial manic episode: a possible role for lithium antioxidant effects. Neurosci Lett 421(1):33–36. https://doi.org/10.1016/j.neulet.2007.05.016

Mah L, Zarate CA Jr, Nugent AC, Singh JB, Manji HK, Drevets WC (2011) Neural mechanisms of antidepressant efficacy of the dopamine receptor agonist pramipexole in treatment of bipolar depression. Int J Neuropsychopharmacol 14(4):545–551. https://doi.org/10.1017/S1461145710001203

Mahmoudi A, Heydari S, Markina YV, Barreto GE, Sahebkar A (2022) Role of statins in regulating molecular pathways following traumatic brain injury: a system pharmacology study. Biomed Pharmacother 153:113304. https://doi.org/10.1016/j.biopha.2022.113304

Matlack K, Tardiff D, Narayan P, Hamamichi S, Caldwell K, Caldwell G, Lindquist S (2014) Clioquinol promotes the degradation of metal-dependent amyloid- (A) oligomers to restore endocytosis and ameliorate A toxicity. Proc Natl Acad Sci USA 111. https://doi.org/10.1073/pnas.1402228111

Mayeux R, Stern Y (2012) Epidemiology of Alzheimer disease. Cold Spring Harb Perspect Med 2(8). https://doi.org/10.1101/cshperspect.a006239

McClean PL, Parthsarathy V, Faivre E, Hölscher C (2011) The diabetes drug liraglutide prevents degenerative processes in a mouse model of Alzheimer's disease. J Neurosci 31(17):6587–6594. PubMed. https://doi.org/10.1523/JNEUROSCI.0529-11.2011

Migdalska-Richards A, Ko WKD, Li Q, Bezard E, Schapira AHV (2017) Oral ambroxol increases brain glucocerebrosidase activity in a nonhuman primate. Synapse (New York, NY) 71(7):e21967. PubMed. https://doi.org/10.1002/syn.21967

Mirshekar MA, Sarkaki A, Farbood Y, Gharib Naseri MK, Badavi M, Mansouri MT, Haghparast A (2018) Neuroprotective effects of gallic acid in a rat model of traumatic brain injury: Behavioral, electrophysiological, and molecular studies. Iran J Basic Med Sci 21(10):1056–1063. PubMed. https://doi.org/10.22038/IJBMS.2018.29639.7165

Mirza N, Sills GJ, Pirmohamed M, Marson AG (2017) Identifying new antiepileptic drugs through genomics-based drug repurposing. Hum Mol Genet 26(3):527–537. https://doi.org/10.1093/hmg/ddw410

Mohamadpour M, Whitney K, Bergold PJ (2019) The importance of therapeutic time window in the treatment of traumatic brain injury. Front Neurosci 13:07–07. PubMed. https://doi.org/10.3389/fnins.2019.00007

Mohammad Sadeghi H, Adeli I, Mousavi T, Daniali M, Nikfar S, Abdollahi M (2021) Drug repurposing for the management of depression: where do we stand currently? Life 11(8). https://doi.org/10.3390/life11080774

Mohammadi E, Nikbakht F, Vazifekhah S, Babae JF, Jogataei MT (2023) Evaluation the cognition-improvement effects of N-acetyl cysteine in experimental temporal lobe epilepsy in rat. Behav Brain Res 440:114263. https://doi.org/10.1016/j.bbr.2022.114263

Mortiboys H, Furmston R, Bronstad G, Aasly J, Elliott C, Bandmann O (2015) UDCA exerts beneficial effect on mitochondrial dysfunction in LRRK2(G2019S) carriers and in vivo. Neurology 85(10):846–852. PubMed. https://doi.org/10.1212/WNL.0000000000001905

Mostert JP, Sijens PE, Oudkerk M, De Keyser J (2006) Fluoxetine increases cerebral white matter NAA/Cr ratio in patients with multiple sclerosis. Neurosci Lett 402(1):22–24. https://doi.org/10.1016/j.neulet.2006.03.042

Mostert JP, Admiraal-Behloul F, Hoogduin JM, Luyendijk J, Heersema DJ, van Buchem MA, De Keyser J (2008) Effects of fluoxetine on disease activity in relapsing multiple sclerosis: a double-blind, placebo-controlled, exploratory study. J Neurol Neurosurg Psychiatr 79(9):1027. https://doi.org/10.1136/jnnp.2007.139345

Mrozek W, Socha J, Sidorowicz K, Skrok A, Syrytczyk A, Piątkowska-Chmiel I, Herbet M (2023) Pathogenesis and treatment of depression: role of diet in prevention and therapy. Nutrition 115:112143. https://doi.org/10.1016/j.nut.2023.112143

Munir S, Takov V (2022) Generalized anxiety disorder (28722900). StatPearls Publishing, Treasure Island (FL). PubMed. http://europepmc.org/abstract/MED/28722900

Muralidar S, Ambi SV, Sekaran S, Thirumalai D, Palaniappan B (2020) Role of tau protein in Alzheimer's disease: the prime pathological player. Int J Biol Macromol 163:1599–1617. https://doi.org/10.1016/j.ijbiomac.2020.07.327

Najm FJ, Madhavan M, Zaremba A, Shick E, Karl RT, Factor DC, Miller TE, Nevin ZS, Kantor C, Sargent A, Quick KL, Schlatzer DM, Tang H, Papoian R, Brimacombe KR, Shen M, Boxer MB, Jadhav A, Robinson AP et al (2015) Drug-based modulation of endogenous stem cells promotes functional remyelination in vivo. Nature 522(7555):216–220. https://doi.org/10.1038/nature14335

Napolitano F, Zhao Y, Moreira VM, Tagliaferri R, Kere J, D'Amato M, Greco D (2013) Drug repositioning: a machine-learning approach through data integration. J Chem 5(1):30–30. PubMed. https://doi.org/10.1186/1758-2946-5-30

Nath N, Khan M, Paintlia MK, Hoda MN, Giri S (2009) Metformin attenuated the autoimmune disease of the central nervous system in animal models of multiple sclerosis1. J Immunol 182(12):8005–8014. https://doi.org/10.4049/jimmunol.0803563

Niculescu AB, Le-Niculescu H, Roseberry K, Wang S, Hart J, Kaur A, Robertson H, Jones T, Strasburger A, Williams A, Kurian SM, Lamb B, Shekhar A, Lahiri DK, Saykin AJ (2020) Blood biomarkers for memory: toward early detection of risk for Alzheimer disease, pharmacogenomics, and repurposed drugs. Mol Psychiatry 25(8):1651–1672. https://doi.org/10.1038/s41380-019-0602-2

Norins LC (2021) Repurposing licensed drugs for use against Alzheimer's disease. J Alzheimer's Dis 81(3):921–932. https://doi.org/10.3233/JAD-210080

Nozaki T, Ura H, Takumi I, Kobayashi S, Maru E, Morita A (2018) The angiotensin II type I receptor antagonist losartan retards amygdala kindling-induced epileptogenesis. Brain Res 1694:121–128. https://doi.org/10.1016/j.brainres.2018.05.027

Nygaard HB, Wagner AF, Bowen GS, Good SP, MacAvoy MG, Strittmatter KA, Kaufman AC, Rosenberg BJ, Sekine-Konno T, Varma P, Chen K, Koleske AJ, Reiman EM, Strittmatter SM, van Dyck CH (2015) A phase Ib multiple ascending dose study of the safety, tolerability, and central nervous system availability of AZD0530 (saracatinib) in Alzheimer's disease. Alzheimer's Res Ther 7(1):35. https://doi.org/10.1186/s13195-015-0119-0

O'Brien JT, Chouliaras L, Sultana J, Taylor J-P, Ballard C, RENEWAL Study Group (2022) RENEWAL: REpurposing study to find NEW compounds with activity for Lewy body dementia-an international Delphi consensus. Alzheimer's Res Ther 14(1):169–169. PubMed. https://doi.org/10.1186/s13195-022-01103-7

Odi R, Invernizzi RW, Gallily T, Bialer M, Perucca E (2021) Fenfluramine repurposing from weight loss to epilepsy: what we do and do not know. Pharmacol Ther 226:107866. https://doi.org/10.1016/j.pharmthera.2021.107866

Ogura J, Yamaguchi H (2022) The effectiveness of antidiabetic drugs in treating dementia: a peek into pharmacological and pharmacokinetic properties. Int J Mol Sci 23(12):6542. PubMed. https://doi.org/10.3390/ijms23126542

Padhi D, Govindaraju T (2022) Mechanistic insights for drug repurposing and the Design of Hybrid Drugs for Alzheimer's disease. J Med Chem 65(10):7088–7105. https://doi.org/10.1021/acs.jmedchem.2c00335

Palacios J, Yildiz A, Young AH, Taylor MJ (2019) Tamoxifen for bipolar disorder: systematic review and meta-analysis. J Psychopharmacol 33(2):177–184. https://doi.org/10.1177/0269881118822167

Palmer AM (2013) Multiple sclerosis and the blood-central nervous system barrier. Cardiovasc Psychiatry Neurol 2013:530356. https://doi.org/10.1155/2013/530356

Panayotis N, Freund P, Marvaldi L, Shalit T, Brandis A, Mehlman T, Michael T, Fainzilber M (2021) β-sitosterol reduces anxiety and synergizes with established anxiolytic drugs in mice. Cell Rep Med 2:100281. https://doi.org/10.1016/j.xcrm.2021.100281

Parsons CG (2019) CNS repurposing—potential new uses for old drugs: examples of screens for Alzheimer's disease, Parkinson's disease and spasticity. Drug repurposing: old molecules, new ways to fast track drug discovery and development for CNS disorders. Neuropharmacology 147:4–10. https://doi.org/10.1016/j.neuropharm.2018.08.027

Paudel YN, Angelopoulou E, Piperi C, Gnatkovsky V, Othman I, Shaikh MF (2020) From the molecular mechanism to pre-clinical results: anti-epileptic effects of Fingolimod. Curr Neuropharmacol 18(11):1126–1137. PubMed. https://doi.org/10.2174/1570159X18666200420125017

Perry T, Lahiri DK, Sambamurti K, Chen D, Mattson MP, Egan JM, Greig NH (2003) Glucagon-like peptide-1 decreases endogenous amyloid-β peptide (Aβ) levels and protects hippocampal neurons from death induced by Aβ and iron. J Neurosci Res 72(5):603–612. https://doi.org/10.1002/jnr.10611

Polis AJ, Fitzgerald PJ, Hale PJ, Watson BO (2019) Rodent ketamine depression-related research: finding patterns in a literature of variability. Behav Brain Res 376:112153. https://doi.org/10.1016/j.bbr.2019.112153

Rebelo R, Polónia B, Santos LL, Vasconcelos MH, Xavier CPR (2021) Drug repurposing opportunities in pancreatic ductal adenocarcinoma. Pharmaceuticals 14(3). https://doi.org/10.3390/ph14030280

Reich DS, Lucchinetti CF, Calabresi PA (2018) Multiple Sclerosis. N Engl J Med 378(2):169–180. https://doi.org/10.1056/NEJMra1401483

Reis DJ, Casteen EJ, Ilardi SS (2019) The antidepressant impact of minocycline in rodents: a systematic review and meta-analysis. Sci Rep 9(1):261. https://doi.org/10.1038/s41598-018-36507-9

Rodriguez-Romaguera J, Sotres-Bayon F, Mueller D, Quirk GJ (2009) Systemic propranolol acts centrally to reduce conditioned fear in rats without impairing extinction. Incentive motivation, conditioning, stress, and neuropsychiatric disorders: a tribute to Jane Stewart. Biol Psychiatry 65(10):887–892. https://doi.org/10.1016/j.biopsych.2009.01.009

Rolan P, Hutchinson M, Johnson K (2009) Ibudilast: a review of its pharmacology, efficacy and safety in respiratory and neurological disease. Expert Opin Pharmacother 10(17):2897–2904. https://doi.org/10.1517/14656560903426189

Romero-Pinel L, Bau L, Matas E, León I, Muñoz-Vendrell A, Arroyo P, Masuet-Aumatell C, Martínez-Yélamos A, Martínez-Yélamos S (2022) The age at onset of relapsing-remitting multiple sclerosis has increased over the last five decades. Mult Scler Relat Disord 68:104103. https://doi.org/10.1016/j.msard.2022.104103

Salat D, Tolosa E (2013) Levodopa in the treatment of Parkinson's disease: current status and new developments. J Parkinsons Dis 3(3):255–269. https://doi.org/10.3233/JPD-130186

Sanders O (2020) Sildenafil for the treatment of Alzheimer's disease: a systematic review. J Alzheimer's Dis Rep 4:91–106. https://doi.org/10.3233/ADR-200166

Santos CO, Caeiro L, Ferro JM, Figueira ML (2011) Mania and stroke: a systematic review. Cerebrovasc Dis 32(1):11–21. https://doi.org/10.1159/000327032

Savva K, Zachariou M, Bourdakou M, Dietis N, Spyrou G (2022) Network-based stage-specific drug repurposing for Alzheimer's disease. Comput Struct Biotechnol J 20. https://doi.org/10.1016/j.csbj.2022.03.013

Schramm C, Katsahian S, Youssov K, Démonet J-F, Krystkowiak P, Supiot F, Verny C, Cleret de Langavant L, Bachoud-Lévi A-C, European Huntington's Disease Initiative Study Group and the Multicentre Intracerebral Grafting in Huntington's Disease Group (2016) How to capitalize on the retest effect in future trials on Huntington's disease. PLoS One 10(12):e0145842. https://doi.org/10.1371/journal.pone.0145842

Serra G, Demontis F, Serra F, De Chiara L, Spoto A, Girardi P, Vidotto G, Serra G (2014) Memantine: new prospective in bipolar disorder treatment. World J Psychiatr 4(4):80–90. PubMed. https://doi.org/10.5498/wjp.v4.i4.80

Shakkour Z, Habashy KJ, Berro M, Takkoush S, Abdelhady S, Koleilat N, Eid AH, Zibara K, Obeid M, Shear D, Mondello S, Wang KK, Kobeissy F (2021) Drug repurposing in neurological disorders: Implications for neurotherapy in traumatic brain injury. Neuroscientist 27(6):620–649. https://doi.org/10.1177/1073858420961078

Singh J, Cleveland J (2018) Comparative effectiveness of allopurinol versus febuxostat for preventing incident dementia in older adults: a propensity-matched analysis. Arthritis Res Ther 20. https://doi.org/10.1186/s13075-018-1663-3

Singh R, Sadiq NM (2022) Cholinesterase inhibitors (31335056). StatPearls Publishing, Treasure Island (FL). PubMed. http://europepmc.org/abstract/MED/31335056

Skibinski G, Finkbeiner S (2011) Drug discovery in Parkinson's disease-update and developments in the use of cellular models. Int J High Throughput Screen 2011(2):15–25. PubMed. https://doi.org/10.2147/IJHTS.S8681

Solaro C, Restivo D, Mancardi GL, Tanganelli P (2007) Oxcarbazepine for treating paroxysmal painful symptoms in multiple sclerosis: a pilot study. Neurol Sci 28(3):156–158. https://doi.org/10.1007/s10072-007-0811-3

Soleimani Zakeri NS, Pashazadeh S, MotieGhader H (2021) Drug repurposing for Alzheimer's disease based on protein-protein interaction network. Biomed Res Int 2021:1–11. https://doi.org/10.1155/2021/1280237

Song Q, Fan C, Wang P, Li Y, Yang M, Yu SY (2018) Hippocampal CA1 βCaMKII mediates neuroinflammatory responses via COX-2/PGE2 signaling pathways in depression. J Neuroinflammation 15(1):338. https://doi.org/10.1186/s12974-018-1377-0

Spain R, Powers K, Murchison C, Heriza E, Winges K, Yadav V, Cameron M, Kim E, Horak F, Simon J, Bourdette D (2017) Lipoic acid in secondary progressive MS. Neurol Neuroimmunol Neuroinflam 4(5):e374. https://doi.org/10.1212/NXI.0000000000000374

Susanna, Amadio Federica, Conte Giorgia, Esposito Giulia, Fiscon Paola, Paci Cinzia, Volonté (2022) Repurposing Histaminergic Drugs in Multiple Sclerosis International Journal of Molecular Sciences 23(11) 6347. https://doi.org/10.3390/ijms23116347

Tafti D, Ehsan M, Xixis KL (n.d.) Continuing education activity

Taheri A, Emami M, Asadipour E, Kasirzadeh S, Rouini M-R, Najafi A, Heshmat R, Abdollahi M, Mojtahedzadeh M (2019) A randomized controlled trial on the efficacy, safety, and pharmacokinetics of metformin in severe traumatic brain injury. J Neurol 266(8):1988–1997. https://doi.org/10.1007/s00415-019-09366-1

Tandon S, Sarkar S (2023) Glipizide ameliorates human poly(Q) mediated neurotoxicity by upregulating insulin signalling in Drosophila disease models. Biochem Biophys Res Commun 645:88–96. https://doi.org/10.1016/j.bbrc.2023.01.022

Tao L, Li D, Liu H, Jiang F, Xu Y, Cao Y, Gao R, Chen G (2018) Neuroprotective effects of metformin on traumatic brain injury in rats associated with NF-κB and MAPK signaling pathway. Brain Res Bull 140:154–161. https://doi.org/10.1016/j.brainresbull.2018.04.008

Terry AV, Gearhart DA, Warner SE, Zhang G, Bartlett MG, Middlemore M-L, Beck WD, Mahadik SP, Waller JL (2007) Oral haloperidol or risperidone treatment in rats: temporal effects on nerve growth factor receptors, cholinergic neurons, and memory performance. Neuroscience 146(3):1316–1332. https://doi.org/10.1016/j.neuroscience.2007.03.003

Tisher A, Salardini A (2019) A comprehensive update on treatment of dementia. Semin Neurol 39:167–178. https://doi.org/10.1055/s-0039-1683408

Titus DJ, Wilson NM, Alcazar O, Calixte DA, Dietrich WD, Gurney ME, Atkins CM (2018) A negative allosteric modulator of PDE4D enhances learning after traumatic brain injury. Neurobiol Learn Mem 148:38–49. PubMed. https://doi.org/10.1016/j.nlm.2017.12.008

Tong H, Zhang X, Meng X, Lu L, Mai D, Qu S (2018) Simvastatin inhibits activation of NADPH oxidase/p38 MAPK pathway and enhances expression of antioxidant protein in Parkinson disease models. Front Mol Neurosci 11:165–165. PubMed. https://doi.org/10.3389/fnmol.2018.00165

Torika N, Asraf K, Apte RN, Fleisher-Berkovich S (2018) Candesartan ameliorates brain inflammation associated with Alzheimer's disease. CNS Neurosci Ther 24(3):231–242. https://doi.org/10.1111/cns.12802

Tzeng N-S, Chung C-H, Lin F-H, Chiang C-P, Yeh C-B, Huang S-Y, Lu R-B, Chang H-A, Kao Y-C, Yeh H-W, Chiang W-S, Chou Y-C, Tsao C-H, Wu Y-F, Chien W-C (2018) Anti-herpetic medications and reduced risk of dementia in patients with herpes simplex virus infections—a Nationwide, Population-Based Cohort Study in Taiwan. Neurotherapeutics 15(2):417–429. https://doi.org/10.1007/s13311-018-0611-x

Veenman L (2020) Raloxifene as treatment for various types of brain injuries and neurodegenerative diseases: a good start. Int J Mol Sci 21:7586. https://doi.org/10.3390/ijms21207586

Vieira DB, Gamarra LF (2016) Getting into the brain: liposome-based strategies for effective drug delivery across the blood–brain barrier. Int J Nanomedicine 11:5381–5414

Volpicelli-Daley L, Brundin P (2018) Prion-like propagation of pathology in Parkinson disease. Handb Clin Neurol 153:321–335. PubMed. https://doi.org/10.1016/B978-0-444-63945-5.00017-9

Walker JE, Giri SN, Margolin SB (2005) A double-blind, randomized, controlled study of oral pirfenidone for treatment of secondary progressive multiple sclerosis. Mult Scler J 11(2):149–158. https://doi.org/10.1191/1352458505ms1134oa

Wallace VCJ, Segerdahl AR, Blackbeard J, Pheby T, Rice ASC (2008) Anxiety-like behaviour is attenuated by gabapentin, morphine and diazepam in a rodent model of HIV anti-retroviral-associated neuropathic pain. Neurosci Lett 448(1):153–156. https://doi.org/10.1016/j.neulet.2008.10.005

Wang J, Ho L, Chen L, Zhao R, Zhao W, Qian X, Humala N, Seror I, Bartholomew S, Rosendorff C, Pasinetti G (2007) Valsartan lowers brain ?-amyloid protein levels and improves spatial learning in a mouse model of Alzheimer disease. J Clin Invest 117:3393–3402. https://doi.org/10.1172/JCI31547

Wang L-L, Pan X-L, Wang Y, Tang H-D, Deng Y-L, Ren R-J, Xu W, Ma J-F, Wang G, Chen S-D (2011) A single nucleotide polymorphism in LRP2 is associated with susceptibility to Alzheimer's disease in the Chinese population. Clin Chim Acta 412(3):268–270. https://doi.org/10.1016/j.cca.2010.10.015

Weintraub MK, Bisson CM, Nouri JN, Vinson BT, Eimerbrink MJ, Kranjac D, Boehm GW, Chumley MJ (2013) Imatinib methanesulfonate reduces hippocampal amyloid-beta and restores cognitive function following repeated endotoxin exposure. Brain Behav Immun 33:24–28. https://doi.org/10.1016/j.bbi.2013.05.002

Wolfgang, Löscher Pavel, Klein (2021) The Pharmacology and Clinical Efficacy of Antiseizure Medications: From Bromide Salts to Cenobamate and Beyond CNS Drugs 35(9):935–963. https://doi.org/10.1007/s40263-021-00827-8

Woodward D, Thorp J, Akosile W, Ong J-S, Gamazon E, Derks E, Gerring Z (2023) Identification of drug repurposing candidates for the treatment of anxiety: a genetic approach. Psychiatry Res 326:115343. https://doi.org/10.1016/j.psychres.2023.115343

Wu J, Vogel T, Gao X, Lin B, Kulwin C, Chen J (2018) Neuroprotective effect of dexmedetomidine in a murine model of traumatic brain injury. Sci Rep 8(1):4935–4935. PubMed. https://doi.org/10.1038/s41598-018-23003-3

Wu H, Wu Z, Shi W, Gao H, Wu H, Bian F, Jia P, Hou Y (2019) Effects of progesterone on glucose uptake in neurons of Alzheimer's disease animals and cell models. Life Sci 238:116979. https://doi.org/10.1016/j.lfs.2019.116979

Yates CC, Charlesworth A, Reese NB, Ishida K, Skinner RD, Garcia-Rill E (2009) Modafinil normalized hyperreflexia after spinal transection in adult rats. Spinal Cord 47(6):481–485. https://doi.org/10.1038/sc.2008.154

Zagórska A, Czopek A, Fryc M, Jaromin A, Boyd BJ (2023) Drug discovery and development targeting dementia. Pharmaceuticals (Basel, Switzerland) 16(2):151. PubMed. https://doi.org/10.3390/ph16020151

Zeng X, Zhu S, Liu X, Zhou Y, Nussinov R, Cheng F (2019) deepDR: a network-based deep learning approach to in silico drug repositioning. Bioinformatics 35(24):5191–5198. https://doi.org/10.1093/bioinformatics/btz418

Zhang L, Zhang L, Li Y, Li L, Melchiorsen JU, Rosenkilde M, Hölscher C (2020) The novel dual GLP-1/GIP receptor agonist DA-CH5 is superior to single GLP-1 receptor agonists in the MPTP model of Parkinson's disease. J Parkinsons Dis 10(2):523–542. https://doi.org/10.3233/JPD-191768

Zhou M, Greenhill S, Huang S, Silva TK, Sano Y, Wu S, Cai Y, Nagaoka Y, Sehgal M, Cai DJ, Lee Y-S, Fox K, Silva AJ (2016) CCR5 is a suppressor for cortical plasticity and hippocampal learning and memory. elife 5:e20985. PubMed. https://doi.org/10.7554/eLife.20985

MIX
Papier aus verantwortungsvollen Quellen
Paper from responsible sources
FSC® C105338

If you have any concerns about our products,
you can contact us on
ProductSafety@springernature.com

In case Publisher is established outside the EU,
the EU authorized representative is:
Springer Nature Customer Service Center GmbH
Europaplatz 3, 69115 Heidelberg, Germany

Printed by Libri Plureos GmbH
in Hamburg, Germany